Electrons and Valence

Electrons and Valence

Development of the Theory, 1900–1925

by Anthony N. Stranges

Texas A&M University Press
COLLEGE STATION

Library of Congress Cataloging in Publication Data

Stranges, Anthony N. (Anthony Nicholas), 1936–
    Electrons and valence.

    Bibliography: p.
    Includes index.
    1. Valence (Theoretical chemistry)   I. Title.
QD469.S73   541.2′24   81-48378
ISBN 0-89096-124-7   AACR2

Manufactured in the United States of America
FIRST EDITION

To Sonya

Contents

List of Illustrations

Preface

Along with the doctrine of atomism, the electron theory of valence ranks as one of the most fundamental developments in the history of modern chemistry. Its development began shortly after J. J. Thomson at Cambridge University discovered the electron in 1897. His valence theory and that of the German chemist Richard Abegg were among the earliest. Abegg and Thomson proposed that the bonds holding atoms in a molecule were electrostatic or polar and that each bond resulted from the complete transfer of an electron from one of the combining atoms to another.

Thomson's visit to Yale University in 1903 to deliver the Silliman Lectures and the publication of his book *The Corpuscular Theory of Matter* in 1907 contributed greatly to the eager acceptance of the polar theory by American chemists. Beginning in 1909, William A. Noyes at the University of Illinois, Harry S. Fry at the University of Cincinnati, K. George Falk at Columbia University, and Julius Stieglitz at the University of Chicago published numerous papers in which they applied the polar theory to the structures and reactions of organic molecules.

Not all chemists were willing to accept the polar theory, especially when dealing with the behavior of organic molecules. The chemical facts, they insisted, were at complete variance with a theory that held that molecules such as methane and carbon tetrachloride contained charged atoms. William C. Bray, Gerald E. K. Branch, and G. N. Lewis at Berkeley in 1913 suggested the additional need of a nonpolar bond. Thomson, arguing from physical evidence, also recognized the shortcomings of the polar theory, and in 1914 he published one of the first electronic interpretations of the nonpolar bond. But only in 1916 did G. N. Lewis succeed in putting forward the accepted electronic mechanism for the nonpolar bond—the shared electron pair. The polar bond thus remained in use for those structures known to consist of ions, but for nonionic or essentially nonpolar compounds the Lewis shared electron pair quickly became the dominant mechanism of bond formation.

This study of the electron theory of valence will show its gradual transition from a purely polar theory to one requiring two kinds of bonds, polar and nonpolar. Many physicists and chemists, such as J. J. Thomson, the Berkeley group of chemists, and Irving Langmuir, made important contributions to the development of the theory, but G. N. Lewis was clearly the central figure. Indeed, his far-reaching idea of the shared electron pair bond, though later given a quantitative interpretation by quantum mechanics in the mid-twenties, remains to this day the foundation of modern valence theory.

In undertaking this study of the electron theory of valence, I wish to thank the History of Science Department, University of Wisconsin, Madison, for awarding me a Ford Foundation Dissertation Fellowship, and David H. Templeton, Department of Chemistry, University of California, Berkeley, for making available G. N. Lewis's unpublished correspondence. I also wish to acknowledge the valuable guidance and assistance of Professor Robert Siegfried of the History of Science Department, Madison. The numerous discussions we had were certainly enlightening and contributed greatly to the ideas developed in this study. To my colleague at Texas A&M University, Professor Kurt Irgolic of the Chemistry Department, I express my appreciation for his reading of the completed manuscript. Dr. William Clark of the Mathematics Department and Mr. David Gill of the Center for Energy and Mineral Resources also assisted in the manuscript's preparation. I thank them for their help.

Electrons and Valence

List of Abbreviated Titles

Amer. Chem. Journal	*American Chemical Journal*
Annalen der Chemie	*Liebig's Annalen der Chemie und Pharmacie*
Annalen der Physik	*Annalen der Physik und Chemie*
B.A.A.S. Report	*British Association for the Advancement of Science Report*
Berichte	*Berichte der deutschen chemischen Gesellschaft*
Comptes rendus	*Comptes rendus hebdomadaires des Séances de l'Académie des Sciences*
J.A.C.S.	*Journal of the American Chemical Society*
J. Chem. Physics	*Journal of Chemical Physics*
J. Chem. Soc.	*Journal of the Chemical Society*
J. Phys. Chem.	*Journal of Physical Chemistry*
J. prakt. Chemie	*Journal für praktische Chemie*
Naturforsch. Ges. Zürich	*Naturforschungen Gesellschaft Zürich*
Phil. Mag.	*Philosophical Magazine*
Phil. Trans.	*Philosophical Transactions of the Royal Society*
Phys. Zeit.	*Physikalische Zeitschrift*
Proc. Nat. Acad. Sci.	*Proceedings of the National Academy of Sciences*
Proc. Roy. Soc.	*Proceedings of the Royal Society*
Trans. Faraday Soc.	*Transactions of the Faraday Society*
Verh. deut. phys. Ges.	*Verhandlungen der deutschen physikalischen Gesellschaft*
Zeit. anorg. Chemie	*Zeitschrift für anorganische und allgemeine Chemie*

4 List of Abbreviated Titles

Zeit. Elektrochemie	*Zeitschrift für Elektrochemie*
Zeit. Physik	*Zeitschrift für Physik*
Zeit. phys. Chemie	*Zeitschrift für physikalische Chemie*

1. The Nineteenth-Century Origin of the Electrochemical Bond and Its Failure in Organic Chemistry

Introduction

In the early years of the nineteenth century, from Humphry Davy onward, evidence for an electrical chemical bond appeared in the literature and found its clearest statement in the dualistic electrochemical theory of Jöns Jacob Berzelius. During the 1830s and 1840s chemists also accumulated an equally impressive body of evidence that not all chemical bonds were electrical. This distinction and realization became conscious and clear with the failure of Berzelius's theory when applied to the molecules of organic chemistry. For those molecules, chemists soon achieved spectacular successes in unraveling their structures without speculating on the nature of the bond.

Electrochemical ideas did not disappear entirely in the nineteenth century. The last two decades witnessed a revival of interest in Berzelius's dualistic theory, resulting chiefly from Hermann von Helmholtz's Faraday Lecture in 1881 and the publication of Svante Arrhenius's electrolytic theory of dissociation in 1887. Their ideas suggested the atomicity of electricity and directed the attention of chemists and physicists to the electrical charge carried by each atom in a molecule. Interest in the electrical behavior of atoms and molecules reached a climax in 1897 with J. J. Thomson's isolation of the electron, the first subatomic particle, and its identification as the fundamental unit of negative charge.

Our discussion on Berzelius and electrochemical dualism begins with a brief look at John Dalton's atomic theory and its success in establishing compositional formulas in the first decades of the nineteenth century.

The Establishment of Compositional and Structural Formulas

John Dalton's theory of indestructible, structureless atoms, each kind differing from the other only in weight, succeeded in bringing order and precision to the study of chemical composition. Earlier atomic theorizing in chemistry had been unsuccessful because it lacked predictive value. But Dalton's assumption that chemical atoms united in small, whole-number ratios gave a simple theoretical explanation for the constant composition of molecules. His 1808 publication on chemical atoms revealed an additional unsuspected relation—the law of multiple proportions.

Dalton (1766–1844) originally introduced his atomic theory to explain the different solubilities of gases in water. His studies on solubilities also showed that each gas dissolved in water as though the others were not present. This behavior resulted, Dalton said, because a natural repulsive force acted only between identical atoms, those in each kind of gas, while a natural attractive force, that of chemical affinity, acted between unlike atoms. Indeed, the rules of combination that Dalton put forward to account for the composition of chemical molecules were a direct consequence of the natural attraction supposedly existing between unlike atoms. They reveal his belief in an axiom of simplicity that guided the union of different atoms:

If there are two bodies A and B, which are disposed to combine, the following is the order in which the combination may take place, beginning with the most simple; namely,

1st. When only one combination of two bodies can be obtained, it must be presumed to be a *binary* one, unless some cause appears to the contrary.

2nd. When two combinations are observed, they must be presumed to be a *binary* and a *ternary*.

3rd. When three combinations are obtained, we may expect one to be a *binary*, and the other two *ternary*.

4th. When four combinations are observed, we should expect one *binary*, two *ternary*, and one *quaternary*, etc.[1]

Dalton's theory of chemical atoms and his arbitrary rules of combination were the subject of much debate in the years following their introduction. They received severe criticism in the writings of Davy (1778–1829) and William H. Wollaston (1766–1828) in England and Claude Louis Berthollet (1748–1822) in France. On the other hand, Berzelius (1779–1848) in Stockholm found Dalton's theory useful, and through his efforts, the value of chemical atoms in establishing compositional formulas gradually gained acceptance.[2]

Berzelius's interpretation of the atomic theory had its roots in Joseph

[1] John Dalton, *A New System of Chemical Philosophy*, vol. 1, pt. 1, pp. 213–14.
[2] Before 1860 the problem of establishing a common set of atomic weights and, hence, of compositional formulas was far from solved.

Louis Gay-Lussac's 1808 "Memoir on the Combination of Gaseous Substances with Each Other."[3] In this memoir, Gay-Lussac (1778–1850) showed that only gaseous volumes of acids and bases combined in small whole-number ratios. The simple laws of chemical combination were evident, he said, only when the reaction took place in the gaseous state. According to Berzelius, Gay-Lussac had founded his law of combining volumes on well-constituted facts. Unlike Dalton, he had not relied on arbitrary rules of combination.[4]

Berzelius maintained that in the union of elementary gases, the number of atoms combining was always proportional to the combining volume of the gas. In this way he determined formulas for the following: water, 2 volumes of hydrogen to 1 volume of oxygen; hydrogen chloride, 1 volume of hydrogen to 1 volume of chlorine; ammonia, 1 volume of nitrogen to 3 volumes of hydrogen. He claimed that "from what we know respecting definite proportions, it follows that it would hold with all bodies at the temperature and pressure at which they would assume the gaseous form." He saw no essential difference between Dalton's theory of atoms and his own theory of volumes. The one represented bodies in solid form; the other, in gaseous form. What Dalton called an *atom*, was a *volume* to Berzelius.[5]

Berzelius did not overlook a serious limitation of his theory of volumes, namely, how to determine the compositional formulas of numerous compounds that did not exist in the gaseous state. He used chemical analogies, assigning similar compositional formulas to compounds that behaved similarly, such as oxides and sulfides, and he introduced a set of arbitrary rules for establishing formulas of compounds when the same element combined with more than one proportion of another element. Thus, if the weight ratio of oxygen that combined with a fixed weight of a second element was $1:2$, Berzelius adopted the formulas EO and EO_2. If the oxygen ratio was $2:3$, possible formulas were EO_2 and EO_3, or EO (E_2O_2) and E_2O_3. For a $3:4$ oxygen ratio, the formulas were either EO_3 and EO_4 or E_2O_3 and EO_2 (E_2O_4). Finally, a $3:5$ oxygen ratio gave EO_3 and EO_5 or E_2O_3 and E_2O_5.

[3] Joseph Louis Gay-Lussac, "Memoir on the Combination of Gaseous Substances with Each Other," *Alembic Club Reprint*, No. 4. See also Maurice P. Crosland, "The Origins of Gay-Lussac's Law of Combining Volumes of Gases," *Annals of Science* 17 (1961), 1–26.

[4] Jöns Jacob Berzelius, "Essay on the Cause of Chemical Proportions, and on Some Circumstances Relating to Them: Together with a Short and Easy Method of Expressing Them," *Thomson's Annals of Philosophy* 2 (December 1813), 450.

[5] Ibid.

A second rule of combination enabled Berzelius to assign formulas to acidic and basic oxides and to the salts formed from their reaction. When an acidic oxide (acid) combined with a basic oxide (base), he assumed the number of oxygen atoms in the acid to be an integral multiple of the number of oxygen atoms in the base, and this number was usually equal to the number of oxygen atoms in the acid:

Acidic Oxide (Acid)	Basic Oxide (Base)	Ratio of Oxygen Atoms in Acid and Base	Number of Oxygen Atoms in Acidic Oxide
SO_3	ZnO	3:1	3
CrO_3	NaO	3:1	3
SO_3	FeO	3:1	3
SO_3	Fe_2O_3	3:3	3

The formulas above represent those Berzelius adopted in 1826. They include the corrections he made in order to bring his formulas into agreement with the law of Dulong and Petit (1819) and Mitscherlich's law of isomorphism (1819). Dulong and Petit's law showed that for many metals the product of its atomic weight and specific heat was nearly constant. It resulted in Berzelius's changing the formulas of the metallic oxides of zinc, copper, and lead from EO_2 to the correct formula EO by halving each metal's atomic weight. He also reduced the formulas of sodium and potassium oxide from EO_2 to EO, though they still gave atomic weights for the metals that were double the values deduced from their specific heats.

Using Mitscherlich's law of isomorphism, which predicted that compounds having the same crystalline form would be similar in chemical composition, and his laws of combination, Berzelius corrected formulas he had assigned previously to several other metallic oxides. According to his second rule of combination, chromic anhydride had the formula CrO_3, and, because the weight ratio of the oxygen atoms in the two known chromium oxides was 2:1, he gave the formula Cr_2O_3 to the second oxide. Since Cr_2O_3 was isomorphous with aluminum oxide and one of the two iron oxides, Berzelius changed his earlier formulas, AlO_3 and FeO_3 to Al_2O_3 and Fe_2O_3. The remaining iron oxide, which originally had the formula FeO_2 because its oxygen ratio compared with the second iron oxide was 2:3, became FeO (Fe_2O_2). And because FeO was isomorphous with the oxides of copper, cobalt, calcium, and lead, Berzelius gave all of them the general formula EO.

By 1840, other chemists, among them Gay-Lussac, Pierre Dulong

(1785–1838), Louis J. Thénard (1777–1857), Eilhardt Mitscherlich (1794–1863), and Antoine J. Balard (1802–1876), had carried out compositional analyses of many metallic compounds and those of the nonmetals nitrogen, phosphorus, sulfur, chlorine, and bromine. The empirical formulas they established, were, for the most part, in agreement with those in use today.

Berzelius and the Electrochemical Dualistic Theory

Dalton had never seriously dealt with the attractive force binding the atoms in a molecule. Berzelius, on the other hand, believed that the force was electrostatic, resulting from the attraction of positive and negative charges that resided in every atom. Indeed, his hypothesis of an electrical force was an idea as old as Dalton's atomic theory. Not only Berzelius, but Davy even earlier and Michael Faraday (1791–1867) later, agreed that the attractions were electrical.

In Davy's theory, the attraction or affinity resulted from opposite charges of electricity that neutral bodies acquired when they came into contact. Conversely, when placed in an electrolytic cell, the body's charged components were attracted to oppositely charged electrodes, released their charges, and returned to the neutral state. In his first Bakerian Lecture before the Royal Society in November 1806, Davy demonstrated the plausibility of his theory when he electrolytically decomposed solutions of acids, alkalis, and salts. The chemical changes Davy observed in the metallic electrodes of the electrolytic cell supported further the intimate relation he believed existed between electricity and chemical combination or decomposition.

Davy never accepted Dalton's theory of ultimate chemical atoms, and consequently he did not link his electrochemical theory to an atomic model. His contemporary Berzelius adopted the doctrine of chemical atoms, and by 1812 Berzelius had given Dalton's structureless atoms an elaborate electrical structure.

Berzelius imagined each atom to consist of one or more electric poles; the atom's positive and negative electricity resided in opposite parts of it like the poles of a magnet. The same atom, therefore, behaved in its reactions with other atoms as either a positively or negatively charged body. But unlike magnetic poles, the electric poles were not of equal strength. The electrolytic decomposition of inorganic salts showed that in metals the positive pole always predominated, as indicated by their invariable appearance at the negative electrode. In nonmetals the negative pole dominated.

Chemical attraction or affinity, according to Berzelius, resulted from the neutralization of opposite electric charges. Two atoms—one positive, the other negative—united to form a binary compound or a compound of the first degree. Berzelius assumed, however, that each kind of atom held a different quantity of electric charge. The quantity of positive charge increased in the order of copper, zinc, aluminum, and calcium, reaching a maximum at potassium, while the quantity of negative charge decreased in the order oxygen, sulfur, nitrogen, and chlorine to a minimum for silicon. Therefore, the union of two atoms often left the resulting binary compound with a residual charge that depended on the difference in strengths of the original charges. This residual charge enabled two binary compounds to unite again, forming a ternary compound or a compound of the second degree, and allowed two ternary compounds to form compounds of even higher degree.

Berzelius could then account for the formation of ternary salts such as $KO.SO_3$ from the binary basic and acid oxides KO^+ and SO_3^-; hydrates, $CuO.SO_3 + 5H_2O$, from the union of a slightly positive salt $CuO.SO_3^+$ with negatively charged water H_2O^-; and alums, $KO.SO_3 + AlO_3.24H_2O$ $[K_2SO_4.Al_2(SO_4)_3.24H_2O]$, from the combination of two positively charged salts with electronegative water.

Faraday's electrolytic studies of 1834 dealt a deadly blow to the theory of residual charges. They revealed that the quantity of electricity required for an element's neutralization and subsequent liberation in electrolysis did not depend at all on the element's degree of electropositive or electronegative character. Instead, equivalent concentrations of ions such as potassium and silver consumed exactly the same quantity of electricity despite the great difference in their electropositive character.

In a series of researches beginning in 1839, John F. Daniell (1790–1845) at King's College, London, presented additional evidence that disproved Berzelius's claim that binary compounds carried residual charges. Daniell showed that in the electrolysis of a salt solution the ions of the metal and the nonmetal, Cu^{++} and $SO_4^=$, rather than the two oxides, CuO^+ and So_3^-, carried the electric charge.[6]

[6]John F. Daniell, "On the Electrolysis of Secondary Compounds," *Phil. Trans.* 129 (1839), 97–112; idem, "Second Letter on the Electrolysis of Secondary Compounds," *Phil.*

Our discussion of Berzelius has now made clear that in his theory he so intimately connected atomism and electrochemical dualism that they had to stand or fall together. His ideas dominated chemical theory for nearly thirty years after their introduction in 1811, because they satisfactorily explained many of the reactions known to inorganic chemists. Indeed, Berzelius found Faraday's law of equivalents unacceptable because he refused to believe that the electric charge that separated a silver atom from an oxygen atom also separated a potassium atom from an oxygen atom, for the first was one of the weakest and the last, one of the strongest combinations known.[7] He rejected Avogadro's hypothesis because it required acknowledging the existence of elementary diatomic gaseous molecules, an acknowledgement incompatible with dualism, though we may wonder why Berzelius never used his argument that the same atom could be positive or negative to explain the formation.

The Rise of Organic Chemistry: The Nonelectrical Bond

In comparison with the considerable progress made in establishing compositional formulas of simple inorganic compounds during the first three decades of the nineteenth century, the analysis of organic compounds was in a rudimentary state. The two main classes of organic compounds, those of vegetable and of animal origin, were known to contain carbon, hydrogen, and sometimes oxygen, nitrogen, and sulfur, though chemists had little understanding of their exact composition. Berzelius had shown in a series of researches on organic acids, starches, and sugars in 1814 and 1815 that chemists could use the atomic theory and the laws of proportion to establish compositional formulas for organic compounds. But due to the much greater complexity of these compounds compared with those of inorganic chemistry, it became increasingly evident that chemists needed some organizing principle in addition to the laws of proportion. The "radicals" that Gay-Lussac introduced in 1815 served as the first simplifying principle in the analysis of organic compounds.

Trans. 130 (1840), 209–24; idem, *An Introduction to the Study of Chemical Philosophy*, pp. 533–35; John F. Daniell and William A. Miller, "Additional Researches on the Electrolysis of Secondary Compounds," *Phil. Trans.* 134 (1844), 1–20. See List of Abbreviations preceding chapter 1.

[7] The confusion existed because Berzelius was unaware of the difference between the quantity factor (the charge) and the intensity factor or voltage (potential).

Gay-Lussac had observed that the gas cyanogen, $(CN)_2$, formed a series of compounds in which the CN group remained intact. Hydrogen, potassium, and iodine reacted with it, giving, respectively, HCN, KCN, and ICN, thus indicating a similarity in the compositional formulas of the cyanides and chlorides: CN, cyanogen radical; Cl, chlorine atom; $(CN)_2$, cyanogen gas; Cl_2, chlorine gas; HCN, hydrocyanic acid; HCl, hydrochloric acid; KCN, potassium cyanide; KCl, potassium chloride; ICN, cyanogen iodide; ICl, iodine chloride.

Twelve years later Jean Baptiste Dumas (1800–1884) noticed the same constancy in behavior of the etherin group, C_2H_4. But the publication that finally established the direction of research in organic chemistry—the search for radicals—was Justus Liebig and Friedrich Wöhler's 1832 paper on benzaldehyde. Their paper demonstrated the simple relation among an aldehyde, an alcohol, an acid, an acyl halide, and a cyanide, showing the conversion of benzaldehyde into each of these compounds. Most important, they recognized that in the conversion the benzoyl radical, C_6H_5CO, remained intact. The relations were benzaldehyde (oil of bitter almonds) or benzoyl hydride, $C_6H_5CO.H$; benzoic acid (benzoyl hydroxide), $C_6H_5CO.OH$; benzoyl chloride, $C_6H_5CO.Cl$; benzoyl cyanide, $C_6H_5CO.CN$; and benzamide, $C_6H_5CO.NH_2$.

Indeed, the simplicity and constancy of these numerical relations in which one atom or atomic group replaced another eventually led to the idea of valence. However, it took more than twenty-five years before chemists realized that these relations were a property common to all atoms.

Following the discovery of the methyl, CH_3, and ethyl, C_2H_5, radicals, Liebig and Dumas in 1837 attempted to define explicitly an organic radical. They pointed out that while the radicals of inorganic chemistry were simple, those of organic chemistry were compound; otherwise, no difference existed. The laws of combination and the laws of reaction were the same in the two branches of chemistry. Their belief that the radicals of organic compounds were equivalent to the atoms of inorganic compounds allowed organic compounds to fit neatly into Berzelius's dualistic scheme, though without assuming an electrical force to hold them together.

The Unitary Theory of Chemical Composition

The same years that had seen the rise of Berzelius's dualistic theory and the radical theory now witnessed the introduction of the unitary theory of chemical composition. Instead of claiming that every compound consisted of distinct components, usually radicals, the unitary theory considered every compound to be a unit structure. Atoms could replace other atoms without seriously altering the properties of the original compound. The theory originated in Gay-Lussac's investigation of cyanogen in 1815, which showed that chlorine replaced hydrogen in HCN giving ClCN, and in a later study in 1828, in which he observed that chlorine replaced hydrogen in the bleaching of oils.

Liebig (1803–1873) and Wöhler (1800–1882) had also demonstrated similar substitutions when they converted benzaldehyde, $C_6H_5CO.H$, to benzoyl chloride, $C_6H_5CO.Cl$, and other derivatives of the benzoyl radical. But the idea of substitution as the basis of the unitary theory originated in France with Dumas in 1834 and Auguste Laurent (1808–1853) the next year. In fact, substitution of electropositive hydrogen by electronegative chlorine led to the eventual abandonment of Berzelius's dualistic theory, which required every compound to consist of oppositely charged components and therefore denied that an electronegative element could replace an electropositive element or vice versa. Electricity was no longer the cause of chemical reaction. Chemists still maintained that a certain affinity brought about reaction, but now the cause of this affinity remained unknown.

Summarizing the rules of substitution, Dumas in 1834 wrote: "When a substance containing hydrogen is submitted to the dehydrogenating action of chlorine, of bromine, of iodine, of oxygen, etc., for each atom of hydrogen which it loses it gains an atom of chlorine, of bromine, of iodine, or half an atom of oxygen."[8]

He cited as examples the conversion of acetic acid (CH_3COOH) to trichloroacetic acid (Cl_3COOH) and the conversion of ethyl alcohol to acetaldehyde and then to trichloroacetaldehyde (chloral):

$$C_2H_5OH \quad \rightarrow \quad CH_3CHO \quad \rightarrow \quad Cl_3CCHO .$$

ethyl alcohol   acetaldehyde   trichloroacetaldehyde

The unitary theory, while maintaining that chlorine could replace hydrogen in a radical, had given no reason why the substitution took place. In 1836 Laurent tried to answer this question with his theory of fundamental

[8] Jean Baptiste Dumas, *Traité de Chimie*, 5:99; idem, "Considerations générales sur la composition théorique des matières organiques," *Journal de Pharmacie* 20 (May 1834), 285; James R. Partington, *A Short History of Chemistry*, p. 241.

and derived nuclei. Organic radicals or fundamental nuclei, he said, had definite geometric arrangements and the arrangement, more than the kind of atom, determined the properties of a molecule. The radical C_8H_{12}, for example, had the shape of a rectangular prism consisting of a carbon atom at each corner and a single hydrogen atom located at the midpoints of the 12 edges. Each of the prism's two narrow sides had an additional position that other atoms such as chlorine, oxygen, or hydrogen could occupy without destroying the fundamental nucleus. Laurent indicated these additions to the fundamental nucleus accordingly:

$$C_8H_{12} + H_2 \quad \text{Hyperhydride}$$
$$C_8H_{12} + Cl_2 \quad \text{Hyperchloride}$$
$$C_8H_{12} + O \quad \text{Aldehyde}$$
$$C_8H_{12} + O_2 \quad \text{Acid.}$$

In the reverse process, if one removed a hydrogen atom from an edge of the prism, Laurent argued that the edge would collapse unless another atom replaced the hydrogen atom. The resulting nuclei he called *derived nuclei* and represented them:

$$C_8(H_{11}Cl) + H_2 \quad \text{or} \quad C_8(H_{11}Cl) + Cl_2 .$$

Here we see Laurent's explanation of why chlorine took hydrogen's place: the fundamental nucleus accepted a chlorine atom in order to retain its geometric form. Liebig and Berzelius ridiculed the theory, but Dumas, aided by his own studies on the chlorination of acetic acid, adopted Laurent's explanation and in 1839 proposed the theory of types: "In organic chemistry there exist certain types which persist even when in place of the hydrogen they contain an equal volume of chlorine, bromine, or iodine is introduced." [9]

The Theory of Types

In a little more than a decade, Dumas's proposal resulted in the appearance of four different types from which chemists hoped to derive and classify all organic compounds. The first of these was the ammonia type,

---

[9] Partington, *A Short History of Chemistry*, p. 251. In its extreme form, the theory of types seemed to suggest that any atom could replace any other atom without essentially altering the substance's properties. But, as held by Laurent, the type theory denied that any other atom could be substituted for carbon, or that chlorine, bromine, and hydrogen were equiv-

whose derivatives August von Hofmann (1818–1892) prepared in 1850 by the successive action of ethyl iodide on ammonia:

$$N \begin{cases} H \\ H \\ H \end{cases} \qquad N \begin{cases} C_2H_5 \\ H \\ H \end{cases} \qquad N \begin{cases} C_2H_5 \\ C_2H_5 \\ H \end{cases} \qquad N \begin{cases} C_2H_5 \\ C_2H_5 \\ C_2H_5 \end{cases}$$

Ammonia   Ethylamine   Diethylamine   Triethylamine .

Alexander Williamson (1824–1904) then developed Laurent's idea of 1846 that alcohol and ether were analogous to water, potassium hydroxide, and potassium oxide:

$$OHH \qquad OC_2H_5H \qquad OC_2H_5C_2H_5 \qquad OHK \qquad OKK .$$

In a series of researches conducted in the years 1850–1853, he formulated the water type:

$$\begin{matrix} H \\ O \\ H \end{matrix} \qquad \begin{matrix} C_2H_5 \\ O \\ H \end{matrix} \qquad \begin{matrix} C_2H_5 \\ O \\ C_2H_5 \end{matrix}$$

Water   Ethyl alcohol   Diethyl ether.

Finally, by 1856 Charles Gerhardt (1816–1856) added the hydrogen chloride type:

$$\begin{cases} H \\ Cl \end{cases} \qquad \begin{cases} C_2H_5 \\ Cl \end{cases} \qquad \begin{cases} C_7H_5O \\ Cl \end{cases}$$

Hydrogen   Ethyl   Benzoyl
chloride   chloride   chloride

and the hydrogen type:

$$\begin{cases} H \\ H \end{cases} \qquad \begin{cases} C_2H_5 \\ H \end{cases} \qquad \begin{cases} C_7H_5O \\ H \end{cases}$$

Hydrogen   Ethyl hydride   Benzaldehyde .
              (ethane)

Though, in retrospect, it is easy to see the idea of valence and structure in these formulas, chemists who believed in the type theory saw instead a

---

alent to oxygen, sulfur, selenium, and tellurium. On the other hand, Laurent recognized that within a group of related elements, such as the nitrogen, oxygen, and halogen groups, one member could replace another, producing little change in the principal properties of the compounds in which it entered.

way to classify or to deduce the constitution of more and more complex compounds by the replacement of certain atoms or groups in each type. Williamson recognized that the group CO was equivalent to two hydrogen atoms, and upon replacing them in a compound it held together two groups that would otherwise fall apart:

$$
\begin{array}{ll}
C_2H_5 & C_2H_5 \\
\quad CO & \quad H_2 \\
N & N \\
\text{Ethyl isocyanate} & \text{Ethylamine .}
\end{array}
$$

But his conclusion was only a more explicit statement of what Laurent's formulas and his own previous work on the constitution of ethers had already implied. Type formulas, for the most part, were empirical formulas useful in the study of double decomposition reactions. Indeed, according to Gerhardt, the types were incapable of isolation, existing only in the imagination.

Edward Frankland (1825–1899), in Manchester, believed that chemists could isolate the radicals of the type formulas and accepted their real existence. He had studied in Robert Bunsen's laboratory in Marburg, where the discovery of apparent radical-containing compounds such as cacodyl, $As(C_2H_3)_2$ [now written as $As_2(CH_3)_4$], and its derivatives had occurred. In fact, Frankland's continuing research on organometallic compounds in 1852 led him to conclude that atoms had a maximum combining power or valence.

The Idea of Valence and Structure

Frankland's 1852 publication "On a New Series of Organic Compounds Containing Metals" showed that though the arsenic atom combined with five equivalents of oxygen, giving AsO_5 [As_2O_5], cacodyl, $As(C_2H_3)_2$ [$As_2(CH_3)_4$], reacted with a maximum of three equivalents of oxygen, forming cacodylic acid, $As(C_2H_3)_2O_3$ [$(CH_3)_2AsO_2H$]. This compound resisted further attack by strong oxidizing agents. He observed a similar relation between AsO_3 [As_2O_3] and $As(C_2H_3)_2O$ [$As_2(CH_3)_3O$], cacodyl oxide, which again indicated that the methyl radical C_2H_3 [CH_3] in combination with arsenic caused a decrease in arsenic's combining power. From these reactions Frankland concluded that the upper and lower values of arsenic's combining power were five and three, respectively. Frankland had, therefore, restricted the number of atoms or radicals with which a specified atom could combine. In other words, he had enunciated a theory of valence.

Of course, Frankland did not limit his new theory of valence to the few organometallic compounds of arsenic, antimony, and tin that he had prepared. In the same 1852 publication, he pointed out that his theory accounted equally well for the formulas of many inorganic compounds:

When the formulae of inorganic compounds are considered, even a superficial observer is struck with the general symmetry of their construction; the compounds of nitrogen, phosphorous, antimony and arsenic especially exhibit the tendency of these elements to form compounds containing 3 or 5 equivs. of other elements, and it is in these proportions that their affinities are best satisfied; thus in the ternal groups we have NO_3, NH_3, PO_3, PCl_3, SbO_3, SbH_3, $SbCl_3$, AsH_3, $AsCl_3$, &c.; and in the five-atom group, NO_5, NH_4O, NH_4I, PO_5, PH_4I, &c. Without offering any hypothesis regarding the cause of this symmetrical grouping of atoms, it is sufficiently evident, from the examples just given, that such a tendency or law prevails, and that, no matter what the characters of the uniting atoms may be, the combining power of the attracting element, if I may be allowed the term, is always satisfied by the same number of these atoms.[10]

The theory of valence, as Frankland envisaged it, was not a theory that brought together types and radicals but one capable of going beyond them. By focusing attention on the atom, it would reveal the general symmetry of molecular construction: "For whilst it is evident that certain types of series of compounds exist, it is equally clear that the nature of the body derived from the original type is essentially dependent upon the electrochemical character of its single atoms, and not merely on the relative position of the atoms."[11] Frankland's 1852 paper did not deal with the structure of the rad-

[10]Edward Frankland, "On a New Series of Organic Compounds Containing Metals," *Phil. Trans.* 142 (1852), 440. Carl Wilhelm Wichelhaus (1842–1927), professor of technological chemistry, University of Berlin, introduced the term *Valenz*, or "valence," in 1868. The British prefer *valency*, but in the United States *valence* is used. Lothar Meyer (1830–95) in *Die modernen Theorien der Chemie und ihre Bedeutung für die chemische Statik* (1864) suggested the adjectives univalent, bivalent, trivalent, quadrivalent, etc. Richard E. Erlenmeyer proposed the hybrids monovalent, divalent, trivalent, and tetravalent in 1860. William Odling introduced the expressions monad, dyad, triad, tetrad. They were used in England for a time but never found general adoption (see Alexander Findlay, *A Hundred Years of Chemistry*, p. 36).

[11]Frankland, "On a New Series of Organic Compounds Containing Metals," p. 441.

icals found in organometallic compounds. Within a few years, other chemists began investigating these radicals trying to discover how the carbon atom held them together.

William Odling (1829–1921) in 1855 and Friedrich August Kekulé (1829–1896) in 1857 clarified the carbon atom's role. They extended Gerhardt's four types, HH, HCl, H_2O, and H_3N, to include the marsh gas or methane type H_4C and, in effect, proposed the tetravalence of the carbon atom. Previously, methane had belonged to the hydrogen type, ${H \atop H}\}$, where the methyl group CH_3 replaced a hydrogen atom.

In addition to an awareness of carbon's tetravalence, chemists had to take one more step to complete the transition from type theory to structural chemistry. This step was accepting the idea of self-linking carbon atoms that Kekulé in Ghent and Archibald Scott Couper (1831–1892) in Paris published nearly simultaneously in 1858. Their proposals, together with Frankland's work on organometallic compounds, constituted the beginning of classic valence theory.

A question that quickly arose upon applying the idea of valence to the study of molecular structure was whether an atom's valence remained invariable. In his first paper on valence (1852) and in subsequent publications, Frankland recognized that each kind of atom had a definite valence and therefore a specific number of bonds, though for some atoms the valence or number of bonds clearly varied. Nitrogen required three bonds for the formation of ammonia, NH_3, and five for ammonium chloride, NH_4Cl. Couper in 1858 proposed for the carbon atom valences of two in carbon monoxide and four in carbon dioxide. Several years later, in 1869, Christian W. Blomstrand (1826–1897) at the University of Lund pointed out the variable valence of chlorine in hydrochloric acid and in its oxyacids, and of sulfur in hydrogen sulfide, sulfur dioxide, and sulfuric acid.

Their view stood in direct contrast with that of Kekulé, who had concluded in 1864 from his examination of simple carbon compounds that an atom's valence was as invariable as its weight. Nitrogen, he said, was trivalent in both ammonia and ammonium chloride, the latter being simply a loose molecular compound of ammonia and hydrogen chloride, $NH_3.HCl$. Indeed, Kedulé's argument that ammonium chloride was a molecular compound long prevailed over Frankland's interpretation of pentavalent nitrogen in ammonium chloride.

Kekulé's belief in the constancy of valence was clearly in error; nevertheless, his determination to maintain the tetravalence of carbon led in the 1860s to the introduction of double and triple bonds in organic compounds. His structures, as well as those of Alexander Crum Brown (1838–1922) in Edinburgh and Richard E. Erlenmeyer (1825–1909) in Munich, represent the earliest application of multiple bonds to the compounds of organic chemistry.[12]

Isomers and Three-Dimensional Chemistry

Structural formulas soon became of great value in interpreting many cases of isomerism. Recognition that there were alternative planar arrangements of the atoms in a molecule enabled Wöhler to resolve the isomerism of ammonium cyanate and urea in 1828. But the isomerism of a second group of compounds—lactic, tartaric, aspartic, and malic acids—remained a mystery when chemists assumed only planar structures. Indeed, chemists had no explanation for the isomerism until 1863, when Johann Wislicenus (1835–1902) in Zurich discovered that lactic acid existed in two optically active forms and showed that structurally both were α-hydroxypropionic acid. Since tartaric, aspartic, and malic acids had two optically active yet structurally identical forms, Wislicenus concluded that for each acid the isomerism resulted from a difference in the spatial arrangement of its atoms.

Chemists also recognized at this time that molecules such as CH_2Cl_2 and CH_2Br_2 did not occur in the two isomeric forms predicted by a planar distribution of the carbon atom's four bonds. Working independently, Jacobus H. van't Hoff (1852–1911) in Holland and Joseph A. Le Bel (1847–1930) in Paris solved this problem in 1874 when they each argued that carbon's bonds were directed in space. The four bonds, they said, pointed to the corners of a tetrahedron, and the optical activity of many carbon compounds resulted from the spatial arrangement of different atoms around a central carbon atom.

Failure of the molecules CH_2Cl_2 and CH_2Br_2 to yield isomers thus became a direct consequence of the tetrahedral model. With van't Hoff's and Le Bel's introduction of the tetrahedral carbon atom, stereochemistry be-

---

[12] For further discussion on the introduction of double and triple bonds in organic compounds see Colin A. Russell, *The History of Valency*, pp. 224–37.

gan. It represented the first attempt to extend the theory of valence beyond its purely numerical character.

Remnants of Nineteenth-Century Electrochemical Dualism

In the period from 1840 to 1900 chemists established highly successful techniques for determining not only atomic or compositional formulas, but structural formulas also. Their successes, however, remained without any electrical explanation of either composition or structure, though there is some evidence that Berzelius's electrochemical ideas did not disappear entirely.

Blomstrand in 1869 tried to reconcile Berzelius's belief that atomic interactions depended on the neutralization of opposite electric polarities with the idea of valence as a direct and specific attraction between atoms.[13] A few years later Vladimir V. Markownikov (1838–1904) in Moscow, recognizing the positive and negative behavior of atoms, put forward his rule for the addition of hydrogen halides and other binary compounds to unsaturated hydrocarbons.[14]

Additional suggestions of negative groups in organic compounds appeared at that time. Victor Meyer (1848–1897) at Stuttgart in 1872 pointed out that nitromethane formed salts with alkalis and concluded that the nitro group was negative.[15] In 1889 Wilhelm Ostwald (1853–1932) found that for organic acids, substituting negative atoms or groups—for example, negative chlorine for positive hydrogen—increased the strength of acids such as acetic acid.[16] These same substituents that increased acid strength also increased reaction velocities.[17]

In 1895 van't Hoff used, perhaps for the first time, the idea of positive and negative ions to explain a reaction involving substances not normally considered to be electrolytes. He tried to show that ozone formation by the

---

[13] Christian W. Blomstrand, *Die Chemie der Jetztzeit*.

[14] Vladimir V. Markownikov, "Sur les lois qui régissent les réactions de l'addition directe," *Comptes rendus* 81 (1875), 670.

[15] Victor Meyer and Otto Stuber, "Über die Nitroverbindungen der Fettreihe," *Berichte* 5 (1872), 399–406.

[16] Wilhelm Ostwald, "Elektrochemische Studien. Das Verdünnungsgesetz," *J. prakt. Chemie* 31 (1885), 433–62; idem, "Über die Affinitätsgrössen organischer Säuren und ihre Beziehungen zur Zusammensetzung und Konstitution derselben," *Zeit. phys. Chemie* 3 (1889), 170–97, 241–88, 369–422.

[17] Wilhelm Ostwald, "Notiz über das elektrische Leitungsvermögen der Säuren," *J. prakt. Chemie* 30 (1884), 93–95.

action of oxygen on moist phosphorus required the presence of oxygen ions.[18] The following year, Julius Stieglitz at the University of Chicago, as a result of his studies on the Beckmann rearrangement and related reactions, proposed the existence of positively charged chloride and bromide ions combined with negatively charged nitrogen in amines, amides, and other compounds.[19]

Berzelius's conception of primary and residual charges clearly resembled early twentieth-century electrochemical ideas, namely the primary and secondary valences of Alfred Werner's coordination theory and Richard Abegg's valences and contravalences. There are also striking similarities between Berzelius's charged atoms and the speculations on intramolecular ions by J. J. Thomson, Walther Kossel, and William A. Noyes.

Werner's Coordination Theory

Kekulé's introduction of the carbon-carbon double bond and the use of triple bonds in the 1860s soon led to elucidation of many unsaturated organic compound structures. The so-called molecular compounds of inorganic chemistry—the hydrates and complex salts, the complex cobalt chlorides and nitrates, for example—presented a different type of unsaturation. In these compounds each of the atoms formed its maximum number of bonds, yet the compounds seemed to possess some residual affinity; otherwise, there was no way of explaining why they combined with other apparently saturated compounds such as ammonia and water. Werner's coordination theory, first put forward in 1891, attempted to show how the residual affinity accounted for the formation of molecular compounds.

According to Werner (1866–1919) in Zurich, chemical affinity, or the atom's property of attraction, radiated from the center of the atom and distributed itself evenly over the atom's surface. When subjected to the "affinity demands" of other atoms, however, an atom concentrated at definite points on its surface enough affinity for the production of one or more valence bonds but left unused a variable amount of affinity. In a coordination compound, Werner envisioned a central atom, usually a metal, having re-

---

[18]Jacobus H. van't Hoff, "Über die Menge und die Natur des Sogen. Ozons, das sich bei langsamer Oxydation des Phosphor bildet," *Ziet. phys. Chemie* 16 (1895), 411–16.

[19]Julius Stieglitz, "On the Beckmann Rearrangement. I. Chlorimidoesters," *Amer. Chem. Journal* 18 (1896), 751–61.

sidual affinity. Direct union with other atoms or ions, using its primary valences, only partially saturated the central atom's affinity; there remained
available a residual affinity for use in bond formation using the central
atom's secondary valences.

Each secondary valence, which Werner represented in his formulas by
a dotted line, enabled the central atom to form an additional bond with an
atom, a neutral molecule such as water or ammonia, or an ion. The last
type of bond required the presence of a second, oppositely charged ion in
order for the coordination compound to remain electrically neutral. In
Werner's structure for the ammonia molecule, the central nitrogen atom
used its three primary valences, represented by solid lines, to bind the
three hydrogen atoms. To add a fourth hydrogen, as in ammonium chloride, the nitrogen atom used its single secondary valence. The chloride ion
balanced electrostatically the positive charge of the ammonium ion.

The atoms, molecules, or ions attached directly to the central atom
were very firmly held. They were in the coordination or first sphere and
were not ionized. The maximum number of groups in this sphere, usually
four or six, was the coordination number. Werner initially employed both
primary and secondary valences in bond formation within the coordination
sphere. But as Nevil Sidgwick (1873–1952) pointed out, Werner later indicated that upon formation of the coordination sphere this distinction van-
ished, and all bonds attaching groups to the central atom were identical.[20]

Ammonium chloride, in which Werner in 1902 assigned nitrogen a valence of four,[21] had the structure:

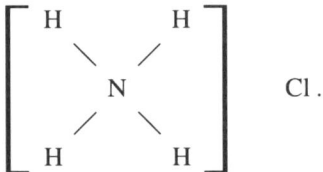

His structure revealed clearly the two different kinds of bonds in ammonium chloride: (1) the un-ionizable, holding the hydrogen atoms to the
nitrogen in the coordination group, NH_4, and (2) the ionizable, uniting the
coordination group and the secondary sphere containing the chloride ion.

[20] Nevil V. Sidgwick, *The Electronic Theory of Valency*, p. 110; Alfred Werner, *New
Ideas on Inorganic Chemistry*, pp. 67–70.
[21] Werner, *New Ideas on Inorganic Chemistry*, pp. 193–200.

We can compare it with the two nineteenth-century structures of ammonium chloride: (1) Kekulé's formula, which showed ammonium chloride to be a molecular compound, $NH_3 \cdot HCl$, and thereby permitted nitrogen to maintain a constant valence of three,[22] and (2) Frankland's structure, in which all five atoms were directly attached to the nitrogen atom, giving nitrogen a valence of five.

G. N. Lewis (1875–1946) at the University of California, Berkeley, was one of the first to recognize the essential correctness of Werner's structural ideas. Lewis's electron formula of ammonium chloride (1916) revived Werner's belief that nitrogen had a valence of four; it showed each hydrogen atom joined to the central nitrogen atom by a nonpolar electron pair bond. A simple electrostatic attraction then held the NH_4 and Cl groups together:

$$H : \overset{..}{\underset{H}{N}} : H + H^+ + : \overset{..}{\underset{..}{Cl}} :^- \rightarrow \left[H : \overset{\overset{H}{|}}{\underset{\underset{H}{|}}{N}} : H \right]^+ + : \overset{..}{\underset{..}{Cl}} :^-$$

The behavior of the cobalt chloride ammines also illustrated the correctness of Werner's coordination theory. In aqueous solution, $CoCl_3 \cdot 6NH_3$ ionized to give three chloride ions, each of which a silver nitrate solution precipitated. On the other hand, the addition of sulfuric acid to the solution did not result in precipitation of the ammonia molecules. Werner and coworker Arturo Miolati (1869–1956) summarized this behavior by the formula:

$$\left[\begin{array}{ccc} NH_3 & NH_3 & NH_3 \\ & \diagdown \; | \; \diagup & \\ & Co & \\ & \diagup \; | \; \diagdown & \\ NH_3 & NH_3 & NH_3 \end{array} \right] Cl_3 .$$

This showed six ammonia molecules in the coordination sphere attached to the central cobalt atom by secondary valences.[23] The three chlorine atoms were in the second sphere and, hence, ionized.

---

[22] A molecular compound was a compound supposedly formed by the union of two or more neutral molecules, in this case the ammonia, NH_3, and the hydrogen chloride, HCl, molecules.

[23] Alfred Werner and Arturo Miolati, "Beiträge zur Konstitution anorganischer Verbindungen," *Zeit. phys. Chemie* 14 (1894), 506.

Werner's formulas for six cobalt compounds are given below:

1. $[Co(NH_3)_6]Cl_3$ 4. $[Co(NH_3)_3(NO_2)_3]$
2. $[Co(NH_3)_5Cl]Cl_2$ 5. $[Co(NH_3)_2(NH_2)_4]K$
3. $[Co(NH_3)_4Cl_2]Cl$ 6. $[Co(NO_2)_6]K_3$.

In each case the cobalt atom displayed a maximum coordination number of six. The fourth compound was a nonelectrolyte because of its empty second sphere. In the first three compounds, the coordination group behaved as a cation; in the last two, as an anion.

When Werner first announced his coordination theory in 1891, it was to protest the Kekulé–van't Hoff conception of directed valence, which the organic chemists had developed. He believed that their ideas could not describe adequately the constitution of inorganic compounds, for those compounds required a broader theoretical foundation. He adopted instead the views of William Lossen (1838–1906) in Königsberg and of Adolf Claus (1840–1900) at Freiburg. Lossen maintained that the valence of an atom was simply a number indicating how many other atoms were present in its combining zone,[24] while Claus argued that valence was not a kind of preexisting force acting in definite units. Such a hypothesis, Claus said, was not only unfounded but also unnatural.[25]

Werner's coordination theory did not receive rapid acceptance. Indeed, it finally gained general support twenty years after its introduction, when Werner predicted and successfully verified by experiment the occurrence of optical activity in the complex *cis*-dichlorobis (ethylenediamine) cobalt (III) chloride. William Ramsay, Samuel Briggs (1880–1935), and John Newton Friend attempted to modify Werner's theory and at the same time develop electron theories of coordination compounds, but only Werner's ideas and terminology were of importance in establishing the modern electron theory of coordination compounds. This was due largely to the efforts of Sidgwick in the 1920s.

The Atomicity of Electricity

With his theory of electric poles and his table of electrochemical affinities, Berzelius by 1812 seemed to be groping toward the idea of a spe-

---

[24] Wilhelm Lossen, "Über die Verteilung der Atome in der Molekül," *Annalen der Chemie* 204 (1880), 327.

[25] Adolf Claus, "Zur Frage nach den Affinitätsgrössen des Kohlenstoffs," *Berichte* 14 (1881), 432.

cific valence and a means of determining the actual amounts of polarized charges possessed by atoms. He knew nothing, however, about the absolute magnitude of the charges or their ratios. This required a clear conception that electricity, like matter, was atomic, and Berzelius never made this deduction.

Faraday provided evidence for electricity's atomic nature in his electrolytic experiments of 1834. But owing to the utter confusion regarding atomic weights and the widespread use of equivalent weights, no one, including Faraday, recognized the whole-number relation existing between the amount of electricity required to neutralize a charged ion and its atomic weight. Faraday made that quite clear in a passage he wrote at that time:

The equivalent weights of bodies are simply those quantities of them which contain equal quantities of electricity . . . it being the ELECTRICITY which determines the equivalent number. . . . Or if we adopt the atomic theory or phraseology then the atoms of bodies . . . have equal quantities of electricity naturally associated with them. But I must confess I am jealous of atoms, it is very difficult to form a clear idea of their nature.[26]

Indeed, the literature of the period (1834–1871) is in sharp contrast with that of the next twenty-five years, which revealed that the atomicity of electricity was a rather popular idea even before the electron's discovery in 1897. Wilhelm Weber (1804–1891) at Göttingen in 1872 suggested the electrical phenomena were due to atoms of electricity.[27] Weber's search for the force that held together the atomic structure of bodies led him to propose the existence of both positive and negative atoms of electricity. He assumed that every ponderable atom had attached to it an electric atom, but did not see the relation between ionic charge and atomic weight implied in Faraday's experiments.

One year later, Clerk Maxwell (1831–1897), discussing electrolysis in *A Treatise on Electricity and Magnetism* (1873), considered the possibility

[26]Michael Faraday, "Experimental Researches in Electricity—Seventh Series," *Phil. Trans.* 124 (1834), 121.

[27]Wilhelm Weber, "Elektrodynamische Massbestimmungen inbesondere über das Prinzip der Erhaltung der Energie," *Leipzig Abhandlung Mathematische-Physikalische* 10 (1871), 1–62 (English translation in *Phil. Mag.* 43 [1872], 1–20, 119–49); idem, *Galvanismus und Elektrodynamik* in *Wilhelm Weber's Werke*, 4:281–85. For earlier discussions by Weber on positive and negative electrons see "Elektrodynamische Massbestimmungen," *Leipzig Abhandlung Jablonowskische Gesellschaft der Wissenschaften* 1 (1846), 209–278, and "Elektrodynamische Massbestimmungen," *Annalen der Physik* 73 (1848), 193–240 (English translation of the 1848 article in Richard Taylor, ed. *Scientific Memoirs*, 5:489–529).

of "molecules of electricity," though Maxwell was not convinced of the fruitfulness of this idea: "For convenience in description we may call this constant molecular charge revealed by Faraday's experiments one molecule of electricity . . . it is extremely improbable that when we come to understand the true nature of electrolysis we shall retain in any form the theory of molecular charges."[28]

The following year, 1874, the Irish physicist G. Johnstone Stoney (1826–1911) delivered a paper before the Belfast meeting of the British Association for the Advancement of Science in which he introduced a fundamental unit quantity of electricity called the "electrine." Stoney had determined this unit quantity from electrolysis (he obtained a value of e, calculated from F/N, equal to 0.3×10^{-10} esu) and found it to be independent of the particular body electrolyzed. He then related the unit quantity of electricity to Faraday's law in a manner that he said gave it precision: "*For each chemical bond which is ruptured within an electrolyte a certain quantity of electricity traverses the electrolyte which is the same in all cases.* This definite quantity of electricity I shall call E_1. If we make this our unit quantity of electricity, we shall probably have made a very important step in our study of molecular phenomena" (emphasis in original).[29]

Stoney had clearly recognized the relation between quantity of electricity and the number of bonds broken. If a current passed successively through a solution of hydrochloric and then sulfuric acid, he pointed out, two "atoms of HCl" will be decomposed for every "one of H_2SO_4," and "the number of bonds separated will be the same in both vessels."[30]

Hermann von Helmholtz's celebrated Faraday Lecture of 1881, "The Modern Development of Faraday's Conception of Electricity," provided the best known and most complete exposition of Faraday's 1834 experiments. Faraday's law, said Helmholtz (1821–1894), showed that "the same definite quantity of either positive or negative electricity moves al-

---

[28] James Clerk Maxwell, "Electrolysis," in *A Treatise on Electricity and Magnetism*, 1:380–81.

[29] G. Johnstone Stoney, "On the Physical Units of Nature," *Phil. Mag.* 11 (1881), 384. Stoney first read this paper before Section A of the Belfast meeting of the B.A.A.S. in August, 1874. He also read it before the Royal Dublin Society on February 16, 1881, and published it in the *Scientific Proceedings of the Royal Dublin Society* 3 (1883), 51–60. Stoney defended his claim to priority in discussing the atomicity of electricity in a letter to the *Philosophical Magazine*: "Of the 'Electron,' or Atom of Electricity," *Phil. Mag.* 38 (1894), 418–20.

[30] G. Johnstone Stoney, "On the Physical Units of Nature," p. 387.

ways with each univalent ion, or with each unit of affinity of a multivalent ion," and "this quantity we may call the electric charge of the atom." It follows, Helmholtz went on, "if we accept the hypothesis that the elementary substances are composed of atoms, we cannot avoid concluding that electricity also, positive as well as negative, is divided into definite elementary portions, which behave like atoms of electricity."[31]

Helmholtz's revival of Berzelius's dualistic theory once again called attention to the identity of the forces of chemical affinity and electricity. "I think the facts leave no doubt," he wrote, "that the very mightiest among the chemical forces are of electric origin. The atoms cling to their electric charges, and opposite electric charges cling to each other."[32] Even for the elementary diatomic gases, H_2, O_2, N_2, and the halogens, Helmholtz thought that the combination of the atoms was probably due to electric neutralization.

Helmholtz maintained further that in atoms if "every unit of affinity is charged with one equivalent either of positive or negative electricity, they can form compounds, being electrically neutral only if every unit charged positively unites under the influence of a mighty electric attraction with another unit charged negatively." This produced compounds having each affinity unit of every atom connected with one and only one affinity unit of another atom and was, Helmholtz declared, "the modern chemical theory of quantivalence."[33]

In calling attention to Faraday's earlier electrolytic studies, Helmholtz as well as Stoney had clearly proposed the electrical nature of chemical affinity. With their conclusion that electricity was atomic, Berzelius's qualitative electrochemical dualism now received a quantitative interpretation.

The Electrolytic Dissociation Theory of Arrhenius

At the same time that belief in the atomicity of electricity was gaining acceptance, Svante Arrhenius (1859–1927) in Uppsala was examining the conductivity of electrolytic solutions. Arrhenius attributed the behavior of these solutions to the presence of ions that resulted from the dissociation of many inorganic compounds when dissolving in water. The dissociation

[31] Hermann von Helmholtz, "The Modern Development of Faraday's Conception of Electricity," *J. Chem. Soc.* 38 (1881), 289–90.
[32] Ibid., p. 302.
[33] Ibid., p. 303.

process, he said, represented a gradual conversion of the compound's inactive, nonconducting molecules into electrolytically active and conducting ions. The degree to which a compound dissociated depended on two factors: (1) the kind of compound and (2) its concentration in solution. For both strong and weak electrolytes, the number of molecules dissociating increased with increasing dilution until the process reached equilibrium. Arrhenius demonstrated a compound's degree of dissociation by measuring its conductivity in aqueous solution at different concentrations.

Upon comparing the conductivities of acid and base solutions with the relative strengths of acids and bases determined thermodynamically by Marcelin Berthelot (1827–1907) in Paris and Julius Thomsen (1826–1909) in Copenhagen, Arrhenius found that those acids and bases having the greatest conductivity were also the most chemically active. Electrically active molecules, he concluded, were identical with chemically active molecules. Arrhenius had reduced chemical and electrical activity to the same cause, the presence of ions. Indeed, the theory of electrolytic dissociation that Arrhenius proposed in 1884 greatly extended the role played by oppositely charged atoms or ions in chemical theory.

It was now an easy matter for Arrhenius to relate his dissociation theory to Germain Hess's theorem of thermoneutrality. This theorem, which Hess (1802–1850) derived experimentally in 1841, showed the heat of neutralization to be identical in reactions of strong acids with strong bases.[34] Hess's theorem was valid, Arrhenius said, because strong acids, strong bases, and salts dissociated nearly completely into their ions:

$$HCl \;\; \rightarrow H^+ + Cl^-,$$
$$NaOH \rightarrow Na^+ + OH^-, \text{ and}$$
$$NaCl \;\; \rightarrow Na^+ + Cl^-.$$

As a result, the quantity of heat liberated in the neutralization process had to be a constant value for equivalent amounts of acid and base because the primary reaction was the formation of water from its ions. Neutralization did not depend on the nature of the acid and base, and, contrary to the belief of most chemists, a salt was no longer the chief product of neutralization:

[34] Germain H. Hess, "Thermochemische Untersuchungen," *Annalen der Physik* 52 (1841), 107.

$$H^+ + Cl^- + Na^+ + OH^- \rightarrow Na^+ Cl^- + HOH$$

or

$$H^+ + OH^- \rightarrow HOH \,.$$

Studies by François Marie Raoult (1830–1901) on the freezing points of solutions[35] in 1885 provided additional support for Arrhenius's dissociation theory. Arrhenius proved that he could calculate the "abnormal" freezing point depression produced by the dissociation of a known concentration of a strong electrolyte using the electrolyte's degree of dissociation. The latter value he calculated from Friedrich Kohlrausch's electrical conductivity measurements.[36]

By 1887 the time appeared ripe for a complete statement on the electrolytic dissociation theory. In volume one of the *Zeitschrift für physikalische Chemie* Arrhenius published the theory in its final form with the title "Über die Dissociation der in Wasser gelösten Stoffe."[37] Three years had passed from the time Arrhenius first introduced his theory in 1884 until he asserted boldly that active molecules dissociated into their ions, and replaced *activity* with the term *electrolytic dissociation*.[38]

The researches of Ostwald and Georg Bredig (1868–1944) at Leipzig verified much of Arrhenius's theory.[39] Using Ostwald's dilution law,[40] they showed that the equilibrium established between the ions and the undissociated molecules of weak acids and bases obeyed the same law of

[35] François-Marie Raoult, "Sur le point de congélation des dissolutions salines," *Annales de Chimie et de Physique* 4 (1885), 401–30.

[36] Friedrich W. Kohlrausch, "Das elektrische leitungsvermogen der Wasserigen Lösungen von Hydraten und Salzen der leichten Metalle, sowie von Kupfervitriol, Zinkvitriol, und Silbersalpeter," *Annalen der Physik* 6 (1879), 145–210.

[37] Svante Arrhenius, "Über die Dissociation der in Wasser gelösten Stoffe," *Zeit. phys. Chemie* 1 (1887), 631–48.

[38] Svante Arrhenius, "Investigations on the Galvanic Conductivity of Electrolytes: I. Determination of the Conductivity of Extremely Dilute Solutions by Means of the Depolariser" and "II. Chemical Theory of Electrolytes," *Bihang till Kongliga Svenska Vetenskaps-Akademiens Handlingar* 8, nos. 13 and 14 (1884).

[39] Georg Bredig, "Über die Affinitätsgrössen der Basen," *Zeit. phys. Chemie* 13 (1894), 289–326; Wilhelm Ostwald, "Über die Dissociationstheorie der Elektrolyte," *Zeit. phys. Chemie* 2 (1888), 270–83, and 3 (1889), 588–602; Ostwald, "Über die Affinitätsgrössen organischer Säuren," pp. 170–97, 241–88, 369–422.

[40] Ostwald, "Über die Affinitätsgrössen organischer Säuren," pp. 170–97, 241–88, 369–422.

equilibrium that held for gaseous dissociation. For solutions of strong electrolytes, on the other hand, Max Planck (1858–1947) in 1888 found that
the law of equilibrium failed completely. No equilibrium between ions and
undissociated molecules, as Arrhenius assumed, existed in these solutions.[41] Indeed, only in the twentieth century did the dissociation of strong
electrolytes receive a successful explanation. The X-ray structure deter-
minations of W. H. and W. L. Bragg, Peter Debye (1884–1966), and Paul
Scherrer (1890–1969) proved that in the solid state the crystal lattice of a
strong electrolyte already consisted of ions.[42] No chemical equilibrium between ions and undissolved molecules existed in those solutions.

The most common criticism of Arrhenius's dissociation theory did not
come from equilibrium studies on strong electrolytes, however. It centered
instead on the chemical and electrical constitution of the ions in solution.
Discussion on their constitution was in fact still taking place in practically
every textbook of general chemistry written in the first decade of the twentieth century, despite most chemists having by then accepted the Arrhenius
theory.[43]

Arrhenius's critics believed that inorganic or polar compounds (as well
as organic or nonpolar compounds) consisted of molecules that upon solution did not dissociate into ions or even atoms but remained in solution as
distinct molecules. They ignored or found unacceptable Arrhenius's postulates that soluble inorganic compounds separated into charged atoms or
ions capable of a stable existence when in solution and that each ion, because of its electric charge, was a species entirely different from the neutral atom. They argued that an atom such as sodium, which reacted ex-
plosively with water, or the poisonous chlorine atom could not possibly
exist in a salt solution as Arrhenius claimed.

Arrhenius's belief that ions existed in solutions of electrolytes was cor-

[41] Max Planck, "Das chemische Gleichgewicht in verdünnten Lösungen," *Annalen der Physik* 34 (1888), 147.

[42] William L. Bragg, "The Analysis of Crystals by the X-Ray Spectrometer," *Proc. Roy. Soc.* A 89 (1913), 468–89; idem, "The Arrangement of Atoms in Crystals," *Phil. Mag.* 40 (1920), 169–89; idem, "The Dimensions of Atoms and Molecules," *Nature* 107 (24 March 1921), 107; Peter Debye and Paul Scherrer, "Atombau," *Annalen der Physik* 19 (1918), 474–83. See also Max Born and Alfred Landé, "Über die Berechnung der Kompressibilität regulärer Kristalle aus der Gittertheorie," *Verh. deut. phys. Ges.* 20 (1918), 210–16; Max Born, "Über die elektrische Natur der Kohäsionskräfte fester Körper," *Verh. deut. phys. Ges.* 21 (1919), 533–38.

[43] Alexander Smith, *General Chemistry for Colleges*, pp. 219–22; John T. Stoddard, *Introduction to General Chemistry*, pp. 160–62.

rect, of course. But he could not account for the ions' stability, and, like Berzelius before him, he left the problem of explaining the relation between the electric charge and the atom carrying the charge unsolved. Indeed, the confusion over the difference between an atom's chemical behavior and that of its ion pointed clearly toward the need for an electron theory of the atom.

J. J. Thomson and the Discovery of the Electron

Though electrolytic and ionization studies had provided sufficient evidence for the existence of electrically charged atoms or ions, they had not demonstrated that the charge could exist independently of the atom. This was the accomplishment of Joseph J. Thomson (1856–1940) at the Cavendish Laboratory, Cambridge, in 1897. Thomson carried to completion the investigations of William Crookes (1832–1919), Julius Plücker (1801–1868), and J. W. Hittorf (1824–1914) on the electrical conductivity of rarefied gases and obtained the first measurements of the charge-to-mass ratio—the e/m ratio—of the fundamental electric charge or electron. He also showed that the electron's properties were independent of the cathode's composition and of the gas present in the discharge tube.[44] Thomson's evidence, later strengthened by the electron's identification with the β ray emanating from radioactive atoms, led to the establishment of the electron as a fundamental constituent of all matter.

The following year (1898) Thomson determined the charge on the electron,[45] and then in 1899 he showed quite conclusively that this charge was the same as the charge on each hydrogen ion produced in the electrolytic decomposition of water.[46] His 1899 investigation took on utmost importance, for it established that the charge carried by an ion in solution was either equal to, or an integral multiple of, the fundamental electron charge. Rather quickly, the theory of electrolytic decomposition, Faraday's electrochemical or chemical equivalents, and the electrolytic dissociation theory acquired an electron interpretation.

Thomson's work in the late 1890s resulted in a convergence of near-

---

[44] J. J. Thomson, "Cathode Rays," *Phil. Mag.* 44 (1897), 293–316.

[45] J. J. Thomson, "On the Charge of Electricity Carried by the Ions Produced by Röntgen Rays," *Phil. Mag.* 46 (1898), 528–45.

[46] J. J. Thomson, "On the Masses of the Ions in Gases at Low Pressures," *Phil. Mag.* 48 (1899), 547–57.

ly 100 years of research on the relation between electricity and matter by both physicists and chemists. This convergence or synthesis became the stimulus that gave rise to the twentieth-century electron conceptions of valence.

Conclusion

The nineteenth century began and ended with the attention of chemists fixed on the electrical constitution of matter. In the intervening years, from about 1830 to 1880, after Berzelius's electrochemical theory failed in organic chemistry, chemists made good progress in establishing the structures of molecules without any electrical theory. Only in the century's last two decades, after Helmholtz, Stoney, and others recognized electricity's atomic structure and Arrhenius published his electrolytic dissociation theory, did chemists turn their attention once again to the electrical constitution of matter. J. J. Thomson's discovery of the electron in 1897 and its rapid establishment as a fundamental particle of all matter then followed, leaving no doubt of matter's electrical constitution.

We shall see that with the electron, chemists accounted successfully for the bond in inorganic or polar compounds. As in Berzelius's theory, the bond was electrostatic. But just as Berzelius's theory could not explain the formation of organic or nonpolar compounds, so were the first electronic theories of valence unsuccessful when applied to organic compounds, for they again assumed an electrostatic bond. Clearly something other than a simple electrostatic attraction was responsible for the bond in these compounds, though the difficulty remained unresolved until the publication of the Lewis valence theory in 1916.

2. The Electron and the First Atomic Models: The Contribution of J. J. Thomson

Introduction

J. J. Thomson's publications in the early twentieth century marked the first serious effort to incorporate the electron into a systematically structured atom. Two major problems confronted Thomson in his work: (1) how to account for the mass of an atom when the electron, the only known subatomic particle, had a mass 1/1000 that of hydrogen, the lightest atom, and (2) how to account for the neutrality of an atom when the only known subatomic particle had a charge of negative one. These were the conditions that any atomic model had to satisfy regardless of whatever else its author intended to accomplish.

Thomson's atom almost alone attempted to deal with the union of atoms at a time when the chief concern of physicists was atomic spectra. Chemical union, according to Thomson, was in every case an electrostatic attraction between oppositely charged atoms that resulted from the complete transfer of an electron from one atom to another. This idea dominated electron theories of valence for almost twenty years before alternative theories based on incomplete electron transfer or electron sharing began to appear.

The Number and Arrangement of Electrons in an Atom

J. J. Thomson's 1897 publication on cathode rays introduced the electron to chemistry. In this and other publications in 1898 and 1899, he had not yet developed an electron structure of the atom,[1] though the experi-

[1] J. J. Thomson, "Cathode Rays," *Phil. Mag.* 44 (1897), 293–316; idem, "On the Charge of Electricity Carried by the Ions Produced by Röntgen Rays," *Phil. Mag.* 46 (1898), 528–45; idem, "On the Masses of the Ions in Gases at Low Pressures," *Phil. Mag.* 48 (1899), 547–57.

mental evidence available strongly suggested the presence of a large number of electrons in every atom. Thomson had shown that the e/m ratio for the electron was large compared with that of the hydrogen ion. When combined with the knowledge that the electron's charge was equal in value to the charge carried by the hydrogen ion in electrolysis, his investigation indicated that the electron's mass was about 1/1000 that of the hydrogen atom. Since an atom, according to Thomson, contained only the nearly weightless ether and a neutralizing amount of nearly weightless positive electricity, there had to be a large number of electrons in every atom. Hydrogen, the lightest of all known atoms, supposedly contained about 1,000 electrons. In the heavier atoms the number of electrons was, of course, larger, probably the same multiple of 1,000 as each atom's mass was of the hydrogen atom's mass.

Even earlier, the numerous bright lines appearing in an atom's emission spectrum had suggested a large number of electrons in an atom. This idea has its origin in Maxwell's electromagnetic theory, which predicted that a rapidly oscillating electric charge would emit light waves. Only after Heinrich Hertz (1857–1894) at Bonn experimentally demonstrated the correctness of Maxwell's prediction in his researches of 1887 and 1888 was there general acceptance that spectral lines resulted from the oscillation of charged particles associated with, or within, the atom.[2] G. Johnstone Stoney's statement in 1891 indicates the acceptance shown the electromagnetic theory of atomic spectra: "Finally, in 1891 . . . I called attention to the fact . . . that the motions going on within each molecule or chemical atom cause these electrons to be waved about in the luminiferous aether, and that in this constrained motion of the electrons the distinctive spectrum of each kind of gas seems to originate; since lines in the spectrum will be furnished by each term of the Fourier's series which represents the special motion of each electron."[3]

[2] Heinrich Hertz, "Über sehr schnelle elektrische Schwingungen" and "Nachtrag zu der Abhandlung über sehr schnelle elektrische Schwingungen," *Annalen der Physik* 31 (1887), 421–88, 543–44; idem, "Über Inductionserscheinungen hervorgerufen durch die elektrischen Vorgänge in Isolatoren," "Über die Ausbreitungsgeschwindigkeit der elektrodynamischen Wirkungen," and "Über elektrodynamische Wellen in Luftraume und deren Reflexion," all in *Annalen der Physik* 34 (1888), 273–85, 551–69, 609–23; idem, "Die Kräfte elektrischer Schwingungen behandelt nach der Maxwell'schen Theorie," *Annalen der Physik* 36 (1889), 1–22. All of these papers appear in English translation in idem, *Electric Waves*.

[3] G. Johnstone Stoney, "Of the 'Electron,' or Atom of Electricity," *Phil. Mag.* 38 (1894), 418. Stoney made the same suggestion in an earlier article, "On the Cause of Double Lines

The electromagnetic theory also predicted that an external magnetic field would affect the charged particles' oscillations or motions and, therefore, the pattern of spectral lines they emitted. Pieter Zeeman (1865–1943), a Dutch physicist, attempted to demonstrate this effect using incandescent sodium vapor, and in 1896 he found that a considerable number of the lines did indeed experience a shifting or broadening in the magnetic field.[4] Zeeman's efforts to explain his experimental results led him to consult his colleague Hendrick A. Lorentz (1853–1928) at Leiden, who believed the spectral lines to be the result of small charged particles revolving in orbits within the atom. An expansion or contraction of these orbits would certainly follow upon applying an external magnetic field, and the change in shape of the orbits, Lorentz said, would produce a change in the wavelengths of the emitted spectral lines—what Zeeman had called a shifting or broadening of the lines.

Lorentz and Zeeman (together with Stoney) accepted the existence of a large number of charged particles within the atom. They also pointed out that the e/m ratio for these particles was the same as the e/m ratio Thomson found for the electron. Zeeman then completed the identification of these particles with Thomson's electrons. He carefully analyzed the direction of the applied magnetic field and of the spectral line shifts and showed that the charge on the light-emitting particles was negative. After learning of Lorentz and Zeeman's results, Thomson wrote:

A reason for believing that there are many more corpuscles [electrons] in the atom than the one or two that can be torn off, is afforded by the Zeeman effect. . . . Now, if there were only one or two of these corpuscles in the atom, we should expect that only one or two lines in the spectrum would show the Zeeman effects. . . . As, however, there are a considerable number of lines in the spectrum which show Zeeman effects . . . we conclude that there are a considerable number of corpuscles in the atom of the substance giving this spectrum.[5]

---

and of Equidistant Satellites in the Spectra of Gases," *Scientific Proceedings of the Royal Dublin Society* 4 (1891), 583. In this article, Stoney introduced the word *electron* to represent each of the negatively charged particles believed to be oscillating within the atom before Thomson had actually isolated the electron.

[4] Pieter Zeeman, "On the Influence of Magnetism on the Nature of the Light Emitted by a Substance," *Phil. Mag.* 43 (1897), 226–39; idem, "Doublets and Triplets in the Spectrum Produced by External Magnetic Forces," *Phil. Mag.* 44 (1897), 55–60, 255–59; idem, "The Effect of Magnetisation on the Nature of Light Emitted by a Substance," *Nature* 55 (11 February 1897), 374.

[5] J. J. Thomson, "On the Masses of the Ions in Gases at Low Pressures," *Phil. Mag.* 48 (1899), 567.

Before 1906, when Thomson showed by three independent methods that the number of electrons in an atom was more nearly equal to the atom's mass, no one questioned the presence of a large number of electrons in an atom. Joseph Larmor (1857–1942) declared in 1897:

Thus Zeeman concludes that the effective mass of a revolving ion, supposed to have the full unitary charge [of the electron] is about 10^{-3} of the mass of the atom. This is about the same as Professor J. J. Thomson's estimates of the masses of the electricity carried in the Cathode Ray. If we took these carriers to be simple electrons, as their constancy under various environments tends to indicate, there would thus be about 10^3 electrons in the molecule.[6]

Similar statements by James Jeans (1877–1946), Lord Kelvin, and Oliver Lodge indicate clearly how familiar Thomson's original hypothesis had become in the decade following the discovery of the electron. Jeans maintained in 1901 that there were roughly 700 electrons in the hydrogen atom and for the other elements a number in proportion to its atomic weight. Kelvin believed that atoms consisted of thousands or millions of "electrions."[7] Lodge suggested the presence of 1,000 electrons in the hydrogen atom, 20,000 or 30,000 in the sodium atom, and 100,000 in the mercury atom.[8]

With the acceptance of a multielectron atom went the problem of finding the arrangement of the electrons in an atom. In 1897 Thomson had very little beyond the evidence suggesting that all atoms apparently contained a large number of identical electrons. For the sake of simplicity, he was content, therefore, to speculate using relatively small numbers of electrons, neglecting for the time the difficulty of accounting for the atom's mass with electrons only. Indeed, in Thomson's initial thoughts on atomic structure, he appealed to several floating-magnet experiments that Alfred M. Mayer, an American physicist at Stevens Institute of Technology, had carried out in 1878.[9]

[6] Joseph Larmor, "On the Theory of the Magnetic Influence on Spectra; and on the Radiation from Moving Ions," *Phil. Mag.* 44 (1897), 503.

[7] James Jeans, "The Mechanism of Radiation," *Phil. Mag.* 2 (1901), 427; Lord Kelvin (William Thomson), "Aepinus Atomized," *Phil. Mag.* 3 (1902), 257–83. Kelvin's term *electrion* came from Faraday's *ion* and Stoney's *electron.* It denoted an atom of negative electricity. He introduced the name *electrion* in the article "Contact Electricity and Electrolysis According to Father Boscovich," *Nature* 56 (27 May 1897), 84.

[8] Oliver Lodge, *Modern Views on Matter.* This was the Romanes Lecture for 1903. See also idem, *Electrons.*

[9] Alfred M. Mayer, "A Note on Experiments with Floating Magnets," *American Journal*

Mayer (1836–1897) had suspended vertically over a container of water a long bar magnet. In the water he then placed a number of needlelike magnets with the same pole up, each stuck through a cork in order to make it float. Both the suspended magnet and the floating magnets were long enough so that only the pole of each nearer the water's surface was active. Mayer assumed the active pole of the suspended magnet to be positive, those of the floating magnets, negative.

When the number of floating magnets did not exceed five, Mayer found that they arranged themselves at the corners of a single polygon or ring: five, at the corners of a pentagon; four, at the corners of a square; three, at the corners of a triangle. Alternatively, when the number exceeded five, the magnets formed concentric polygons or rings. Six magnets did not arrange themselves at the corners of a hexagon, rather they split into two groups consisting of one in the middle surrounded by five at the corners of a pentagon. As the number of floating magnets increased, Mayer observed a splitting into three and even four concentric rings of magnets. A few illustrations of Mayer's arrangements follow:

1	2	3	4	5
1,5	2,6	3,7	4,8	5,9
1,6	2,7	3,8	4,9	
1,5,9	2,7,10	3,7,10	4,8,12	5,9,12
1,6,9	2,8,10	3,7,11	4,9,13	5,9,13
1,6,10		3,8,11		
1,6,11		3,8,12		
		3,8,13		
1,5,9,12	2,7,10,15	3,7,12,13	4,9,13,14	
1,6,10,13	2,7,12,14	3,7,12,14	4,9,14,15	
		3,7,13,14		
		3,7,13,15		

In this example, 1,6,10,13 indicates one magnet in the middle, then successive rings of 6, 10, and 13 magnets.

The important feature of Mayer's experiments, according to Thomson,

of Science 15 (1878), 276–77; idem, "Floating Magnets," *Nature* 17 (18 April 1878), 487–88; idem, "Note on Floating Magnets," *American Journal of Science* 15 (1878), 477–78; idem, "Floating Magnets," *Nature* 18 (4 July 1878), 258–60.

was that successive rings of magnets formed naturally, and these rings might offer an explanation of the properties of atoms and the periodic law:

If we regard the system of magnets as a model of the atom, the number of magnets being proportional to the atomic weight . . . any property conferred by three magnets forming a system by themselves, would occur with atomic weights 3, 10, and 11; 20, 21, 22, 23, and 24; 35, 36, 37, and 39 [*sic*]; in fact, we should have something quite analogous to the periodic law, the first series corresponding to the arrangements of the magnets in a single group, the second series to the arrangements in two groups, the third series in three groups, and so on.[10]

Thomson's ideas regarding the nature and distribution of the positive charge in an atom were, on the other hand, far less definite. They appear in the description of the atom he gave in 1899:

I regard the atom as containing a large number of smaller bodies which I will call corpuscles, these corpuscles are equal to each other. . . . In the normal atom, this assemblage of corpuscles forms a system which is electrically neutral. Though the individual corpuscles behave like negative ions, yet when they are assembled in a neutral atom the negative effect is balanced by something which causes the space through which the corpuscles are spread to act as if it had a charge of positive electricity equal in amount to the sum of the negative charges of the corpuscles. . . . The detached corpuscles behave like negative ions, each carrying a constant negative charge which we shall call for brevity the unit charge; while the part of the atom left behind behaves like a positive ion with the unit positive charge and a mass large compared with that of the negative ion.[11]

Thomson's Early Attempt to Account for the Chemical Union of Atoms

Thomson's early publications on the electron mention only briefly the role the electron played in the union of atoms. In Thomson's atom a Faraday tube of electrostatic force held each electron in the atom, but whenever an electron with its Faraday tube migrated from one atom and attached itself to another, compound formation resulted between the two atoms. In the formation of hydrogen chloride from its elements, for example, the hydrogen atom lost an electron to chlorine and acquired a unit positive

---

[10] J. J. Thomson, "Cathode Rays," *Phil. Mag.* 44 (1897), 314. The sum of the magnets 3, 7, 13, 15, which Thomson gave as 39, should have been 38.

[11] Thomson, "On the Masses of the Ions," p. 564.

charge; chlorine by gaining the electron now had a unit negative charge. The electrostatic attractive force of the oppositely charged atoms held the binary compound together: H^+Cl^-.

Faraday's laws of electrolysis, as Helmholtz and Stoney had already pointed out, provided an experimental method of relating an atom's electrical charge to its chemical valence. They showed that the unit charge assigned to the hydrogen and the chlorine atom in hydrogen chloride agreed with the quantity of electricity—1 Faraday—required to liberate one gram-atomic weight of each element from an aqueous solution of hydrogen chloride. Electrolysis also established the sign of each charge, H^+ and Cl^-. In both magnitude and sign, each charge was in agreement, therefore, with the univalence of the hydrogen and chlorine atom determined by chemical analysis.

At the same time that Thomson proposed his theoretical interpretation of the electrostatic bond, Wilhelm Wien (1864–1928) in Aachen, Germany, was analyzing the behavior of positively charged gaseous ions in discharge tubes. Wien found for each kind of positive ion an e/m ratio, which suggested it had a mass comparable with that of an ordinary atom. Indeed, the maximum e/m ratio calculated for a positive ion using the smallest known m value was the e/m ratio for the hydrogen ion; and this still gave a mass ratio for the hydrogen ion and the electron of 1000 to 1. Wien's analyses revealed that no positive ion had a mass as small as that of the negatively charged electrons present in gases at low pressures.[12] They showed further that the positive charge did not exist independently of matter or in the "free state." His study supported Thomson's belief that a positive ion, unlike the negative electron, was not readily transferred from one atom to another and thus was not involved in the formation of an electrostatic bond.

That an electrostatic bond involved only the transfer of a negative electron was by far the most widely held electrical theory of valence. Oliver Lodge, John Newton Friend, Samuel Briggs, and William Ramsay in England and William A. Noyes and G. N. Lewis in the United States developed valence theories based on this hypothesis. Except for the theories of the German chemists Walther Nernst, Richard Abegg, and Guido Bod-

---

[12] Wilhelm Wien, "Untersuchungen über die elektrische Entladung in verdünnten Gasen," *Annalen der Physik* 65 (1898), 440–52; Thomson, "On the Masses of the Ions," p. 564.

länder,[13] every electron theory of valence that has appeared since 1897 has used only the negative electron to account for chemical union.

Lodge's 1902 discourse to the Institution of Electrical Engineers[14] and his 1903 Romanes Lecture, *Modern Views on Matter*, were expositions of Thomson's electrostatic valence theory. In each publication Lodge (1851–1940) discussed the atom's structure and how the atom entered into chemical union: "It becomes a reasonable hypothesis to surmise that the whole of the atom may be built up of positive and negative electrons interleaved together, and of nothing else; an active or charged ion having one negative electron in excess or defect, but the neutral atom having an exact number of pairs. The oppositely charged electrons are to be thought of on this hypothesis as flying about inside the atom, as a few thousand specks."[15]

His inference that positive electrons were present in an atom was not a suggestion that positive electrons took part in bond formation. A neutral atom required some kind of positive charge to balance the electron's negative charge, and Lodge was fulfilling this requirement with a positive electron. Indeed, at that time the chief defect in the electrical theory of matter was its inability to account for the nature and distribution of an atom's positive electricity. "The positive electron, if it exists," Lodge remarked, "has never yet been isolated from the rest of an atom of matter . . . has never been found detached from a mass less than the hydrogen atom; whereas the negative electron is constantly and freely encountered flying about alone, its mass being little more than the thousandth part of an atom of hydrogen."[16]

Electrostatic attraction, according to Lodge, resulted then from either attaching a negative electron to an atom or detaching it, giving, respectively, negatively and positively charged ions. Each ion now had a center of force with which it attached itself to an oppositely charged ion and entered into chemical combination with it. When such pairing took place, the excess charge of one ion compensated for the other's deficiency, forming a neutral molecule.[17]

---

[13] Their theories required the transfer of positive as well as negative electrons, though Abegg and Bodländer unlike Nernst held this view for a short time.

[14] This discourse was published in 1906 as Lodge's *Electrons*.

[15] Lodge, *Modern Views on Matter*, p. 11.

[16] Ibid., p. 12.

[17] Ibid., p. 6; Lodge, *Electrons*, p. 52.

Lodge's suggestion that positive and negative charges were interleaved or coupled in the interior of an atom (together with Thomson's positive sphere atom) was a fairly common hypothesis in the period preceding Ernest Rutherford's introduction of the nuclear atom in 1911. Both G. N. Lewis's cubic atom of 1902 and Philipp Lenard's dynamid theory of 1903 employed a similar conception of internal structure.[18]

Lenard (1862–1947), professor of physics at Kiel, arrived at his theory after investigating the behavior of cathode rays. He found that swiftly moving cathode rays penetrated a thin aluminum window; they passed freely through a layer of several thousand atoms. His demonstration led him to conclude that the greater part of an atom must be empty space; it was, in fact, the first demonstration of an atom's almost total emptiness.

Lenard also noticed a small amount of cathode ray absorption, which he found to be proportional to the amount of matter traversed. To account for the absorption, Lenard proposed that every atom contained tiny impenetrable centers of force called dynamids, of radius 3×10^{-12} cm—a value not differing greatly from the presently accepted radius of the nucleus, 10^{-13} cm. Each dynamid consisted of a coupled positive and negative electron and the number of dynamids contained in an atom was proportional to the atom's weight. Lenard thought the hydrogen atom probably contained one dynamid, but he never expanded his theory to include spatial arrangements of the dynamids for atoms of higher atomic weight, nor did he attempt to use it to develop an electron interpretation of chemical union.

In the same year that Lenard put forward his dynamid theory, Johannes Stark (1874–1957), *Privatdozent* at Göttingen, suggested that chemical union resulted from a sharing of electrons by the atoms in a molecule. Instead of an electrostatic attraction resulting from the complete transfer of an electron and its Faraday tube of force, Stark believed that many lines of force emanated from an electron to each of the bonded atoms. In a simple binary compound, each electron attached itself with a larger number of

---

[18] Philipp Lenard, "Über die Absorption von Kathodenstrahlen verschiedener Geschwindigkeit," *Annalen der Physik* 12 (1903), 714–44; Edward N. da C. Andrade (*The Structure of the Atom*, p. 4, and "The Birth of the Nuclear Atom," *Scientific American* 195 [November 1956], 93–104), also discussed Lenard's dynamid theory. Lewis had not published his cubic atom in 1902, but it was well known to his colleagues. His atom will be examined in chapter 7. Hantaro Nagaoka published his theory of the saturnian atom in "Kinetics of a System of Particles Illustrating the Line and Band Spectrum and the Phenomena of Radioactivity," *Phil. Mag.* 7 (1904), 445–55 (read before the Physico-Mathematical Society, Tokyo, December 5, 1903).

lines of force to the more negative atom, and with a smaller number, to the more positive atom.[19]

Stark continued to regard the electron as the "bond of union" in his later publications.[20] However, he was not sure how many electrons constituted the bond. Sometimes it was one; at other times, two. Nevertheless, his theory of the chemical bond—an electron or electrons held in common by two atoms—was probably the earliest application of electron sharing, introduced at a time when purely electrostatic theories of chemical union were dominant.

Thomson's Positive Sphere Atomic Model and Its Relation to Chemical Periodicity

In the years after 1897, the negative electron clearly had begun to find a place in theories of chemical union. Thomson, however, made no further attempt to develop the electron's role in chemical union after his brief discussions on electrostatic attraction in 1897 and 1899. Then in 1902 an opportunity arose when Yale University asked Thomson to deliver the Silliman Lectures. He gave the lectures in May 1903, choosing as his themes the constitution of matter and the nature of electricity. They were published the following year in a volume entitled *Electricity and Matter*. Thomson treated these same themes in the article "On the Structure of the Atom," which appeared in the March 1904 issue of *Philosophical Magazine*.[21]

In these publications, Thomson adopted and greatly developed mathematically Lord Kelvin's positive sphere model of the atom. Kelvin's atom consisted of negative electrons moving in a sphere of positive electricity. One electron was in stable equilibrium at the center of the sphere while higher numbers of electrons achieved stability by taking on such geometrical patterns as an equilateral triangle, a tetrahedron, an octahedron, and a cube. Mayer's concentric rings of floating magnets, Thomson pointed out, seemed to offer a visual demonstration of how the electrons might arrange themselves in such an atom.

[19] Johannes Stark, *Dissoziierung und Umwandlung chemischer Atome*, pp. 3–8.

[20] Chapter 6 of this volume contains a detailed examination of Stark's electron theory of valence.

[21] J. J. Thomson, *Electricity and Matter*; idem, "On the Structure of the Atom: An Investigation of the Stability and Periods of Oscillation of a Number of Corpuscles Arranged at Equal Intervals around the Circumference of a Circle; with Application of the Results to the Theory of Atomic Structure," *Phil. Mag.* 7 (1904), 237–65.

Kelvin (1824–1907) had described his atom in a 1901 paper "Aepinus Atomized," but he did not think that merely assuming a different number of electrons in an atom accounted for the different properties of chemical atoms. In fact, Kelvin was an avowed opponent of the electrical theory of matter applied to chemistry and believed in a Boscovichian law of force as the ultimate explanation of an atom's behavior:

We might be tempted to assume that all chemical action is electric, and that all varieties of chemical substance are to be explained by the numbers of electrions required to neutralize an atom or a set of atoms; but we can feel no satisfaction in this idea when we consider the great and wild variety of quality and affinities manifested by the different substances or the different 'chemical elements;' and as we are assuming the electrions to be all alike, we must fall back on Father Boscovich, and require him to explain the difference of quality of different chemical substances by different laws of force between the different atoms.[22]

Thomson, on the other hand, was optimistic that Kelvin's atom could explain theoretically the behavior of chemical atoms. He saw in it a possible solution to a problem he had considered earlier, namely, the arrangement of the elements in the periodic table: "The properties conferred on the atom by this ring structure are analogous in many respects to those possessed by the atoms of the chemical elements, and . . . in particular the properties of the atom will depend upon its atomic weight in a way very analogous to that expressed by the periodic law."[23]

Thomson's mathematical development of Kelvin's atom began with his proving the stability of an atom having its electrons arranged in several concentric rings in a single plane. This arrangement did not imply that atoms actually had planar structures. It was only a temporary and simplifying assumption that Thomson introduced because of the greater analytical and geometrical difficulties resulting from a distribution of electrons in three-dimensional shells rather than in planar rings. Nevertheless, Thomson believed the same kind of properties would be associated with shells as with rings.

He argued, for example, that the position of the bright lines in the emission spectrum of a chemical element probably depended on a particular configuration of electrons recurring in different atoms, regardless of

[22] Kelvin, "Aepinus Atomized," pp. 257–83, quote on p. 272. This is the article Kelvin contributed to a jubilee volume presented to Professor Johannes Bosscha (1831–1911), November 1901. Bosscha was director of the Polytechnic Institute in Delft, Holland, and secretary of the Dutch Scientific Society.

[23] Thomson, "Structure of the Atom," p. 255.

whether the electrons were in rings or in shells. Those elements which contained such configuration were members of the same periodic family, and their spectra should form a homologous series of lines with the distance between the lines of the doublets and triplets increasing with the elements' atomic weight.[24]

The spectroscopic investigations of Johannes R. Rydberg (1854–1919) in Sweden and Carl Runge (1856–1927) and Heinrich Kayser (1853–1946) in Germany supported Thomson's arguments. They showed that for the alkali metals and alkaline earths the distance between lines forming a doublet was nearly proportional to the square of each element's atomic weight. In other periodic groups, this distance varied as some power of the element's atomic weight.[25]

Thomson called attention to a second relation between his electron configuration and the periodic properties of atoms. He pointed out that there were certain configurations that changed very abruptly upon addition of a single electron. Mayer's floating magnets illustrated this behavior. Five magnets formed a single group, while six magnets formed two groups; similarly, fourteen magnets formed two groups, while fifteen, three; twenty-seven magnets formed three groups, but twenty-eight, four.

The abrupt change in an atom's electron configuration suggested a possible explanation for the sudden change in properties periodically observed upon arranging the chemical elements in order of increasing atomic weight. As Thomson noted, an extreme difference in properties existed between the nearly adjacent elements fluorine and sodium. Yet, beginning with sodium, there followed a more or less continuous variation in properties up to chlorine where a second discontinuity occurred. Another significant break took place between bromine and rubidium.[26] Thomson's early effort to give an electron interpretation to the periodic table, though purely speculative—he had no idea of the actual number of electrons in the different atoms—nevertheless, was in its time quite remarkable.

In his 1904 paper, Thomson worked out mathematically for the Kelvin atom configurations of electrons increasing in number from 1 to 100. But

[24] Thomson, *Electricity and Matter*, p. 121.

[25] Johannes R. Rydberg, "Über den Bau der Linienspektren der chemischen Grundstoffe," *Zeit. phys. Chemie* 5 (1890), 227–32; Carl Runge and Heinrich Kayser, "Über die Spektren der Alkalien," *Annalen der Physik* 41 (1890), 302–20. See also, Albert E. Garrett, *The Periodic Law*, p. 209.

[26] Thomson, *Electricity and Matter*, pp. 122–23.

to follow how he related these configurations to the periodic arrangement and valences of the chemical elements, we need to consider, as Thomson did, only one particular series of configurations.[27] This series contained nine hypothetical atoms that Thomson compared with the helium-neon period, or the neon-argon period. The number of electrons increased singularly from 59 to 67, though each atom always contained 20 electrons in its outer ring. They are shown below along with a few members of the two series that immediately preceded and followed the 59–67 series and contained 19 and 21 electrons in their outermost rings.

19	19	19	19	20	20	20	20.	Outer ring
16	16	16	16	16	16	16	17	
12	12	12	13	13	13	13	13	
7	8	8	8	8	8	9	9	
1	1	2	2	2	3	3	3	Inner ring
55	56	57	58	59	60	61	62	Total number of electrons

20	20	20	20	20	21	21	Outer ring
17	17	17	17	17	17	17	
13	13	14	14	15	15	15	
10	10	10	10	10	10	11	
3	4	4	5	5	5	5	Inner ring
63	64	65	66	67	68	69	Total number of electrons

From this arrangement we see that 59 was the smallest total number of electrons for the series having an outer ring of 20. Thomson had calculated that the outer ring's stability was least when the number of inner electrons was a minimum and increased as the number of inner electrons increased. The 59-atom, therefore, had just a sufficient number of electrons inside its outer ring to make this ring stable.

In forming the atoms 59 through 67, the addition of inner electrons made it increasingly difficult for the outer ring to lose an electron. The successive atoms in this series were, as a result, more electronegative. Indeed, Thomson argued that the increasing stability of the outer ring led to a corresponding increase in the electronegative character of the atoms until there were 67 electrons in an atom. At this point, the stability of the outer

[27] Thomson, "Structure of the Atom," pp. 258–63.

ring would reach a maximum. The addition of one more electron now started a new series of atoms each containing 21 electrons in the outer ring.

To illustrate the different valences of the atoms 59 through 67, Thomson assumed that atoms 60, 61, 62, and 63 were able to lose 1, 2, 3, and 4 electrons, respectively, and form positive valences without disrupting their outer ring of 20. To take a specific example, if the 61-atom, which lost a maximum of 2 electrons, were to lose 3, it would revert to the 58-atom. But the 58-atom being the last and most stable member of the previous series, was, according to Thomson, an atom that had a strong tendency to attract and retain electrons. This tendency prevented the 61-atom from losing more than two electrons. A similar explanation held for the formation of other positive valences in this series.

Negative valences followed whenever the stability of the outer ring increased because of the addition of inner electrons. The 64-, 65-, and 66-atoms, for example, could hold one or more electrons on their surface without breaking up the outer ring. In this way, they acquired negative valences of 3, 2, and 1.

Thomson's comparison of the total electron number of his atoms with the valence of the elements in periods 2 and 3 of the periodic table is summarized below:

Number of electrons	59	60	61	62	63	64	65	66	67
Atom	He	Li	Be	B	C	N	O	F	Ne
	Ne	Na	Mg	Al	Si	P	S	Cl	Ar
Valence	0	+1	+2	+3	+4	−3	−2	−1	0

Atoms with 59 or 67 electrons represented a stable configuration of electrons and were analogous to the stable structures of the rare gases. They had, therefore, zero valence. The valences of the intermediate atoms were either negative or positive, depending on the atom's ability to gain or lose electrons and acquire the configuration of a rare gas—that is, a configuration of 59 or 67 electrons. In this respect, Thomson's belief that his stable electron configurations corresponded to the rare gas structures and that other atoms tended toward these structures resembled G. N. Lewis's and Richard Abegg's ideas on valence and periodicity.

Two years later Thomson again discussed valence and periodicity in a course of lectures that he gave at the Royal Institution in the spring of 1906 and in an article appearing in the *Philosophical Magazine* for June 1906. He published the lectures in a single volume, *The Corpuscular Theory of Matter*, the following year.[28] Thomson's discussion was essentially the same as previously, except now he incorporated Abegg's suggestion that every atom had both a positive and a negative valence (valence and contravalence), the sum of its maximum positive and negative valence always being eight. Abegg's "rule of eight" seemed to fit nicely into Thomson's valence scheme, at least for the one series that contained 20 electrons in its outer ring.

Number of electrons	59	60	61	62	63	64·	65	66	67
Valence	+0	+1	+2	+3	+4	−3	−2	−1	−0
Contravalence	−8	−7	−6	−5	−4	+5	+6	+7	+8
		Electropositive atoms					Electronegative atoms		
Corresponding	He	Li	Be	B	C	N	O	F	Ne
chemical atom	Ne	Na	Mg	Al	Si	P	S	Cl	Ar

Thomson, therefore, was quite aware that he could never reconcile Abegg's rule with most of his calculated electron configurations precisely because of the variable number of atoms present in each series:

Number of Electrons in Outermost Ring of Each Atom in the Series	Total Number of Atoms in the Series
18	5
19	4
20	9
21	10
22	8
23	7
24	8

He readily admitted that any agreement between his valences and the positive and negative valences that chemists assigned to atoms was purely accidental.[29] Nevertheless, Thomson believed in the eventual establishment of

[28] J. J. Thomson, "On the Number of Corpuscles in an Atom," *Phil. Mag.* 11 (1906), 769–81; idem, *The Corpuscular Theory of Matter*.
[29] Thomson, *Corpuscular Theory*, p. 118.

a periodic table arranged according to the number of electrons in each kind of chemical atom.

Arrhenius's Criticism of the Thomson Atom

About the time Thomson's ideas on valence and periodicity were becoming well known, Svante Arrhenius took the opportunity to discuss them in a course of lectures he gave upon visiting the University of California at Berkeley in the summer of 1904. The electron theory of valence had not greatly impressed Arrhenius, who believed it was a formal conception that led to no new results.[30] He found Thomson's electron interpretation of the periodic table interesting but criticized it too, rather severely. Its major shortcoming was that Thomson attributed practically all the weight of an atom to the number of electrons it contained.

For example, the weights of the sodium, magnesium, and aluminum atoms increased in the order given because each atom contained one electron more than the preceding one. But in forming the Na^+, Mg^{+2}, and Al^{+3} ions, each ion now contained the same number of electrons and, therefore, had the same atomic weight. The equal weights of these three ions that Thomson's electron configurations predicted (and we can easily see that the same argument applied to other ions containing the same number of electrons, N^{-3}, O^{-2}, and F^{-1}) was, according to Arrhenius, a conclusion fatal to Thomson's entire theory. In the chemist's periodic table, the atomic weight increased with increasing positive valences in every series; in Thomson's theory, it did not. This shortcoming, Arrhenius continued,

is only a specially striking instance of a more general one; the systems in Thomson's series differ from each other by one electron, so that the difference between two consecutive atomic weights is constant. This does not agree with the more complicated behavior of the natural elements. In the series 2 (sodium through chlorine) this difference varies between 1.05 and 3.39, i.e., in the proportion 1:3.2, in a somewhat irregular manner. In other series this variation is of the same order.[31]

A second shortcoming of Thomson's electron configurations appeared in the number of atoms in each series having zero valence. Both the 59-atom and the 67-atom, the first and last members of the series with twenty electrons in the outer ring, had zero valence. If we were to continue the analogy, the 58-atom, as the last member of the previous series and the 68-

[30] Svante Arrhenius, *Theories of Chemistry*, p. 86.
[31] Ibid., p. 100.

atom, the first member of the following series, also should have zero valence. Thus, in every series there were two atoms belonging to Group 0. Neon could correspond to either the 58- or 59-atom and argon to the 67- or 68-atom. Thomson's configurations, as Arrhenius pointed out, resulted in a duplication of atoms in Group 0 that did not exist.

Arrhenius's final criticism of Thomson's electron configurations questioned Thomson's inconsistencies regarding the number of electrons in an atom and in the difference in the number of electrons when passing from one atom to the next in the periodic table. Experiments, including Thomson's, had now shown the mass of an electron to be anywhere from $1/1700$ to about $1/2000$ of a hydrogen atom.[32] Thus, the neon atom ($A = 19.9$) probably contained 39,000 electrons, sodium ($A = 23.05$), 46,100 electrons, magnesium ($A = 24.1$), 43,200 electrons, and other large numbers for the remaining atoms.

Arrhenius argued that since the differences in atomic weight between adjacent atoms were not at all constant, neither were the differences in the number of electrons constant as Thomson's configurations indicated. He also argued that the difference in valence between adjacent atoms was not one or two electrons but a large number. "By this amendment," he concluded "the Thomson scheme loses much of its simplicity, and at the same time much of its theoretical value."[33]

In defense of Thomson we should note that he never claimed his electron configurations were those of *actual* atoms. They provided only "a theory which enables us to picture a kind of model atom and to interpret chemical and physical results in terms of such a model . . . even though the models are crude."[34] His hypothesis that the ions of adjacent, or nearly adjacent, atoms in the periodic table often contained the same number of electrons was correct. Indeed, Thomson could have avoided most of Arrhenius's rather obvious criticisms had he dealt at all with the problem of

---

[32] The following deal with changes in the value of the mass of the electron: Walther Kauffmann, "Die magnetische Ablenkbarkeit der Kathodenstrahlen und ihre Abhängigkeit vom Entladungspotential," *Annalen der Physik* 62 (1897), 598; Emil J. Wiechert, "I. Über das Wesen der Elektrizität. II. Experimentelles über Kathodenstrahlen," *Annalen der Physik* 21 (1897), 443–44; Thomson, "Masses of the Ions," pp. 547–67; Thomson, "Number of Corpuscles," pp. 769–81; Lenard, "Über die Absorption von Kathodenstrahlen verschiedener Geschwindigkeit," pp. 714–44; August Becker, "Messungen an Kathodenstrahlen," *Annalen der Physik* 17 (1905), 381–480. See also Robert Millikan, *The Electron.*

[33] Arrhenius, *Theories of Chemistry*, p. 101.

[34] Thomson, *Corpuscular Theory*, pp. v–vi.

accounting for the mass of an atom with only a limited number of electrons at his disposal.

Despite some of the erroneous conclusions that followed from his speculative electron configurations, Thomson still had given a simple explanation of chemical union based on the loss or gain of electrons. With his configurations, he also accounted for the variable valence of an atom such as copper or iron: it was due to the loss of one, two, or even three electrons, the number lost depending on the chemical reaction. Not explained was the reason why an atom's valence varied in many well-known cases by intervals of two. Tin had valences of $+2$ and $+4$; nitrogen, $+3$ and $+5$; while chlorine and iodine formed halides and oxyacids whose valences ranged in intervals of two from -1 to $+7$.

Thomson's account of variable valence can be compared with a second theory that tried to explain why an atom's valence often varied by intervals of two. This theory, the theory of latent or residual valence, had as its basic premise the idea that an atom "called out" one or more pairs of oppositely charged valences whenever it needed them.[35]

In one variation of the theory, Leopold Spiegel (1865–1927) in Berlin renamed the latent pairs "neutral affinities." He believed that when they were not externally saturated, they neutralized each other within the atom and as a result exerted no effect on an atom's electrochemical behavior.[36] The platinum atom, for example, had four active valences, and two pairs of neutral affinities:

This gave for potassium hexachloroplatinate (II), K_2PtCl_6, the structural formula:

---

[35] John Newton Friend later gave the theory of latent or residual valence an electron interpretation. See chapter 3.

[36] Leopold Spiegel, "Über Neutralaffinitäten," *Zeit. anorg. Chemie* 29 (1902), 365–70.

Arrhenius in 1904 also used latent pairs of valences, which he called "electrical double valences," to describe the formation of ammonium chloride from ammonia and hydrogen chloride. In the ammonia molecule, each of the three hydrogen atoms was positive and was held electrostatically by three negative valences of nitrogen; in hydrogen chloride, hydrogen was again positive and electrostatically bound to a negative chlorine. To form ammonium chloride, the nitrogen atom "called out" a pair of latent valences of opposite sign that then united with the positive hydrogen and negative chlorine of hydrogen chloride.[37] This was illustrated:

$$
\begin{array}{l}
\text{H} + - \\
\text{H} + - \\
\text{H} + -
\end{array}
\mathbf{N}
\begin{array}{l}
+ - \text{Cl} \\
\\
- + \text{H}
\end{array}
$$

Thomson's Evidence for a Smaller Number of Electrons in an Atom

In 1906 Thomson supplied the experimental evidence that permitted the introduction of an atom with far fewer electrons than ever imagined just a few years earlier. From three independent methods—(1) the scattering of X rays by gases, (2) the absorption of β rays, and (3) the dispersion of light by gases—he argued that the number of electrons in an atom of an element was approximately equal to the element's chemically determined atomic weight.[38]

The first of these methods, the scattering of X rays by gases, required establishing the ratio of the energy in a beam of primary X rays passing through a gas to that of the secondary X rays emitted or scattered by the gas. Thomson had shown in his text *Conduction of Electricity through Gases*[39] that for a medium containing N electrons per cubic centimeter, this ratio was

$$
\frac{E_s}{E_p} = \frac{8\pi}{3} \frac{Ne^4}{m^2},
$$

where E_s/E_p = ratio of secondary to primary radiation emitted per unit time and unit volume; N = number of electrons per cubic centimeter of air

[37] Arrhenius, *Theories of Chemistry*, pp. 83–84.
[38] Thomson, "Number of Corpuscles," pp. 769–74; idem, *Corpuscular Theory*, pp. 142–53.
[39] J. J. Thomson, *Conduction of Electricity through Gases*, p. 326.

at 0°C; e = electron charge, 1.10×10^{-20} emu; m = electron mass; and $e/m = 1.7 \times 10^7$ emu/gram. For the scattering in air,[40] it had the value

$$\frac{E_s}{E_p} = \frac{8\pi}{3} \frac{Ne^4}{m^2}$$

$$= 2.4 \times 10^{-4}.$$

Substituting known values of e and e/m reduced the equation to $Ne = 10$. Now if n equaled the number of molecules in a cubic centimeter of air at 0°C, then

$$ne = 2.8 \times 10^{19} \times 1.10 \times 10^{-20} \text{ emu}$$

or

$$ne = .4,$$

and since $Ne = 10$,

$$\frac{N}{n} = \frac{\text{number of electrons/cm}^3 \text{ of air}}{\text{number of molecules/cm}^3 \text{ of air}}$$

$$\frac{N}{n} = \frac{10}{.4}$$

$$\frac{N}{n} = \frac{25}{1}.$$

According to Thomson, there were, therefore, 25 electrons in each "molecule" of air.[41]

To complete his argument Thomson pointed out that the major component of air, nitrogen gas, had a scattering ratio nearly equal to that of air. This resulted, he said, because the number of electrons in a "molecule" of air was probably nearly equal to the number in a nitrogen molecule. And since the number of electrons in a "molecule" of air was nearly equal to the weight of a nitrogen molecule, $N_2 = 28$, Thomson's X-ray scattering seemed to indicate that perhaps the nitrogen molecule contained 28 elec-

[40] Charles G. Barkla, "Energy of Secondary Röntgen Radiation," *Phil. Mag.* 7 (1904), 543–60.

[41] The value of ne is actually 0.31, not Thomson's value of 0.4. This error does not change Thomson's general conclusion.

trons. This was, of course, the same number as its molecular weight. Thus the number of electrons in a nitrogen atom, N, was equal to the atomic weight of nitrogen, or 14. Generalizing the results of his experiment, Thomson concluded that the number of electrons in every kind of atom equaled the weight of that atom.[42]

Thomson developed his second method, the absorption of β rays from an equation for the coefficient of absorption of β rays, λ, that he derived in his 1906 paper:

$$\lambda = \frac{4\pi N e^4}{m^2}\; \frac{V_0^{\,4}}{V^4}\, \log\left(\frac{1}{2}\,\frac{aV^2 m}{V_0^{\,2}e^2} - 1\right),$$

where N = number of electrons per cubic centimeter, V = velocity of β rays, V_0 = velocity of light, e = electron charge, m = mass of electron, and a = a length comparable with the distance between the electrons in an atom.[43]

He substituted δ, the density of the absorbing substance; M, the mass of the atom containing the unknown number of electrons, n; and the experimentally determined relation, $\delta = (N/n)M$, to get for the coefficient of absorption:

$$\lambda = \delta \cdot \frac{4\pi e^2}{m^2}\; \frac{en}{M}\; e\; \frac{V_0^{\,4}}{V^4}\, \log\left(\frac{1}{2}\,\frac{aV^2 m}{V_0^{\,2}e^2} - 1\right).$$

Knowing that λ/δ was approximately constant regardless of the absorbing substance and assuming that the logarithmic term varied only very slightly, Thomson next inserted known values for the velocity of β rays from uranium, $V = 1.6 \times 10^{10}$ cm/sec, and for the ratio λ/δ. For silver or copper, $\lambda/\delta = 7$. This gave the expression[44]

$$\frac{ne}{M} = \frac{1.4 \times 10^4}{\log\left(\dfrac{mV^2}{V_0^{\,2}}\dfrac{a}{e^2} - 1\right)}.$$

To eliminate the factor 10^4 he substituted the e/M ratio for a hydrogen atom, or $e/M' = 10^4$, which reduced the equation to

[42] Thomson, *Corpuscular Theory*, p. 145; idem, "Number of Corpuscles," p. 773.
[43] The derivation is also in Thomson's *Conduction of Electricity through Gases*, p. 377.
[44] Thomson gave no reason for the disappearance of the 1/2 factor in the logarithmic term.

$$n = \frac{M}{M'} \frac{1.4}{\log \left(\dfrac{mV^2}{V_0^2} \dfrac{a}{e^2} - 1 \right)}.$$

Finally, because the value of a, the distance between the electrons in an atom, though unknown had to be small (ca. 10^{-13} cm to 10^{-14} cm), Thomson assumed that the logarithmic term was also small (between 1 and 10) and could not alter the order of the right-hand side. Thus

$$n \simeq \frac{M}{M'},$$

and because M, the mass of the hydrogen atom, equaled 1, the number of electrons in an atom was of the same order as its atomic weight.

In his final method for determining the number of electrons in an atom, namely, the dispersion of light by gases, Thomson intended to use an equation that he had derived for the refractive index of a monatomic gas:

$$\frac{\mu^2 - 1}{\mu^2 + 2} = P_0 + P_0^2 \frac{M}{E} \frac{m}{e} \frac{1}{N(M + nm)} \frac{3\pi}{\lambda^2},$$

where λ = wavelength of light used, P_0 = frequency of light, μ = refractive index of the light, M = mass of the sphere of positive electrification, E = the entire charge on the positive sphere of electricity (emu), N = number of atoms in a unit volume of the gas, e = electron charge (emu), m = electron mass, and n = number of electrons in the atom.

At that time (1906) there were few experimental measurements of the refractive index for monatomic gases with which Thomson could compare his derived equation. However, studies by Lord Rayleigh (J. W. Strutt, 1842–1919) indicated that the refractive index for these gases was of the same order as that for diatomic gases.[45] Rayleigh's study permitted Thomson to use an equation that Eduard Ketteler (1836–1900) at Bonn had obtained upon measuring the refractive index for the diatomic gas hydrogen.[46] That equation was

[45] Thomson, *Corpuscular Theory*, p. 153; idem, "Number of Corpuscles," p. 771; Lord Rayleigh (John William Strutt), "On Some Physical Properties of Argon and Helium," *Proc. Roy. Soc.* 59 (1896), 198–208.

[46] Eduard Ketteler, "Über die Dispersion des Lichts in den Gasen," *Annalen der Physik* 124 (1865), 390–406; idem, "On the Dispersion of Light in Gases," *Phil. Mag.* 32 (1866),

$$\frac{\mu^2 - 1}{\mu^2 + 2} = \frac{1}{3}\left(2.8014 \times 10^{-4} + \frac{2 \times 10^{-14}}{\lambda^2}\right).$$

Equating his own equation for the refractive index with Ketteler's gave

$$\frac{M}{E}\frac{m}{e}\frac{1}{N(M + nm)} = 6 \times 10^{-8}.$$

Upon substituting values for m/e ($1/1.7 \times 10^7$) and Ne (0.8), he reduced this to

$$\frac{M}{(M + nm)}\frac{e}{E} \simeq 1.$$

Because $E = ne$, the final equation was

$$\frac{M}{M + nm}\frac{1}{n} \simeq 1.$$

Thomson concluded from this equation that in a hydrogen atom n, the number of electrons in the atom, did not differ much from 1, which was the hydrogen atom's weight.

In arguing that the number of electrons in an element's atom did not differ greatly from the element's atomic weight, Thomson admitted that his evidence was rather indirect and his data not very numerous. He had, however, employed three different methods that dealt with widely separated physical phenomena, and though no one of them was conclusive in itself, the evidence became very strong upon considering that such different methods led to the same result.[47]

Nonetheless, at the time of its publication 1906–7, Thomson's conclusion was incompatible with the generally held assumption that even the lighter atoms had to contain at least several hundred electrons. How else could physicists account for the large number of lines in the elements' spectra. Indeed, Charles Barkla's experiments on the scattering of X rays by gaseous molecules appeared to confirm this assumption. For these reasons George A. Schott (1868–1927) in 1907 described as unwork-

---

336–45. See also idem, "Constanz des Refraktions," *Annalen der Physik* 30 (1887), 285–316; and idem, *Theoretische Optik*.

[47] Thomson, "Number of Corpuscles," p. 769.

able Thomson's hypothesis that the hydrogen atom contained only one electron.[48]

The Distribution of the Mass in an Atom

Thomson's final method, in addition to providing a means of estimating n, the number of electrons in an atom, led to a new conclusion regarding the distribution of mass in an atom. According to the equation

$$\frac{M}{M + nm}\, \frac{1}{n} \simeq 1 \, ,$$

the mass M associated with the positive electricity could not be small compared with nm, the total mass of the electrons. In other words, as early as 1906 Thomson had shown that if the hydrogen atom contained only one electron with mass about $1/1700$ the mass of the hydrogen atom, then the hydrogen atom's mass was due chiefly to its positively charged component.[49] Indeed, Thomson believed that for all atoms the mass of "the carrier of unit positive charge" was large compared with that of "the carrier of unit negative charge" if one assumed that the charge components accounted for the atom's entire mass.[50]

This interpretation thus differed from what Thomson had written earlier in his 1904 text *Electricity and Matter*. There the atom's total mass resulted from the tubes of negative electric force, the electrons, as they moved through the ether. Having reduced the number of electrons in an atom to an insignificant fraction of what he once supposed, Thomson therefore associated a large amount of mass with the positive charge or charges in the atom.[51]

Thomson's studies, supported by those of Paul Drude (1863–1906) on the refraction of light by atoms, also led him to postulate in each atom the

[48] Barkla, "Energy of Secondary Röntgen Radiation," pp. 543–60; "The Constitution of the Atom," *Chemical News* 96 (23 August 1907), 94–95; G. A. Schott, "On the Electron Theory of Matter and on Radiation," *Phil. Mag.* 13 (1907), 189–213.

[49] Thomson, *Corpuscular Theory*, pp. 153, 162. The figures used in the calculations are from pp. 10, 15–16. For the electron, $e/m = 1.7 \times 10^7$ emu/g. For hydrogen (from acidic electrolysis), $E/M = 10^4 e/m$, and $e/m = 1700 E/M$. Since $E = e = 3.1 \times 10^{-10}$ esu $= 10^{-20}$ emu, then $M = 1700m$, and $m = 6 \times 10^{-28}$ g.

[50] Thomson, "Number of Corpuscles," p. 774. *The Corpuscular Theory*, pp. 162–63, contains essentially the same discussion.

[51] Thomson, *Electricity and Matter*, pp. 49–52; Thomson, *Corpuscular Theory*, p. 163.

existence of easily moveable electrons equal in number to the atom's valence. He imagined these electrons embedded in a low-density sphere or shell of positive electricity that surrounded a denser positive core containing the remaining electrons. Thomson's atom thus had a "crowded center, surrounded by a rarified atmosphere through which a few corpuscles [electrons] are scattered, the positive electricity in the atmosphere being equivalent to the negative charge on the corpuscles scattered through it." [52]

This new model of atomic structure received some unexpected additional support in 1911. Barkla (1877–1944), upon revising his earlier calculations, showed that for the lighter elements the number of electrons in an atom was equal to one-half the element's atomic weight. His experiments demonstrated that Thomson's conclusion regarding the number of electrons in an atom was not too low but, on the contrary, was actually too high. Ernest Rutherford's 1911 publication that introduced the nuclear model of the atom led to the same result. But it remained for Henry Gwyn-Jeffreys Moseley (1887–1915) in 1913–14 to establish beyond doubt with X-ray spectral analysis the number of electrons in an atom. For many elements this number happened to equal one-half the atomic weight, and for all the elements it was exactly equal to the number of positive charges contained in the atom's nucleus. The latter, Moseley pointed out, was equal to the numerical position occupied by the element in the periodic table. [53]

When Rutherford (1871–1937) introduced the nuclear atom in 1911, he did not mention the results of Thomson's 1906 experiments. In particular, he did not comment on Thomson's important conclusions that the hydrogen atom almost certainly contained only one electron and that the carrier of positive charge accounted for most of the atom's mass. Thomson's atomic model was obviously compatible with Rutherford's nuclear atom, the chief difference between the two being the radius of the positive sphere. According to Thomson, it was of the order 10^{-8} cm; in Rutherford's atom, 10^{-13} cm. The α-scattering experiments of Hans Geiger

[52] Paul Drude, "Optische Eigenschaften und Elektronentheorie," *Annalen der Physik* 14 (1904), 677–725, 936–61; Thomson, *Corpuscular Theory*, pp. 155, 166. Drude, like Thomson, based his discussion on the number of electrons in an atom on the optical dispersion of transparent bodies in the ultraviolet region of the spectrum.

[53] Charles G. Barkla, "Note on the Energy of Scattered X-Radiation," *Phil. Mag.* 21 (1911), 648–52; Ernest Rutherford, "The Scattering of α and β Particles by Matter and the Structure of the Atom," *Phil. Mag.* 21 (1911), 669–88; Henry Gwyn-Jeffreys Moseley, "The High Frequency Spectra of the Elements," pt. 1, *Phil. Mag.* 26 (1913), 1024–34, and pt. 2, *Phil. Mag.* 27 (1914), 703–13.

(1882–1945), Ernest Marsden (1889–1970), and Rutherford at Manchester in the period 1909–1913, ultimately resolved the problem in favor of the nuclear atom.[54]

Thomson's Development of the Electrostatic Theory of Valence

In his 1897 publication on cathode rays, Thomson had proposed that a Faraday tube of electrostatic force held each electron in an atom. He also assumed that an electrostatic attraction, between the two atoms of a binary compound, for example, occurred whenever a stray tube migrated from one of the atoms and attached itself to the other. The Faraday tube physically connected the resulting, oppositely charged atoms, and the ether carried along with it provided the medium through which the electrostatic force could act. Further development of these ideas appeared in Thomson's *Electricity and Matter*, in which he clearly defined the Faraday tube as follows: the origin of the Faraday tube corresponds to a unit of positive electricity, its termination, to a unit of negative electricity, the electron.[55]

Such monovalent atoms as sodium, hydrogen, or chlorine, when combined in a molecule, carried either a single positive or negative charge because each was either the beginning or the end of a Faraday tube and had either donated or received an electron. Similarly, atoms having higher valences were either positive or negative in a molecule depending on their being the origin or termination of two or more Faraday tubes.

Whether an atom donated or received electrons depended on the velocity of its electrons. In some atoms Thomson believed the velocities were so great that one or more electrons escaped, leaving a positively charged atom. In others, the velocities were small and permitted no electron escape. These atoms accepted one or more electrons, becoming negatively charged atoms sufficiently stable to overcome the repulsive force due to the presence of the extra electrons.[56]

Thomson regarded the tubes of force as nearly identical with the chemists' bonds, differing from them only in having direction:

If we interpret the "bond" of the chemist as indicating a unit Faraday tube, connecting charged atoms in the molecule, the structural formulae of the chemist can

---

[54]Rutherford, "The Scattering of α and β Particles," pp. 669–88; Hans Geiger and Ernest Marsden, "On a Diffuse Reflexion of the α-Particles," *Proc. Roy. Soc.* A 82 (1909), 495–500; Hans Geiger and Ernest Marsden, "The Laws of Deflexion of α-Particles through Large Angles," *Phil. Mag.* 27 (1913), 604–23.

[55]Thomson, *Electricity and Matter*, pp. 14–15.

[56]Ibid., pp. 132–33.

be at once translated into the electrical theory . . . but the symbol indicating a bond on the chemical theory is not regarded as having direction, no difference is made on this theory between one end of a bond and the other. On the electrical theory, however, there is a difference between the ends, as one corresponds to a positive, the other to a negative charge.[57]

He illustrated his electron theory of valence by describing the bonding in the hydrocarbons ethane, C_2H_6, and ethylene, C_2H_4. They were good choices for they enabled Thomson to show the great similarity between the organic chemists' structural formulas with nonelectrical valence bonds and his new electrical valence formulas, thereby enhancing their chance of acceptance by chemists.

In these molecules, as in every molecule, the sign of the charge on each atom that indicated the direction of the Faraday tube depended on the ease with which an electron or electrons left one atom and were acquired by another. In his discussion in *Electricity and Valence*, Thomson assumed that in both ethane and ethylene, hydrogen was more negative than carbon and assigned a negative charge to each hydrogen atom, a positive charge to the two carbon atoms. Three Faraday tubes, therefore, went from a single carbon atom, leaving each with a charge of $+3$, and ended on three different hydrogen atoms:

$$
\begin{array}{ccc}
\text{H} & & \text{H} \\
\nwarrow & & \nearrow \\
\text{H} \leftarrow \text{C}_1 - \text{C}_2 \rightarrow & \text{H}. \\
\swarrow & & \searrow \\
\text{H} & & \text{H}
\end{array}
$$

But in addition to the Faraday tubes connecting the hydrogen atoms, there was also a single tube holding together the carbon atoms. As a result, one of the carbon atoms (C_1) gained an additional positive charge, giving it a total charge of $+4$, and left the remaining carbon atom (C_2) with a net charge of $+2$ (3 positive plus 1 negative charge).

$$
\begin{array}{ccc}
\text{H} & & \text{H} \\
\nwarrow & & \nearrow \\
\text{H} \leftarrow \text{C}_1 \rightarrow \text{C}_2 \rightarrow & \text{H} \\
\swarrow & & \searrow \\
\text{H} & & \text{H}
\end{array}
$$

---

[57] Ibid., p. 134.

Ethane, according to Thomson, was not the symmetrical molecule it appeared to be when chemists used an ordinary graphic formula to indicate its structure. Thomson had adopted the established structural formulas of organic chemistry, but he interpreted valence, or better electrovalence, as the net charge remaining on an atom in a molecule after taking into account the number of electrons gained and lost by that atom. The electrovalence of an atom was then not necessarily equal to the number of bonds it formed. On the other hand, in the organic chemists' nonelectrical structural formulas, no such difference existed between the two carbon atoms. Each formed four bonds and consequently had a valence of four. The number of bonds an atom formed always equaled its valence.

A similar situation existed for ethylene, C_2H_4. Along with symmetrical structure (2), Thomson's theory predicted the existence of an asymmetrical structure (1):

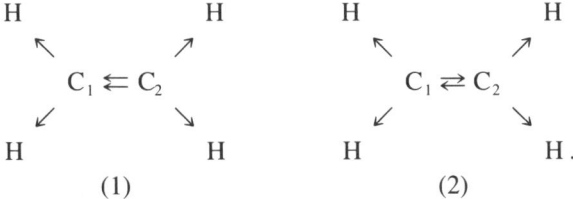

The two structures differed only in the magnitude of the charges carried by the carbon atoms. In (1), C_1 had a charge of 0; C_2, a charge of $+4$. The C—C bond did not affect the net charge on each carbon in structure (2); each carbon carried a charge of $+2$. Later H. S. Fry, in developing Thomson's electrostatic valence theory, called the two structures of ethylene, *electromers*, that is, electron isomers.

The Physicists' Reception of Thomson's Atom and Electrostatic Theory of Valence

Unlike Arrhenius, who had severely criticized Thomson's attempt to give the atom and the periodic table an electronic interpretation, the physicists Oliver Lodge at Birmingham and Norman R. Campbell (1880–1949) at Cambridge spoke favorably of Thomson's atomic theory. Lodge, in reviewing Thomson's *Electricity and Matter* in *Nature*, saw it as a distinct contribution to a mathematical chemistry and called it "steps toward a new *Principia*." In his 1906 edition of *Electrons*, in which he examined the nature and properties of electrons, Lodge again praised Thomson's atomic theory. It is "capable of carrying us a long way towards a rational theory of Mendeleeff's series of the chemical elements, together with some of their chemical—especially their electro-chemical—properties." [58]

Lodge did not regard Thomson's atom as the only model of atomic structure, however. Thomson's atom, especially the idea of positive electricity existing as a permeable sphere, lacked confirmatory experimental evidence. Lodge offered an atom with positive and negative charges in separate alternating layers or planes. It resembled Joseph Larmor's proposal in 1894 that positive and negative electric charges were mirror images of each other. But both the Lodge and Larmor atoms were incompatible with experiments on positive rays that showed that positive electricity never seemed to exist separately from particles of matter similar in weight to ordinary atoms. It was probably with this difficulty in mind that Lodge wrote: "The relations of positive electricity constitute in fact the main outstanding problem of Physics at the present time, and until they can be probed, further progress towards understanding the constitution of an atom must remain in a state of suspended animation." [59]

Norman R. Campbell echoed Lodge's opinion regarding the uncertainty surrounding the positive electricity in an atom. In his well-known *Modern Electrical Theory* (1907) Campbell remarked that though no evidence existed to indicate the manner of positive charge distribution, it was nevertheless best "to adopt a hypothesis which renders most easy the task of deducing the properties of the atom from its structure." [60]

Campbell was referring to Thomson's atom, which he maintained had an immense advantage over all other atomic models chiefly because it explained both chemical and physical phenomena using a single particle, the negative electron. He agreed with Thomson that positive and negative valences resulted from the loss and gain of electrons and that all chemical combination was electrostatic.

As a proponent of the electron theory of matter, Campbell also assumed that the ultimate explanation of the periodic table would come from

[58] Oliver Lodge, Review of *Electricity and Matter* by J. J. Thomson, in *Nature* 70 (26 May 1904), 73; Lodge, *Electrons*, p. 149.

[59] Joseph Larmor, "A Dynamical Theory of the Electric and Luminiferous Medium," *Phil. Trans.* 185 (1894), 719–822; also idem, "On the Theory of Moving Electrons and Electric Charges," *Phil. Mag.* 42 (1896), 201; Wien, "Untersuchungen über die elektrische Entladung," pp. 440–52; Lodge, *Electrons*, pp. 147–48.

[60] Norman R. Campbell, *Modern Electrical Theory*, p. 232.

a knowledge of the atoms' electron configurations. He believed, with Thomson, that the periodic recurrence of electronic patterns in the atoms accounted for the periodicity of chemical properties exhibited in the periodic table.

In Thomson's table, adding to an existing electron configuration a single electron and a corresponding amount of positive charge resulted in the formation of a new atom. Campbell, on the other hand, thought that atoms of nearly identical atomic weight might have the same electron configuration and thus the same chemical properties. Those substances which chemists regarded as elements might actually be mixtures of a large number of very similar but not identical atoms. "Such a suggestion may appear rather startling," Campbell remarked, "but I do not think any valid evidence can be urged against it." Campbell's statement, like that of William Crookes nearly twenty years before, seemed to be another early speculation on the existence of isotopes,[61] which Frederick Soddy (1877–1956) at Glasgow and Kasimir Fajans (1887–1975) at Karlsruhe proposed independently in 1913 from studies on radioactivity.[62]

Unlike Crookes, Campbell had indeed based his speculations on developments in the field of radioactivity which suggested the existence of many short-lived elements for which no place remained in the periodic table. Nevertheless, Campbell did not find the difference between his and Thomson's interpretation of the periodic table serious enough to reject Thomson's atomic theory. In the group of similar atoms possibly only one was stable and capable of permanent existence.

Lines of Force and Residual Valence

In *Electricity and Matter* (1904), Thomson did not discuss the bonding in molecules in any greater detail than given above. He simply assumed

[61] Ibid., pp. 262, 263; William Crookes, "Elements and Meta-Elements," presidential address, *J. Chem. Soc.* 53 (1888), 490; also Crookes, "Spectroscopic Researches on the Rare Earths," *J. Chem. Soc.* 55 (1889), 257.

[62] Frederick Soddy, "The Radio-Elements and the Periodic Law," *Chemical News* 107 (28 February 1913), 97–99. Soddy introduced the term *isotopes* in the last paragraph of his 1913 paper "Intra-Atomic Charge," *Nature* 92 (4 December 1913), 399–400. Fajans had suggested the term *pleiad* (Kasimir Fajans, "Die Stellung der Radioelemente im periodischen System," *Phys. Zeit.* 14 [1913], 136–42; English translation by Alfred Romer, "The Placing of the Radioelements in the Periodic System," in *Radiochemistry and the Discovery of Isotopes*, p. 211). See also Kasimir Fajans, "On a Relation between the Nature of a Radioactive Transformation and the Electrochemical Behavior of the Radioelement Involved," in *Radiochemistry and Discovery of Isotopes*, pp. 198–206.

that the bond between a pair of atoms was an electrostatic tube of force, regardless of whether the atoms were the same, as in the diatomic gaseous elements oxygen, hydrogen, or nitrogen, or different, as in the hydrocarbons methane, ethane, or ethylene. His electron interpretation of the chemists' valence bond tended to harden the idea of simple numerical valence relations existing among atoms. Indeed, his interpretation came at a time when many chemists were already abandoning those relations, as seen particularly in Thiele's theory of partial valence and Werner's coordination theory or theory of principal and secondary valences.

Yet there is evidence that Thomson, too, had recognized the need for a valence theory that did more than account for the structures of simple molecules—one that could deal with the still unsolved problem of residual valence. He believed, for example, that the conditions under which combination took place influenced to some degree an atom's valence and that an atom had available valences in addition to its ordinary valences.[63] An atom that normally combined with *n* atoms might combine under certain conditions with more than *n* atoms, though it held the original *n* atoms more firmly than the others. Oliver Lodge in 1904 modified Thomson's interpretation of the Faraday tube and provided a solution to the problem of residual valence.

In a letter to *Nature*, Lodge pointed out that he knew of no evidence for assuming each electron to have only one solitary line of force assigned to it; rather, the electron might have a large number of lines emanating from it.[64] Thomson's Faraday tube, according to Lodge, was actually a Faraday bundle, and unlike the electron charge, its lines of force were divisible.

Lodge suggested the following description of residual valence: "When opposite charges have paired off in solitude, every one of these lines start[s] from one and terminate[s] on the other constituent of the pair, and the bundle or field of lines constitutes a full chemical 'bond;' but bring other charges or other pairs into the neighborhood, and a few threads or feelers are at once available for partial adhesion in cross directions also. . . ."[65]

While the majority of the lines of force formed an ordinary valence bond, Lodge believed that a few strands were operating elsewhere. In fact, the subdivision of force might go on to any extent, "giving rise to molecular combination and linking molecules into complex aggregates, so that a

---

[63] Thomson, *Electricity and Matter*, p. 132.
[64] Lodge, "Residual Affinity," *Nature* 70 (23 June 1904), 176.
[65] Ibid.

quite gradual change of valency is conceivably possible, the number of wandering lines being sometimes equal to, or even greater than, the number of faithful lines—though this would usually represent an unstable condition not likely to persist." [66]

In concluding his letter on residual valence, Lodge invited comments on it and immediately received a reply from Edward Frankland's son Percy. Percy Frankland (1858–1946) found Lodge's theory of residual valence highly suggestive and claimed that chemists who were endeavoring to interpret chemical phenomena with the physicist's electron theory would welcome it. Indeed, he argued that dispersion of the lines of force constituting a Faraday bundle provided a theoretical explanation for the color changes accompanying the attachment of water molecules to another molecule as water of crystallization.

Choosing copper sulfate for illustration, Frankland noted that the direct union of the copper atom and the sulfate group with two Faraday bundles gave a colorless compound, anhydrous copper sulfate, $CuSO_4$. But dividing the lines of these bundles and attaching to them five molecules of water, severed the copper and sulfate union. This action permitted the blue color characteristic of the copper ion to appear. [67]

Frankland proposed a similar explanation for the solubility of all electrolytes in water. Consider his description for sodium chloride. In the solid state a Faraday bundle connected each sodium and chlorine atom, resulting in that compound's great stability. Adding water, however, deflected some of the lines of the bundle in a way that permitted each atom to attach itself to the water molecules. Further addition of water then totally diverted the strands of the Faraday bundle, which had formerly united the sodium and chlorine atoms, abolishing entirely their union: "The Faraday bundle starting with its positive extremity on the sodium atom will terminate at its negative end by means of a plurality of strands on a number of water molecules, and similarly the Faraday bundle emanating by its negative extremity from the chlorine atom will terminate at its positive end in a plurality of strands also on a number of water molecules." [68]

Frankland thought that the union between the oxygen and hydrogen atoms in a water molecule was only slightly weakened because the sodium and chlorine atoms diverted but a small fraction of the strands in the bundle

[66] Ibid.
[67] Percy Frankland, "Residual Affinity," *Nature* 70 (7 July 1904) 222–23.
[68] Ibid.

connecting the oxygen and hydrogen atoms. In Frankland's theory, therefore, hydration of the resulting ions did not merely accompany, it actually conditioned electrolytic dissociation.

The Conflict between the Electrostatic Theory and the Arrhenius Theory of Dissociation

Lodge and Frankland clearly differed with Thomson in their interpretation and use of lines of force emanating from an electron. But each of them believed that a chemical bond always resulted from the transfer of an electron from one atom to another, giving rise to oppositely charged ions in a compound. This view of complete ionization as a necessary consequence of chemical combination was in conflict with the Arrhenius dissociation theory, in which a compound dissociated into ions only upon solution.

As early as 1905, Edward C. C. Baly (1871–1948) and Cecil H. Desch (1874–1958) in London tried to reconcile these two conflicting opinions. Their explanation depended on the ability of the solvent to lengthen the connecting Faraday bundles or tubes.[69] Baly and Desch accepted Thomson's idea that ions were present in every compound, held there by Faraday tubes, but they argued that when the length of the tube fell below a certain critical value, as in a solid or a highly concentrated solution, the oppositely charged ions were so close to each other that the compound appeared to be undissociated.

Conversely, in a more dilute solution, where the average length of the tube was equal to or greater than the critical length, a few interchanges of the ions took place and the compound partially dissociated. Addition of more solvent caused further lengthening of the Faraday tubes and greater separation of the positive and negative ions eventually leading to complete dissociation.

Actually, comparison of the theories of Arrhenius and of Thomson and his followers shows that later advances confirmed parts of both. The X-ray analyses of numerous compounds by the Braggs and Debye and Scherrer, as well as the investigations of Max Born (1882–1970) and Alfred Landé (1888–1975) carried out over a decade later, demonstrated that ions (intermolecular ions) did indeed exist in the undissolved crystalline state of inorganic salts, as Thomson believed, and were not produced in the solution

[69]Edward C. C. Baly and Cecil H. Desch, "The Ultra-violet Absorption Spectra of Certain Enol-keto Tautomerides," *J. Chem. Soc.* 87 (1905), 784.

process.[70] But their work provided no evidence for the existence of ions in other types of compounds such as the hydrocarbon solids.

In these and other non-ionizing compounds, the "ions," according to Thomson, were of a special kind. They were more firmly bound in compounds and not set free upon solution, as were the ions of salts and other dissociated compounds.

Arrhenius's dissociation theory, on the other hand, did not require the presence of either special ions or intermolecular ions in any chemical compound. The ions resulted only upon solution of compounds such as acids, bases, and salts, and their number increased with increasing dissociation of the solute. His theory accounted for the electrolytic behavior of many compounds when in solution, though it could not disprove the presence of ions in these compounds before solution.

While Thomson went to one extreme in assuming that ions of one kind or another existed in all compounds, Arrhenius went to the other. He denied the presence of ions in any compound, even in solutions of salts like sodium chloride, in which dilution had little effect on its electrical conductivity. In fact, the inability of Arrhenius's theory to deal adequately with the anomalous behavior of salt solutions and, in general, with solutions of strong electrolytes, was its chief failure. This shortcoming was corrected only after X-ray analysis revealed the presence of ions in crystals of salts,[71] and investigators such as Niels Bjerrum (1879–1958) in Denmark, Debye and Erich Hückel at Zurich, and Lars Onsager (1903–76) in Norway modified some of Arrhenius's original premises in the 1920s in order to take into account the effects of interionic forces.[72]

Kauffmann's Electronic Interpretation of Residual or Partial Valence

Frankland, like Lodge, had seen that the divisibility of lines of force made possible an indefinite number of different degrees of chemical union.

[70]Peter Debye and Paul Scherrer, "Atombau," *Annalen der Physik* 19 (1918), 474–83; Max Born, "Über die elektrische Natur der Kohäsionskräfte fester Körper," *Verh. deut. phys. Ges.* 21 (1919), 533–38; Max Born and Alfred Landé, "Über die Berechnung der Kompressibilität regulärer Kristalle aus der Gittertheorie," *Verh. deut. phys. Ges.* 20 (1918), 210–16.

[71]Debye and Scherrer, "Atombau," pp. 474–83.

[72]Niels J. Bjerrum, *Selected Papers*; Peter Debye and Erich Hückel, "Zur Theorie der Elektrolyte: I. Gefrierpunktserniedrigung und verwandte Erscheinungen," *Phys. Zeit.* 24 (1923), 185–206; Debye, "Zur Theorie der Elektrolyte: II. Das Grenzgesetz für die elektrische Leitfähigkeit," *Phys. Zeit.* 24 (1923), 305–25; Debye, "Report on Conductivity of

The chemical bond, in which all the lines connected a pair of atoms, was only one degree—actually the most extreme degree—of chemical union. Residual valence was a necessary consequence of this divisibility and not of the electron's charge.

Johannes Thiele (1865–1918) in his theory of partial valence (1899) also proposed the divisibility of valence. He employed this idea to account for the apparent residual valence of many atoms as well as the addition reactions of unsaturated and conjugated hydrocarbons such as 1,3-butadiene.[73] But neither Thiele nor Frankland, in his brief article, provided an electronic interpretation of residual valence, though Frankland's belief in the divisibility of the electron's lines of force easily lent itself to one.

An electron interpretation of Thiele's theory was not long in coming, however. In a 1908 publication Hugo Kauffmann (1870–?) in Stuttgart argued that accepting the electron theory of valence actually demanded the divisibility of valence, that is, divisibility of the lines of force associated with an electron.[74] His electron structures showed a single electron dividing its lines of force among three atoms, A_1, A_2, and A_3, as well as a chemical bond resulting when two atoms, A_1 and A_2 or A_2 and A_3, shared a pair of electrons:

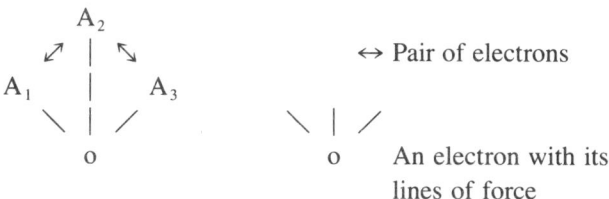

↔ Pair of electrons

o   An electron with its lines of force

His structures, Kauffmann claimed, were entirely analogous to Thiele's notation

$$A_1 \leftrightharpoons A_2 \leftrightharpoons A_3 \ldots$$

in which the curved lines indicated that one valence of A_2 had divided itself between the two atoms A_1 and A_3.

Strong Electrolytes in Dilute Solutions," *Trans. Faraday Soc.* 23 (1927), 334–40; Lars Onsager, "Zur Theorie der Elektrolyte," *Phys. Zeit.* 27 (1926), 388–92, and 28 (1927), 277–98.

[73] Johannes Thiele, "Zur Kenntnis der ungesättigten Verbindungen: I. Theorie der ungesättigten und aromatischen Verbindungen," *Annalen der Chemie* 306 (1899), 87–142.

[74] Hugo Kauffmann, "Elektronen und Valenzlehre," *Phys. Zeit.* 9 (1908), 311–14; idem, *Die Valenzlehre*, pp. 334–42, 539–41.

Thus, in Thiele's formula for ethylene, the dotted lines indicated both ethylene's unsaturated character and its ability to undergo additional reactions.[75] He illustrated this:

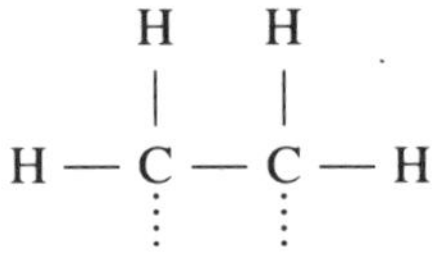

$$
\begin{array}{ccc}
\text{H} & & \text{H} \\
| & & | \\
\text{H} - \text{C} & - & \text{C} - \text{H} \\
\vdots & & \vdots
\end{array}
$$

But in Kauffmann's interpretation each dash was equivalent to a complete Faraday bundle, and each dotted line to a divided bundle; or each dash represented a single valence bond, and each dotted line a partial valence. Hence, the obvious similarity between Thiele's formula for benzene and its electron representation by Kauffmann:

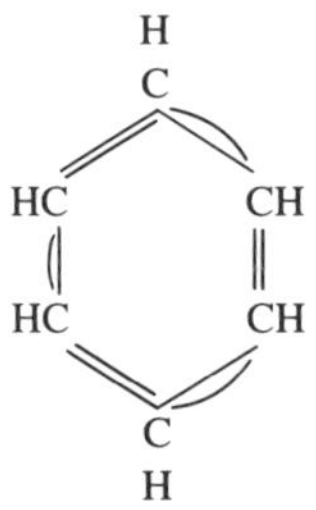

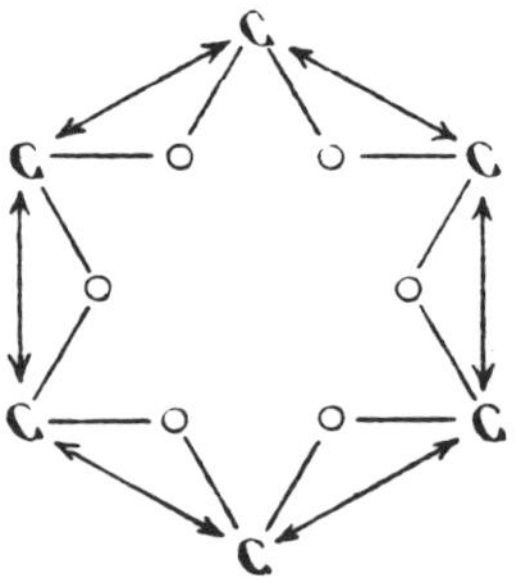

In this same publication, Kauffmann attempted to correct what seemed to him a serious weakness of the organic chemists' structural formulas. Every chemist knew, he pointed out, that a saturated hydrocarbon such as ethane was far less reactive than its unsaturated homologs ethylene and acetylene. Yet, the structural formulas given to each suggested exactly the opposite behavior. The triple-bonded carbon atoms in acetylene and the

---

[75] We might also note at this time another obvious resemblance, this one between Thiele's ethylene formula and the electron dot formula that G. N. Lewis used to show the reactivity of ethylene:

$$
\begin{array}{cc}
\text{H} & \text{H} \\
\text{H} : \text{C} : \text{C} : \text{H}
\end{array}
$$

In Lewis's formula each of Thiele's partial valences was equivalent to an unpaired electron, which Lewis represented by a single dot and considered to be equal to one-half a normal valence bond. See G. N. Lewis, *Valence and the Structure of Atoms and Molecules*, pp. 89, 124.

double-bonded carbons in ethylene appeared more firmly united and, consequently, less reactive than the single-bonded carbon atoms in ethane.

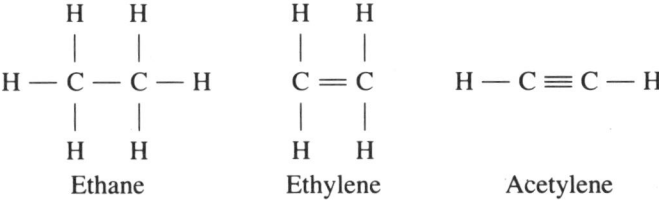

Ethane   Ethylene   Acetylene

Adolf von Baeyer (1835–1917) in 1885 had already tried to overcome this difficulty with his strain theory.[76] He accepted the presence of a double and a triple carbon-carbon bond in ethylene and acetylene, respectively, but he attributed their chemical reactivity to a distortion of these bonds from the normal tetrahedral direction found in ethane. Kauffmann proposed instead that the greater reactivity of double- and triple-bonded molecules was due to a repulsion of the numerous and closely packed lines of force from the line joining the centers of the two carbon atoms. These lines now extended like tentacles farther into space, permitting other atoms or molecules to come within their sphere of influence, and made multiple-bonded molecules more reactive than those with a single bond. At the same time, Kauffmann's proposal accounted for the state of strain set up in double- and triple-bonded molecules, just as Baeyer had postulated.

Conclusion

With the discovery of the electron and its establishment as a fundamental constituent of all matter, there arose an obvious need to incorporate the electron into an atomic model. J. J. Thomson's positive sphere model was the first serious attempt to develop an electrical atom. However, Thomson could not account for the atom's mass when the only known subatomic particle, the electron, had a mass $1/1700$ that of a hydrogen atom without assuming an immense number of electrons in every atom. Nor could he account for the neutrality of an atom containing only negative electrons without introducing a neutralizing sphere of positive electricity. The theory of the nuclear atom that Ernest Rutherford put forward in 1911 treated successfully the difficulties of Thomson's atom.

[76] Adolf von Baeyer, "Über Polyacetylenverbindungen," *Berichte* 18 (1885), 2277.

Thomson's electrostatic theory of valence gained the support of most chemists and physicists for nearly the first two decades of the twentieth century simply because they knew no other way of explaining the union of atoms. His discovery of the electron sparked efforts to regard all chemical union as the result of electron transfer, though there was frequently no evidence of the oppositely charged atoms that the transfer supposedly produced. This assumption eventually led to replacement of a purely electrostatic valence theory by the Lewis theory of electron pair sharing, at least for those compounds clearly not held together by the attractive force of oppositely charged atoms.

3. The Chemists' Reception of the Electron: 1897–1909

Introduction: The General State of Valence Theory

We have seen that those physicists who showed an interest in an electron theory of valence—Oliver Lodge and Norman R. Campbell—for example, looked favorably upon Thomson's electrostatic valence theory. There was, on the other hand, no agreement among them regarding the atom's internal structure, particularly the nature and distribution of the positive charge. Indeed, physicists had proposed at least four different internally structured atomic models—not to say anything of the various interpretations given to the Kelvin-Thomson positive sphere atom.[1]

In 1903 Lodge suggested that the atom, instead of having its negative electrons imbedded in a neutralizing sphere of positive electricity as in the Kelvin-Thomson model, might consist of alternating layers of positive and negative electrons. In that same year, Lenard published his dynamid theory of the atom, in which each dynamid consisted of a positive and a negative charge coupled together. Then in 1904, Hantaro Nagaoka (1865–1950) in Tokyo announced his Saturnian atom. A few years later, in 1909, an arrangement of the charges in an atom identical to Lodge's appeared in Albert E. Garrett's publication *The Periodic Law*.[2] Clearly, speculation on the atom's internal structure seemed to be the order of the day among physi-

---

[1] The book by George K. T. Conn and Henry D. Turner, *The Evolution of the Nuclear Atom* (pp. 91–130), contains a discussion of the atoms of J. Jeans, Rayleigh, Kelvin, H. Nagaoka, and G. A. Schott.

[2] Oliver Lodge, *Modern Views on Matter*, p. 11 (the same model is described in Lodge's *Electrons*, pp. 148–49); Philipp Lenard, "Über die Absorption von Kathodenstrahlen verschiedener Geschwindigkeit," *Annalen der Physik* 12 (1903), 714–44; Hantaro Nagaoka, "Kinetics of a System of Particles Illustrating the Line and Band Spectrum and the Phenomena of Radioactivity," *Phil. Mag.* 7 (1904), 445–55; Albert E. Garrett, *The Periodic Law*, p. 283.

cists in the years before Ernest Rutherford introduced the nuclear atom in 1911.

These structured atomic models aroused little interest among chemists. They were highly mathematical, and, with the exception of J. J. Thomson's model, physicists had developed them solely to account for radiation and atomic spectra, rather than to solve the chemists' problems of periodicity and bonding. No doubt chemists found little reason for paying attention to the physicists' structured atoms in a second factor. They had already developed a valence theory that satisfactorily accounted for the number and arrangement of atoms in molecules without using a structured atom. The physicists were still arguing among themselves about the number and arrangement of positive and negative charges within the atom.

Examination of the major chemical journals in the decade following the electron's discovery and its establishment as a fundamental constituent of all atoms in fact shows that electron models made little impact on the chemists' treatment of valence. The publications that dealt with valence and the structures of organic and inorganic molecules were almost always descriptive and empirical. Whenever questions on theory arose, as in Thiele's theory of partial valence (1899)[3] or Moses Gomberg's hypothesis of trivalent carbon in the organic radical triphenylmethyl (1900),[4] no one (except Hugo Kauffmann) attempted to give them an electron interpretation.

Chemistry's highly descriptive and empirical state in the early years of the twentieth century, of course, did not satisfy all chemists. In Europe a new school of theoretical chemists led by Svante Arrhenius in Sweden and Jacobus H. van't Hoff, Walther Nernst, and Wilhelm Ostwald in Germany had begun to move chemistry away from its older empirical, descriptive state. Though Arrhenius's major investigation was the theory of dissociation and van't Hoff, Nernst, and Ostwald engaged mainly in studying chemical thermodynamics, each of them showed an interest in the problems of valence. Indeed, Richard Abegg, a product of the new school, set

[3] Johannes Thiele, "Zue Kenntnis der ungesättigten Verbindungen: I. Theorie der ungesättigten und aromatischen Verbindungen," *Annalen der Chemie* 306 (1899), 87–142.

[4] Moses Gomberg, "An Instance of Trivalent Carbon: Triphenylmethyl," *J.A.C.S.* 22 (1900), 757–71; idem, "Triphenylmethyl, ein Fall von dreiwertigem Kohlenstoff," *Berichte* 33 (1900), 3150–63; idem, "Über das Triphenylmethyl," *Berichte* 34 (1901), 2726–33, and 35 (1902), 2397–408; idem, "Über Triphenylmethyl," *Berichte* 36 (1903), 376–88, and 37 (1904), 1626–44; idem, "Über Triphenylmethyl Acetat," *Berichte* 36 (1903), 3924–30; idem, "On Trivalent Carbon," *Amer. Chem. Journal* 25 (1901), 317–35.

out to provide a theoretical basis for the study of valence with the newly discovered electron. In his role as editor of the *Zeitschrift für Elektrochemie* (1901–10) and founder of the *Handbuch der anorganischen Chemie* (1905), he intended to elevate chemistry from its descriptive position to a rational science.

William A. Noyes's address, delivered at the International Congress of Arts and Sciences in St. Louis in 1904, was another early attempt to convince chemists that the time had now come for chemistry to move beyond its descriptive and empirical nature. Noyes (1857–1941) pointed out that the chemical journals presented too much evidence that the "ever-increasing army of nascent doctors" apparently had nothing better to do than prepare new compounds. He urged that "chemists should not be content with rounding out organic chemistry as a descriptive science nor even with adding to the number of empirical rules which enable us to predict certain classes of phenomena. We must instead, place before ourselves the much higher ideal of gaining a clear insight into the nature of atoms and molecules and of the forces or motions which are the real reason for the phenomena which we study."[5]

Contemporary textbooks of general chemistry also reflected chemistry's descriptive and empirical condition, in particular that the study of valence meant learning a collection of numerical rules.[6] Hydrogen, which never combined with more than one atom of another element, had a valence of one and was the reference with which to compare the numerical combining power of other atoms. Chemists took the bond symbolism used at first by organic chemists, the familiar dash, and applied it indiscriminately to the bonding in all compounds. Seldom did they ever discuss the nature of the chemical bond or attempt to discover what kind of force held the atoms in a molecule other than to say that a force of affinity was responsible for chemical union.

But just as one finds in the scientific journals occasional publications by chemists, such as Abegg and Noyes, which dealt with electron theories

[5]William A. Noyes, "Present Problems of Organic Chemistry," *Science* 20 (14 October 1904), 501.

[6]F. W. Clarke and L. M. Dennis, *General Chemistry*, pp. 102–7; William McPherson and William Edwards Henderson, *An Elementary Study of Chemistry*, pp. 116–21 (first edition published in 1905). Clarke was chief chemist, U.S. Geological Survey; Dennis was professor of chemistry at Cornell University. Both McPherson and Henderson were at Ohio State University.

of the atom and valence, so one finds these same topics discussed in a few chemistry textbooks used during the first decade of the twentieth century. Two popular American textbooks that mentioned an electron theory of the atom and valence were Alexander Smith's *General Chemistry for Colleges* (1908) and Louis Kahlenberg's *Outlines of Chemistry* (1909).[7]

While Smith's treatment was merely descriptive, Kahlenberg (1870–1941), like Arrhenius, wanted chemists to adopt a wait-and-see attitude before applying the electron theory to chemistry. His discussion was so brief that the following quotation presents everything Kahlenberg wrote on the electron in his *Outlines*:

The electron theory considers electricity itself to be material in character and to consist of corpuscles or electrons that weigh about 0.0005 as much as a hydrogen atom. This theory has developed from a study of radium rays and the discharge of electricity through rarefied gases. The electrons are considered to be negative electricity itself. Positive electrons appear to be much more difficult to isolate. J. J. Thomson has recently attempted to construct a theory that the atoms of the various elements are composed entirely of electrons, and has shown that, on the basis of such an assumption, the properties of the elements would exhibit periodicity as indicated by the periodic system. The electron theory has not yet been tested as to its value in the study of chemical changes.

Indeed, in a 1915 edition of his book, Kahlenberg still expressed doubt about the fruitfulness of the electron theory in chemistry, remarking that it had "thus far not proved to be of special value."[8]

George S. Newth's *A Textbook of Inorganic Chemistry* (1905) and Wilhelm Ostwald's *Outlines of General Chemistry* (1908) are additional examples of textbooks that examined the electron's role in valence theory. Considering his book's publication date (1905), Newth presented a remarkable description of the electrostatic bond and the Faraday bundle as the equivalent of the chemist's valence bond. Newth, at the Royal College of Science in London, obviously was aware of the Lodge–Percy Frankland articles published in *Nature* in 1904. These authors had interpreted the

[7] Alexander Smith, *General Chemistry for Colleges*; Louis Kahlenberg, *Outlines of Chemistry*. Smith (1865–1922) was professor of chemistry at the University of Chicago; Kahlenberg was chairman of the chemistry department at the University of Wisconsin. See also William Simon and Daniel Base, *Manual of Chemistry*, p. 87; John T. Stoddard, *Introduction to General Chemistry*.

[8] Kahlenberg, *Outlines of Chemistry*, p. 432; idem, *Outlines of Chemistry*, 2d ed., p. 454.

chemical bond to be a single bundle of lines of force and had also suggested that residual valence resulted whenever some of the lines connecting the atoms had their ends free for additional bonding. On the electron theory of valence, Newth wrote:

The trend of modern thought . . . lies in the direction of an electrical interpretation of valency. . . . Stated in briefest outline, this chemical "bond" or unit of affinity, which formerly has been regarded in the light of a single line of force—a fraction of a bond being considered as altogether inadmissible—is now regarded as a *bundle* of lines of force (a Faraday bundle). Under appropriate conditions, such as the proximity of suitable molecules or ions, it is conceived that some strands of the bundle may become loosened from one of the attached atoms and thus become available for attraction by similar wandering strands from other molecules. Obviously, therefore, this view admits of practically an unbroken gradation in degrees of chemical affinity. Instead, therefore, of *residual* affinity, we now have varying fractions of the total bundle of lines of force which in its entirety constitutes the chemical "bond;" the two conceptions are not very widely different.[9]

Ostwald's discussion on the electron theory of valence in his *Outlines* amounted to about one and a half pages.[10] According to Ostwald, the theory's simplest application was to the ions of electrolytes, which he believed, following Nernst, to be compounds of chemical atoms and electrons. Each electron added to an ion equaled one unit of valence. There was no need, Ostwald said, to be concerned with an internal structure of the atom, other than to assume an atom had a few electrons available for bond formation.[11]

For those compounds not ionizing in solution, Ostwald argued that a unit of valence was not necessarily equal to an electron. The compounds NO and CO, for instance, reacted to give NO_2 and CO_2, and, according to the electron theory, each of them must have possessed a free electron or free valence. Yet their obvious lack of electrical properties, Ostwald pointed out, stood in direct contrast to any theory that claimed they possessed free electrical charges. As Ostwald's reasoning indicates, he had equated the behavior of an atom having an additional electron—an ion, which would of course have the ability to conduct an electric current—

[9] George S. Newth, *A Textbook of Inorganic Chemistry*, pp. 67–68.

[10] Wilhelm Ostwald, *Outlines of General Chemistry*, trans. William W. Taylor, pp. 539–40.

[11] This concern only with an atom's outermost electrons was the approach chemists usually followed in their treatment of atomic structure and valence.

with a neutral atom having a free electron, that is, an electron still available for additional bond formation. The erroneous assumption in Ostwald's argument is sufficiently evident and requires no further comment.

The discussion above has made clear that the electron in the decade after its discovery had not become essential to the average chemist's theory of valence. But we turn now to the work of the few exceptions, those theoretically minded chemists who provide an entirely different picture. Here we find a swift assimilation of the electron into valence theory that was remarkable. Indeed, Nernst at Göttingen included an electron interpretation of valence in the 1898 edition of his *Theoretische Chemie*. This was only a year after Thomson's discovery of the electron. Then in 1899 Richard Abegg and Guido Bodländer put forward a theory of chemical affinity based on measurements of the attractive force between an atom and an electron.[12] A few years later, Abegg, William A. Noyes, and others attempted to develop an electron theory of valence.

Nernst's Electron Theory of Valence: The Application of the Positive Electron in Valence Theory

The electron theory of valence, by definition, cannot predate Thomson's discovery of the electron in 1897. Yet we must not overlook the influence that Hermann von Helmholtz's 1881 Faraday Lecture had in directing the attention of chemists once again to electrical valences. Helmholtz had shown that if Faraday's "electrochemical equivalent" was the quantity of electrical charge required to liberate one gram-atomic weight of an element, then the quantities required by the different elements were in the same ratio to each other as the valences of the elements. Thus hydrogen and chlorine, each having a valence of one (H = +1 and Cl = −1) required, respectively, the addition and the removal of 96,500 coulombs (one Faraday) to liberate one gram-atomic weight of the element. Zinc, with a valence of two (Zn = +2) gained 2 × 96,500 coulombs or two Faradays in liberating one gram-atomic weight of zinc. A similar relation held for other ions in solution.

In his Faraday lecture Helmholtz also proposed that an electrostatic force held the atoms in all compounds, even those which gave no evidence

[12] Walther Nernst, *Theoretische Chemie*, pp. 346–47; Richard Abegg and Guido Bodländer, "Die Elektroaffinität, ein neues Prinzip der chemischen Systematik," *Zeit. anorg. Chemie* 20 (1899), 453–99.

of ions in solution. To account for each atom's charge, Helmholtz assumed that electricity, like matter, consisted of atoms, and he postulated the existence of both a positive and a negative atom of electricity. A positive ion was a compound of a chemical atom with a positive atom or atoms of electricity; a negative ion, a compound of a chemical atom with one or more negative atoms of electricity.

Nernst (1864–1941) accepted Helmholtz's electrical explanation of valence, pointing out that his own studies on the variation of osmotic pressure in electrolytic solutions gave identical results for the magnitude of the ionic charges.[13] With the discovery of the negative electron, it was a simple matter for him to translate Helmholtz's positive and negative atoms of electricity into positive and negative electrons and to develop in 1898 an electrostatic theory of valence requiring both kinds of electrons. Nernst was thus one of the first chemists to incorporate the electron into valence theory, though in his theory he assumed a positive in addition to the negative electron, $\oplus$ and $\ominus$.

In accordance with Helmholtz's belief that electricity, like the chemical elements, was atomic, Nernst treated electricity as an elementary substance. It then followed from the laws of definite and multiple proportions that electrons could combine with chemical atoms only in whole-number ratios, forming ions that, Nernst said, behaved as saturated chemical compounds. Hydrogen ion, $H\oplus$, arose upon substituting a positive electron for the chlorine atom in a molecule of hydrochloric acid, HCl. Similarly, substitution of a negative electron for the hydrogen atom in HCl produced the compound $Cl\ominus$. Substitution of two negative electrons for the two hydrogen atoms in the sulfuric acid molecule, H_2SO_4, gave the compound $SO_4{\ominus\atop\ominus}$ and so on. Both atoms and radicals, according to Nernst, had different chemical affinities for the positive and negative electrons. Those that had a strong tendency to combine with positive electrons yielded positively charged ions, but an affinity for negative electrons characterized those atoms or radicals that formed negatively charged ions.[14]

While Thomson's extensive researches on gases in discharge tubes had confirmed beyond doubt the reality of the negative electron apart from an

[13] Walther Nernst, "Die elektrolytische Zersetzung wässiger Lösungen," *Berichte* 30 (1897), 1547–63; idem, *Theoretical Chemistry*, trans. Robert A. Lehfeldt (1904), pp. 350–79, 390–400.

[14] Nernst, *Theoretische Chemie*, pp. 346–47; idem, *Theoretical Chemistry* (1904), pp. 390–92. These same ideas are found in George S. Newth, *A Textbook of Inorganic Chemistry*, p. 105.

atom, no comparable experimental evidence supported the existence of a free positive electron. Nernst acknowledged that no one had yet proved the existence of free positive electrons through the study of gaseous discharge; but this, he claimed, was probably because the positive electron had a much greater affinity for ordinary atoms and radicals than its negative counterpart, making its isolation much more difficult. Nernst, therefore, never doubted the eventual isolation of the positive electron.[15]

Actually, evidence allegedly supporting the separate existence of positive electrons began to appear in 1906. In one of these investigations Jean Becquerel (1878–1953) in Paris examined the spark spectrum of yttrium. Becquerel found that a certain group of lines exhibited the ordinary Zeeman effect, that is, a displacement of the spectral lines upon placing the light source in a magnetic field, but that another group gave an inverse effect. If the negative electrons in the yttrium atom produced the first displacement, then surely, Becquerel argued, the presence of positive electrons must have caused the inverse effect. Becquerel also thought he had obtained positive electrons for a very brief moment after passing an electrical discharge through a vacuum tube. However, he claimed they disappeared rapidly due to recombination with negative electrons, forming hydrogen gas, which always seemed to appear in the vacuum tube after the discharge had occurred.[16]

In another experiment, Robert W. Wood (1868–1955) at Johns Hopkins University investigated the influence of a magnetic field on the spectrum of sodium. Wood discovered a magnetic rotation or shifting of some of sodium's D lines in one direction and of others in the opposite direction, which once again suggested the possibility of positive and negative electrons in an atom. Zeeman had shown that the negatively charged electrons accounted for the "positive rotation" of sodium's D lines; Wood thus claimed that the reverse direction of rotation, the "negative rotation" of the sodium D lines, clearly favored the presence of positive electrons.[17]

Though magnetic deflection of spectral lines was the only evidence directly supporting the positive electron, the positive electron's popularity was fairly widespread. But not everyone assumed, as Becquerel and Wood

[15] Nernst, *Theoretical Chemistry* (1904), p. 393.

[16] Jean Becquerel, "Sur un phénomène attribuable à des électrons positifs, dans le spectre d'étincelle de l'yttrium," *Comptes rendus* 146 (1908), 683–85.

[17] Robert W. Wood, "On the Existence of Positive Electrons in the Sodium Atom," *Phil. Mag.* 15 (1908), 274–79. Becquerel's and Wood's observations require a quantum mechanical interpretation, including electron spin.

did, that it would be a mirror image of the negative electron. Rutherford, while accepting the existence of a positive electron, argued that the reason why no one had yet isolated it was "simply because everybody expected that it would have properties exactly opposite to those of the negative electron, or in other words, that it would be the mirror image of the negative electron." The positive electron, Rutherford suggested, might "have entirely different properties, and so elude the vigilance of those who were seeking it." [18]

The real existence of what Nernst and others called the positive electron was resolved only after 1911, when Rutherford's nuclear atom became the accepted model of atomic structure. Nernst, however, continued to maintain his belief in a positive electron that was a mirror image of the negative electron until the publication of his 1923 edition of *Theoretical Chemistry*, in which he wrote: "It was, however, a great surprise to learn that the positive electron is identical with the hydrogen ion—not, that is to say, with the hydrogen ion present in aqueous solutions, which may be hydrated, but rather with the positive nucleus of the hydrogen atom, which results from the removal from the latter of the rotating electron." [19]

In concluding the discussion on Nernst's valence theory, we should note that his idea of an ion behaving as a saturated chemical compound found some support in Alfred Werner's study of platinum-ammonia coordination compounds. Werner had shown that one or more chloride ions could replace an identical number of ammonia molecules in the coordination sphere of the central platinum atom, though the replacement always brought about a measurable change in the coordination compound's conductivity. Nernst's idea also formed the basis of William Ramsay's valence theory, as seen from the title of Ramsay's 1908 publication "The Electron as an Element." Further, there is considerable similarity between Nernst's saturation idea and the 1916 electron theories of G. N. Lewis and the German physicist Walther Kossel (1888–1956) in Munich. [20]

[18] "The Construction of the Atom," *Chemical News* 96 (23 August 1907), 94.

[19] Walther Nernst, *Theoretical Chemistry*, trans. L. W. Codd (1923), p. 464. This statement does not appear in the seventh German edition of Nernst's *Theoretical Chemistry*, trans. H. T. Tizard (1916), p. 431.

[20] Alfred Werner, "Beiträge zur Konstitution anorganischer Verbindungen," *Zeit. anorg. Chemie* 3 (1893), 267–330, and 8 (1895), 153–97; Alfred Werner and Arturo Miolati, "Beiträge zur Konstitution anorganischer Verbindungen," *Zeit. phys. Chemie* 12 (1893), 35–55; William Ramsay, "The Electron as an Element," *J. Chem. Soc.* 93 (1908), 774–88; G. N. Lewis, "The Atom and the Molecule," *J.A.C.S.* 38 (1916), 762–85; Walther Kossel, "Über Molekülbildung als Frage des Atombaus," *Annalen der Physik* 49 (1916), 229–362.

Lewis and Kossel argued that atoms gained or lost electrons in order to acquire the electron structure of one of the rare gases, which, with respect to chemical reactivity, appeared to be saturated chemical compounds. Kossel required only electron transfer from one atom to another to achieve saturation in his electrostatic theory of valence. But Lewis, in addition to electron transfer, introduced at that time an idea that has since remained the very foundation of modern valence theory—a pair of electrons shared between two atoms, as in the elementary diatomic gases or other un-ionized molecules, led to saturation.

At the time Nernst was discussing the electron's role in valence theory, chemists had no conception of electron sharing to account for the obvious differences between polar and nonpolar compounds. Nernst maintained that as well as an electrostatic or polar attraction atoms exercised a chemical attraction on each other. Two atoms of the same element, such as hydrogen or nitrogen, could unite and form a stable diatomic molecule without any apparent involvement of their electrons. Chemical attraction also accounted for the formation of the nonmetallic compounds, among them iodine chloride and phosphorous sulfide, and for the numerous organic compounds. Nernst did not rule out the possibility that in nonpolar reactions, electrical forces operated in the background, though this, he believed, was a question for chemists of the future to answer.[21]

Nernst's theory of valence thus required a positive and a negative electron to explain the formation of polar compounds (the simple salts) and the ions present in solutions of acids, bases, and salts. At the same time, his theory retained the notion of a chemical or atomic affinity to describe the bonding in the nonpolar molecules of both inorganic and organic chemistry.

Abegg's Theory of Normal and Contravalences

Richard Abegg (1869–1910) had been a *Privatdozent* and an assistant to Nernst at Göttingen and then professor of chemistry at Breslau. Like Nernst, he was one of the first chemists to apply ideas about the electron to valence theory. In his 1899 publication with Guido Bodländer (1855–1904), Abegg suggested that chemists should replace the older idea of a chemical affinity existing among atoms with the new theory based on the affinity of atoms for electrons. Because they could not measure elec-

[21] Nernst, *Theoretical Chemistry* (1923), pp. 472–73.

troaffinities directly, Abegg and Bodländer urged instead the adoption of electromotive force (EMF) potentials, which followed the same order as the electroaffinities.[22]

Their table of potentials thus resembled those in use today. It listed values of an atom's affinity for negative electricity, that is, a measure of the force that bound an electron to an atom to form an anion. Included as well were measurements of an atom's affinity for positive electricity, or the force with which a cation held its positive electron, opposing the formation of a neutral atom. Abegg and Bodländer were following Nernst in using both positive and negative electrons in their theory. But as they pointed out, this hypothesis did not affect their conclusion, for they believed no essential difference resulted in the affinity measurements using either a unitary or a dualistic theory of electricity.

An important feature of Abegg and Bodländer's work was their recognition that electroaffinity accounted for all chemical combinations including those combinations supposedly due to Werner's secondary valences and to molecular attractions. They traced a parallel between the electroaffinity of an atom or a radical and its tendency to form complex ions. For the ion $Fe(CN)_6^{-3}$, and others involving iron, copper, or silver combined with chloride, bromide, iodide, and cyanide, they showed that in many cases the ease of forming a coordination compound increased as the negative atom's electroaffinity increased. Because of observations such as these, Abegg and Bodländer ruled out a constant valence for each element. They proposed instead a maximum valence, which an element did not always attain in its compounds but varied according to the element combining with it. Indeed, with their theory of electroaffinity, Abegg and Bodländer believed that they could express quantitatively the elements' well-known change from electropositive to electronegative character when moving from left to right across the periodic table. Their theory also accounted for the increase in electropositive character in a particular group of elements with increasing atomic weight.

In later publications on valence in 1902, 1904, and 1907, Abegg continued to call attention to the importance of electroaffinity, though now he

[22] Abegg and Bodländer, "Die Elektroaffinität," pp. 454–55. Criticism of Abegg and Bodländer's paper came from James Locke of Yale University in "Electro-Affinity as a Basis for the Systematization of Inorganic Compounds," *Amer. Chem. Journal* 27 (1902), 105–17. For Abegg and Bodländer's reply, see their article "Electro-Affinity as a Basis for the Systematization of Inorganic Compounds," *Amer. Chem. Journal* 28 (1902), 220–28.

argued that electroaffinity was the tendency of an atom to gain or lose only negative electrons.[23] An atom's valence—or better, its electrovalence—was the number of negative electrons gained or lost to insure stability. All bonds whether in atomic, molecular, or coordination compounds were, therefore, electrostatic, or polar. Hence, chemists should draw no hard and fast lines among these compounds. Such distinctions were misleading and unnecessary because an electron theory of valence explained the formation of all compounds and by so doing solved the problem of variable valence.

According to Abegg, the total number of electrovalences, positive and negative, equaled the number of positions on an atom to which electrons could attach or remove themselves. This number had a maximum value of eight for all atoms, but the actual electrovalence of an atom in its combinations depended on two factors: (1) the atom's position in the periodic table and (2) the electropolarity difference between the atoms, which indicated the electroaffinity difference of the combining atoms.[24]

Summarizing his hypothesis, which included a statement of his famous "rule of eight" and a description of the two kinds of electrovalence, Abegg wrote:

Every element possesses a positive as well as a negative maximum valence whose sum is always eight. The maximum positive valence is the same as the number of the periodic group to which the element belongs. Whether an element manifests its positive or its negative electrovalence depends on the differences of polarity [and the electroaffinities] of the elements with which it combines. The manifestation of one kind of valence appears greatly to hinder, but not completely to suspend that of the other kind. We shall call those valences which are less in number (<4), and therefore stronger than the others, the *normal valences*, and those which are more

---

[23]Richard Abegg, "Versuch einer Theorie der Valenz und der Molekülarverbindungen," *Christiania Videnskabs-Selskabet Skrifter* 12 (1902); idem, "Die Valenz und das periodische System: Versuch einer Theorie der Molekülverbindungen," *Zeit. anorg. Chemie* 39 (1904), 330–80; idem, "Valency," *B.A.A.S. Report* 77 (1907), 481.

[24]The significance attached to the number eight did not originate with Abegg. Dmitri Mendeleev at St. Petersburg, Russia, on investigating the relation between periodicity and valence, had called attention to the hydride and oxide formulas of atoms in Groups IV through VII, pointing out that the valence sum of each atom's hydride and oxide was eight. For the hydride and oxide of a Group V element, RH_3 and R_2O_5, R had a valence of 3 in RH_3 and a valence of 5 in R_2O_5. Translated into electrical terms, Mendeleev had shown that the sum of the negative and positive valences of these atoms was eight (Mendeleev, "Essai d'un système des éléments d'après leurs poids atomiques et propriétés chimiques," *Journal of the Russian Physical and Chemical Society* 1 [1869], 60–77). Abegg gave this well-known relation its first published electron interpretation more than thirty years later. At about the same time, G. N. Lewis had developed, but not published, an electron theory of valence based on the distribution of eight electrons at the corners of a cubic atom (see chapter 7).

in number than the others, and weaker because of contrary polarity, the *contravalences*, of the elements. Thus, Cl possesses one negative normal valence and seven positive contravalences, and analogously Ag, one positive normal valence and seven (hypothetical) contravalences. The manifestation of the maximum valence is not necessary. . . . The contravalences, in particular, are seldom completely employed. Increasing atomic weight [within a periodic group] facilitates their manifestation.[25]

The following table taken from Abegg's 1904 paper[26] shows clearly his theory of normal and contravalences and his "rule of eight":

Groups	I	II	III	IV	V	VI	VII
Normal valence	+1	+2	+3		−3	−2	−1
Contravalence	(−7)	(−6)	(−5)	±4	+5	+6	+7

Abegg's contravalences resembled Berzelius's hypothesis that a residual charge remained in an atom even after compound formation. In his valence theory the residual charge became the residual contravalences remaining in an atom after satisfying its normal valences.[27] Abegg was also restating in electronic terms Berzelius's belief that every atom consisted of one or more electric poles that permitted the positive and negative electricity to reside in opposite parts of an atom like the poles of a magnet. Accordingly, every atom had both a positive and a negative electric character, or was amphoteric.

Abegg argued that even in compounds that chemists usually assumed were nonpolar, like many organic compounds, a polarity would result because of the electroaffinity difference of the combining atoms. However, the polarity might be small enough to be negligible. This was the reason, he pointed out, why hydrochloric acid, HCl, did not react with methane, $H_3C.H$, whereas methyl alcohol, $H_3C.OH$, being more polar because of the negative hydroxyl group, reacted with the acid to give methyl chloride, $H_3C.Cl$, and water.[28] Indeed, the greater the compound's polarity, the greater its reactivity.

[25] Abegg, "Die Valenz und das periodische System," pp. 343–44.

[26] Ibid., p. 344.

[27] We can make exactly the same case for Werner's secondary valences. Werner at one time spoke of "a fundamental difference between the two types of valency" but later believed they were "identical," that is, both his primary and secondary valences were electrochemical (Alfred Werner, *New Ideas on Inorganic Chemistry*, pp. 67–70).

[28] Ibid., pp. 375, 380–81. His suggestion that some compounds were more polar than others did not imply the idea of an incomplete transfer or the sharing of an electron. In refer-

In methyl alcohol and methyl chloride, the carbon atom used a positive valence to combine with the negative groups, Cl and OH. Alternatively, when a carbon atom in a methyl group combined with zinc to give zinc dimethyl, H_3C—Zn—CH_3, it used one of its negative valences to form the bond. It was, therefore, not unlikely, Abegg believed, that in the hydrocarbon ethane, H_3C—CH_3, one of the carbon atoms carried a positive and the other a negative charge.[29]

Abegg's Rules for Determining the Electropolarity of Atoms

The behavior of the carbon atom in methyl alcohol and zinc dimethyl was a clear example of an atom's using, depending on the circumstances of combination, either a positive or a negative valence. On the basis of this and other similar experimental observations, Abegg proposed in 1904 the following four rules for establishing the electropolarity of an atom in a compound. These rules enabled him to determine whether an atom lost or gained electrons and was exercising its positive or negative valence.[30]

First, if a compound ionized in a solvent, electrolytic decomposition revealed the polar character of its constituents. This method provided the chief evidence for the charge on an atom. Abegg also used Paul Walden's studies on the specific conductivity of nonaqueous solutions and those of Henri Moissan (1852–1907) on metallic hydrides. Walden (1863–1957) at the University of Riga investigated solutions of the halogens bromine and iodine and of the interhalogen compounds iodine chloride and iodine bromide in liquid sulfur dioxide, ammonia, and ether. His results established the presence of the anions I^{+1}, I^{+3}, Br^{+1}, Br^{+3} and, together with the unquestioned existence of the anions of these elements, Br^-, I^-, Cl^-, supported Abegg's idea that the bond in the elementary diatomic molecules was polar.[31] In Paris, Moissan's experiments on the conductivity of the metallic hydrides of sodium, potassium, rubidium, and cesium led Abegg to

---

ring to polarity differences, Abegg meant only that the differences in the strengths of his measured electroaffinities were sometimes large and sometimes small, producing compounds possessing different degrees of polarity.

[29] Ibid.

[30] Ibid., pp. 338–44.

[31] Paul Walden, "Über abnorme Elektrolyte," *Zeit. phys. Chemie* 43 (1903), 385–464; idem, "Über einige anorganische Lösungs und Ionisierungsmittel," *Zeit. anorg. Chemie* 25 (1900), 225. J. J. Thomson used Walden's results in his later investigations on electrovalence. See J. J. Thomson, *The Corpuscular Theory of Matter*.

suspect the existence of the hydride ion H^- and to represent the hydrogen molecule as H^+H^-.[32] Abegg assumed the atoms to be materially the same but different in polarity.

Second, if a compound hydrolyzed, its positive component attached itself to the hydroxyl group, and the other component, having attached itself to the hydrogen ion, was negative. Thus, if the compound XY hydrolyzed accordingly,

$$XY + HOH = XH + YOH,$$

X used a negative and Y a positive valence in the original compound, even though XY might not ionize in the usual sense. The hydrolysis of B_2S_3 and P_2S_3 indicated the presence of B^{+3} and P^{+3}, while other hydrolysis reactions gave evidence for As^{+3}, As^{-3}, N^{-3}, and Sb^{-3}.

Harry Shipley Fry (1878–1949) at the University of Cincinnati developed an electron theory of valence a few years later and assumed as Abegg had that the ions existed prior to hydrolysis. Indeed, Fry argued that hydrolysis reactions provided evidence for a new kind of isomerism called *electronic isomerism* or *electromerism*. A compound XY could exist in two forms, X^+Y^- and X^-Y^+, which differed only in the location of the electrons, though Fry emphasized that in general only one electromer was stable. The other, if it existed at all, was present in minute concentration.[33]

Third, the relative positions of two atoms in the periodic table indicated their polar character in a binary compound. For compounds consisting of a metal and a nonmetal, including those not ionizing in aqueous solution, such as PbS or HgS, Abegg merely stated what chemists already accepted, namely, that the metallic groups of elements were positive and the nonmetallic groups, negative. But in a compound like ICl, in which both atoms belonged to the same periodic group, he argued that the atom having the greater atomic weight, in this case iodine, was always positive because it could use its positive contravalence more easily. This rule followed from the fact that fluorine formed no oxygen compounds, whereas

[32] Henri Moissan, "Sur la non-conductibilité électrique des hydrures métalliques," *Comptes rendus* 136 (1903), 591–92; idem, "Étude de la combinaison de l'acide carbonique et de l'hydrure de potassium," *Comptes rendus* 136 (1903), 723–27.

[33] Harry S. Fry, "Interpretations of Some Stereochemical Problems in Terms of the Electronic Conception of Positive and Negative Valences: Part IV. The Simultaneous Formation of Ortho-, Meta- and Para-Substituted Derivatives of Benzene," *J.A.C.S.* 37 (1915), 863–83; idem, "The Electronic Conception of Positive and Negative Valences," *J.A.C.S.* 37 (1915), 2368–73.

the oxygen compounds of the other halogens generally increased in stability with increasing atomic weight. The halogens' behavior seemed to indicate a greater availability of positive contravalences as the atomic weight increased within a group.

Fourth, the positive valence of an atom varied up to a maximum, but its negative valence was invariable. In the compounds SO_2 and SF_6, or ICl and ICl_3, the valence of the sulfur and the iodine atoms varied, and they were, therefore, using their positive contravalences in these compounds.

The Valence of Nitrogen and the Structure of Ammonium Chloride

The valence of nitrogen in ammonium chloride and the structure of the ammonium chloride molecule had been the subject of much debate among chemists since the doctrine of valence first appeared in the 1850s.[34] Ammonium chloride dissociated on heating to give ammonia and hydrogen chloride and sublimed when perfectly dry without decomposition, but upon solution it yielded ammonium and chloride ions.

Kekulé, in keeping with his belief that an atom's valence was invariable, considered ammonium chloride, NH_4Cl, to be a molecular compound of ammonia and hydrogen chloride, $NH_3.HCl$. This interpretation permitted the nitrogen atom to maintain a constant valence of three in both ammonia and ammonium chloride and received experimental support from ammonium chloride's dissociation on heating to give ammonia and hydrogen chloride.

According to Edward Frankland, an atom's valence varied. Frankland maintained that ammonium chloride was an atomic compound having all five atoms, four hydrogen and one chlorine, joined directly to the central nitrogen atom, giving nitrogen a valence of five and not three as in Kekulé's theory. Indeed, nitrogen's pentavalence seemed confirmed when Herbert B. Baker (1862–1935) showed in 1894 that ammonium chloride, if perfectly dry, sublimed without decomposition.[35] However, neither Kekulé's nor Frankland's structure could account for ammonium chloride's dissociation upon solution into ammonium, NH_4^+, and chloride, Cl^-, ions.

[34] For examples of the continuing debate on ammonium chloride's structure see John Newton Friend, "Electrochemical Conceptions of Valency," *J. Chem. Soc.* 119 (1921), 1040–47; and J. D. Main Smith, "Friend's Theory of Valency," *Chemical News* 124 (17 February 1922), 84–86.

[35] Herbert B. Baker, "Influence of Moisture on Chemical Change," *J. Chem. Soc.* 65 (1894), 612.

Werner's coordination theory presented yet a third alternative, for it revealed an unsuspected relation between the structures of the many so-called molecular compounds and their electrical conductivity when in solution. In Arrhenius's dissociation theory, the electrical conductivities of solutions of equal concentration varied directly with the number of ions present in the solution. Werner's introduction of coordination formulas showed the number of ions resulting upon solution of these compounds. Applied to the problem of ammonium chloride, Werner's formula indicated the presence of two groups, the coordination group NH_4 and a chlorine atom, held together electrostatically:

$$\begin{bmatrix} H & & H \\ & \diagdown & \diagup \\ & N & \\ & \diagup & \diagdown \\ H & & H \end{bmatrix}^{+} \quad Cl^{-}.$$

Upon solution the electrostatic bond broke, resulting in the formation of ammonium and chloride ions.[36]

In this respect, Werner's formula was a definite improvement over those of Kekulé and Frankland. His formula also showed clearly that the nitrogen atom had a valence of four since directly attached to it were the four hydrogen atoms comprising the coordination or first sphere. The chlorine atom remained outside the coordination sphere. It formed the second sphere and did not attach itself to any one atom within the coordination sphere but belonged to the NH_4 group as a whole.

Abegg's theory of normal and contravalence when applied to the ammonium chloride structure showed that ammonium chloride was an atomic compound and that the nitrogen atom was pentavalent. Nitrogen's normal valences held three of the hydrogen atoms, while contravalences of opposite sign bound the fourth hydrogen atom (which used its negative contravalence) and the chlorine atom.[37] This was illustrated:

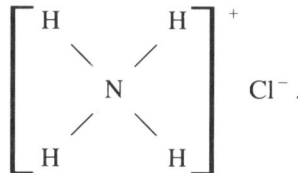

$$\begin{matrix} H + - & & + - H \\ H + - & \mathbf{N} & + - H \\ H + - & & + - Cl. \end{matrix}$$

Since weak contravalences held the chlorine and fourth hydrogen atom, Abegg's formula provided an explanation for ammonium chloride's

[36] Werner, *Inorganic Chemistry*, pp. 193–94.

[37] Abegg, "Die Valenz und das periodische System," p. 372.

dissociation on heating. But as Arrhenius and John Newton Friend pointed out,[38] the difference in binding the fourth hydrogen suggested the existence of two isomers when an alkyl group, CH_3, C_2H_5, C_6H_5, replaced a hydrogen atom:

$$
\begin{matrix} H + - \\ H + - \\ R + - \end{matrix} \mathbf{N} \begin{matrix} + - H \\ \\ + - Cl \end{matrix} \quad \text{and} \quad \begin{matrix} H + - \\ H + - \\ H + - \end{matrix} \mathbf{N} \begin{matrix} + - R \\ \\ + - Cl \end{matrix},
$$

where R = CH_3, C_2H_5, or C_6H_5.

In a paper on substituted ammonium derivatives published thirty years earlier in 1876, Victor Meyer at Zurich had shown the improbability of such isomers' existing.[39] Meyer demonstrated that in the two reactions,

$$(CH_3)_2NH + 2\ C_2H_5I = (CH_3)_2N(C_2H_5)_2I + HI,$$

$$(C_2H_5)_2NH + 2\ CH_3I = (C_2H_5)_2N(CH_3)_2I + HI,$$

the same ammonium derivative resulted, which indicated clearly the identity of nitrogen's four valences. At the time, chemists used his demonstration to refute Kekulé's suggestion that ammonium compounds were molecular associations. Of course the same argument was equally valid when applied to Abegg's use of normal and contravalences to describe the structure of ammonium chloride. Indeed, as a result of Meyer's studies, which showed the equivalency of four of nitrogen's valences, Arrhenius and Friend proposed that Abegg's contravalences in ammonium chloride, which they called *neutral* or *latent* valences, were of opposite sign to one another.[40] By doing so they insured the identity of four of nitrogen's valences:

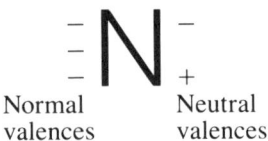

Normal   Neutral
valences   valences

[38] Arrhenius, *Theories of Chemistry*, pp. 81–84; John Newton Friend, *The Theory of Valency*, p. 144.

[39] Victor Meyer and Marco T. Lecco, "Über die Konstitution der Ammoniumverbindungen und des Salmiaks," *Berichte* 8 (1875), 233–42; idem, "Untersuchungen über die Konstitution der Ammoniumverbindungen und des Salmiaks," *Annalen der Chemie* 180 (1876), 173–91.

[40] See chapter 2, "Arrhenius's Criticism of the Thomson Atom," and Werner's structure of ammonium chloride, chapter 1, "Werner's Coordination Theory."

Abegg also applied his theory of normal and contravalences to the structures of complex cations and anions. The ions $[Ag_3I]^{++}$ and $[AgI_2]^-$ he illustrated as follows:

$$+Ag(-)(+)$$
$$\qquad\qquad I- +Ag \qquad \text{and} \qquad -I(+)(-)Ag+ -I$$
$$+Ag(-)(+)$$
$$\qquad [Ag_3I]^{++} \qquad\qquad\qquad [AgI_2]^-,$$

where ordinary $+$ and $-$ signs represent the normal polar valences and $(+)$ and $(-)$ the polar contravalences of the silver and iodine atoms.[41]

An Evaluation of Abegg's Valence Theory

Abegg believed that all chemical bonds, whether in simple molecules or complex ions, were electrostatic or polar. In the light of modern electron valence theory this is, of course, incorrect. Chemists have replaced his structure, as well as the other electrostatic structures of ammonium chloride mentioned above, with a model that G. N. Lewis introduced in 1916. In the Lewis model an electron pair bond connected each of the four hydrogen atoms to the central nitrogen atom, forming the positively charged ammonium ion; an electrostatic attraction held the negative chloride ion to the positive ammonium ion.

Actually, the Lewis structure for ammonium chloride was an electronic interpretation of Werner's coordination model. The nitrogen atom was tetravalent in each case, and each required the presence of two different kinds of bonds. Werner used coordination (un-ionizable) bonds to hold the hydrogen atoms to the nitrogen in the ammonium group, and an electrostatic bond between the oppositely charged ions, NH_4^+ and Cl^-.[42] Lewis introduced electron pair bonds to connect nitrogen with the four hydrogen atoms in the ammonium group, and again, an electrostatic bond between the two ions.

Neither did the modern electron interpretation of complex ions or coordination compounds stem from Abegg's theory of polar normal and contravalences. Rather, it originated in Sidgwick's application of the Lewis shared electron pair to Werner's coordination models. Nonetheless, at the

[41] Abegg, "Die Valenz und das periodische System," p. 362.

[42] Werner seems to have believed that all chemical bonds ultimately would be electrical (*Inorganic Chemistry*, pp. 67–70).

time of its publication, Abegg's theory seemed quite promising, chiefly because chemists and physicists increasingly believed that in the last analysis all affinity forces resulted from electron gain or loss. Abegg's electropolar theory eliminated the need to postulate other kinds of affinity existing between atoms, or between molecules as in the molecular compounds and associated compounds like water, hydrofluoric acid, and nitrogen dioxide. Nor were indefinite crystallographic forces required to explain the formation of alums and double chlorides.

In commenting on the vast extent to which Abegg applied his valence theory, the British chemist Pattison Muir (1848–1931) in 1906 wrote: "To bring the constitution of all compounds under one category, to establish the electroamphoteric character of the elements, and to connect the exhibitions of that character with the atomic weights of the elements, is to make a most important and seminal contribution to the final answer one day to be given to the question—What happens when homogeneous substances interact?"[43]

There is little doubt of the importance that those chemists and physicists who were developing electron theories of valence attached to Abegg's rule of eight. We have already seen Thomson using Abegg's rule in trying to draw an analogy between the gain or loss of electrons by his electron rings and the actual valences of the chemical elements. Ramsay also used this rule when he assumed that eight electrons were the maximum number present in an atom's valence ring. With this assumption he proceeded to develop electronic models of the cobalt-ammonia coordination compounds.

The significance that Abegg saw in the number eight (it represented the maximum number of electron positions on an atom's surface) and his belief that all chemical bonds were polar foreshadowed many of the ideas found in Walther Kossel's 1916 valence theory. Kossel pointed out that the atomic numbers of the first three rare gases differed by eight electrons: He = 2, Ne = 10, and Ar = 18; and because these gases did not enter into chemical combination with any other element, he assigned a special stability to their electron configurations, writing them as He 2, Ne 2-8, and Ar 2-8-8. Kossel also proposed that an atom's valence was equal to the number of electrons it lost or gained in attempting to attain an electron structure identical to that of the rare gas nearest it in the periodic table. This was essentially the same idea Abegg tried to illustrate with his theory that the

---

[43] M. M. Pattison Muir, *A History of Chemical Theories and Laws*, p. 545.

maximum sum of an atom's positive and negative valences (normal and contravalences) was eight, though, of course, he did not know the number of electrons in an atom or their manner of distribution.

It appears unlikely that Abegg influenced G. N. Lewis's development of the cubic atom, which Lewis published in 1916, nearly simultaneously with Kossel's paper on valence. For in 1902, Lewis had already discussed, but not published, an atomic model that had the electrons situated individually at the corners of concentric cubes. In that same year Lewis arrived at an electrostatic theory of valence based on the number of electrons either lost or gained by an atom to give an outermost completed cube of eight electrons.

The Kinetic Valence Theory of William A. Noyes

In his 1904 address to the International Congress of Arts and Sciences, William A. Noyes remarked that it had become customary among chemists to think of the electron's involvement only in those reactions that occurred in solution. Yet, if we accept the electron theory of matter, he pointed out, "It is evident that the electrons must be present in the molecule of an electrolyte no matter in what manner it is formed, [and] it is just a step further to the conclusion that the electrons are involved in every combination or separation of atoms and, indeed, may be the chief factor in chemical combination." [44]

As Abegg and Thomson had done, Noyes was claiming that the forces binding the atoms in all compounds were electrical. He expressed this same idea a few years later, in 1909: "But if we assume that the forces which hold atoms together in electrolytes are electrical it is difficult to escape from the conclusion that the forces are electrical in the molecules of non-electrolytes also, for the two classes pass over into each other so gradually that it is very hard to believe that after the line is passed we are dealing with a radically different kind of atomic force." [45]

According to Noyes, organic reactions were ionic, and acceptance of this idea would enable chemists to understand many reactions not easily understood otherwise, for example, the reaction of phenol with phosphorus pentachloride to give chlorobenzene and phenyl phosphate. By assuming

[44] Noyes, "Present Problems of Organic Chemistry," p. 498.
[45] William A. Noyes, "Molecular Rearrangements," *J.A.C.S.* 31 (1909), 1370.

that phenol ionized slightly to phenyl and hydroxyl ions but mainly to hydrogen and phenoxy ions, chemists would find the formation of the products followed directly.[46] Expressed in an equation, it appeared:

$$3\ C_6H_5O^- + 3H^+ + C_6H_5^+ + O^{-2} + H^+ + P^{+5} + 5Cl^-$$
$$= C_6H_5Cl + (C_6H_5)_3PO_4 + 4HCl\,.$$

Even if chemists and physicists accepted the identity of atomic and electrical forces, Noyes argued that a true understanding of these forces still remained unknown. He had difficulty conceiving of an attraction between atoms acting at a distance without a medium and as an alternative suggested that the attraction might be kinetic. The motions of the electrons in an atom enabled the atoms to act on each other.[47]

Noyes never made clear what kind of electron motion he had in mind, but because of it, the electrons in certain atoms acquired a positive charge, and in other atoms, a negative charge. Hence, whenever combination took place between two atoms (or groups of atoms) it was because the electrons in these atoms had motions that produced positive and negative charges. When the combined atoms separated, the electrons retained, lost, or even reversed their motions, so upon separation either atom might become positive or negative due partly to its own behavior and partly to the behavior of the other reactants. This conception, said Noyes, was similar to the behavior of a magnetic pole: it could attract a pole of opposite kind, induce the formation of a pole of opposite kind, or reverse the polarity of another magnet.

Noyes's kinetic valence theory thus had the advantage of providing a theoretical explanation for an atom's apparent ability to acquire either a positive or a negative charge, as chemists and physicists often had to assume in order to account for the existence of the diatomic elementary gases and the interhalogen compounds. It was also simpler than J. J. Thomson's electrostatic theory for, unlike Thomson, Noyes assumed no inherent attraction between negative electrons and the positive component of an atom.[48]

In later publications Noyes abandoned the idea that an electron, depending on its motion, was sometimes positive and sometimes negative.

[46] Ibid.

[47] Noyes, "Present Problems of Organic Chemistry," p. 499.

[48] Ibid.; Noyes, "Molecular Rearrangements," p. 1371.

He continued to hold, however, that a kinetic or dynamic hypothesis requiring only negative electrons was still necessary to explain the electron's role in the combination of atoms. Indeed, in a 1917 paper Noyes suggested that a single electron revolving in an orbit that included two atoms might hold the atoms together. He applied this idea to the structure of the carbon atom in methane. The carbon atom did not have its positive charge concentrated in a single nucleus, as in Rutherford's atom; the charge was instead distributed tetrahedrally.[49] This arrangement accounted for the localization of the electron bonds in particular regions of the carbon atom as required by the structures of organic chemistry:

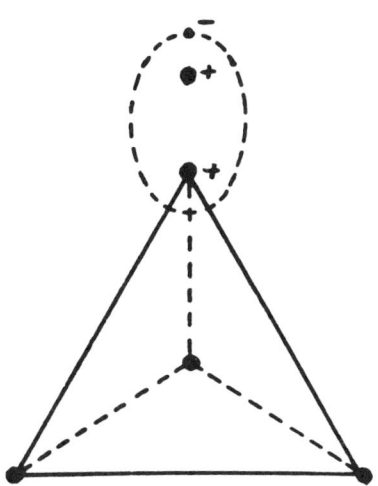

William Ramsay's Theory of Valence: The Electron as an Element

Two ideas, Abegg's rule of eight—the belief that an atom in a molecule could surround itself with a maximum of eight electrons—and Nernst's proposal that electricity was an element, were central to Ramsay's electron theory of valence. But while Nernst believed in two elements of electricity, using positive and negative electrons or "atoms of electricity" to account for an atom's positive and negative valences, Ramsay adopted a one-element theory of electricity, requiring only the negative electron in his va-

[49] William A. Noyes, "A Kinetic Hypothesis to Explain the Function of Electrons in the Chemical Combination of Atoms," *J.A.C.S.* 39 (1917), 879–82.

lence theory. According to Ramsay, positive valence resulted from the loss of negative electrons, and conversely, negative valence from their gain.

Ramsay (1852–1916) at University College, London, first published his theory in a small volume called *Modern Chemistry: Theoretical Chemistry* (1907), and in the following year, it was the subject of his presidential address to the London Chemical Society: "Electrons are atoms of the chemical element, electricity; they possess mass; they form compounds with other elements; they are known in the free state, that is, as molecules; they serve as the 'bonds of union' between atom and atom. The electron may be assigned the symbol 'E.' "[50]

To illustrate his electron bond of union, Ramsay chose the reaction of metallic sodium with chlorine gas, $ENa + Cl = NaECl$, where he used Cl, the atomic symbol for chlorine, rather than Cl_2, the molecular form in which chlorine naturally occurred, merely for the sake of simplicity. The symbols ENa and $NaECl$, representing sodium metal and sodium chloride, require further clarification.

ENa, metallic sodium, was no longer an element. According to Ramsay's interpretation of the electron as an atom of the element electricity, ENa was a compound of "sodion," Na^+, and an electron, E.[51] Following Abegg, Ramsay believed that an atom's highest positive valence determined the number of detachable electrons in the atom. For sodium, which had a single valence and formed an ion with a charge of $+1$, he had only to indicate the presence of one electron ENa. To the metals barium and aluminum he assigned the symbols E_2Ba and E_3Al, for barium and aluminum lost two and three electrons, respectively, in forming their metallic ions, Ba^{++} and Al^{+++}. Since the iron atom gave two metallic ions having

---

[50] William Ramsay, *Modern Chemistry: Theoretical Chemistry*; idem, *Modern Chemistry: Systematic Chemistry*; idem, "The Electron as an Element," *J. Chem. Soc.* 93 (1908), 778.

[51] Ramsay's atom required only negative electricity or electrons, which he considered to be a particular form of matter, and positive electricity, "which was matter deprived of negative electricity—that is minus this electric matter" (*Essays, Biographical and Chemical*, p. 176). Ramsay also discussed the idea that atoms of the heavier elements had evolved from the simpler atoms in his 1909 presidential address, "Elements and Electrons," *J. Chem. Soc.* 95 (1909), 624–37. His hypothesis resembled Thomson's earlier description of the evolution of the elements given in *Electricity and Matter*. The generic difference between elements, Ramsay wrote in his address, "is due to their gain or loss of electrons; not of such supplementary electrons as convert an element into an ion but of electrons more closely associated with the atom, constituents of the atoms, as it were" (p. 625). As a chemist, however, Ramsay's chief concern lay not with an atom's internal or constitutional electrons but with the "satellites" or detachable electrons that determined the atom's valence.

charges of $+2$ and $+3$, $\mathbf{E}_3$Fe represented the neutral iron atom, $\mathbf{E}$Fe indicated Fe^{++}, and Fe the Fe^{+++} ion.

With the term NaECl, Ramsay wanted to show that the sodium atom had transferred its single electron, $\mathbf{E}$, to chlorine at the moment of combination and that the transference resulted in a bond of union between the two ions in sodium chloride. The bond was an electrostatic bond, and the solid salt differed from its solution only because the solution permitted movement of the sodium and chloride ions.[52]

In attempting to present a mental picture of what occurred in the reaction, Ramsay offered the following fanciful analogy:

An electron is an amoeba-like structure, and that ENa may be conceived as an orange of sodium surrounded by a rind of electron, that on combination, the rind separates from the orange and forms a layer or cushion between the Na and Cl, and that on solution the electron attaches itself to the chlorine in some similar fashion forming an ion of chlorine. It will be noticed that E fills the place usually occupied by a bond; thus Na—Cl. It happens providentially that the bond and the negative sign are practically the same. Na—Cl may be supposed to ionize thus: Na($-$Cl), the negative charge or electron remaining with the chlorine.[53]

Because sodium chloride's formation involved the transfer of a single electron from sodium to chlorine, Ramsay's description of the reaction did not include the seven latent valence electrons that he believed were associated with the chlorine atom. Abegg also maintained that the chlorine atom had seven valence electrons, and for this reason, he assigned it a maximum contravalence of $+7$ in addition to its normal valence of -1. Both Ramsay and Abegg argued that the presence of seven valence electrons in the chlorine atom followed from chlorine's ability to form a well-known series of oxyacids, $HClO$, $HClO_2$, $HClO_3$, and $HClO_4$, and their corresponding salts. Indeed, accepting that oxygen was divalent and hydrogen monovalent in each oxyacid, chlorine had valences of 1, 3, 5, and 7 or what chemists considered the same thing at that time, chlorine formed 1, 3, 5, and 7 bonds, respectively:

$$
H{-}O{-}Cl \qquad \overset{\displaystyle O}{\overset{\|}{Cl}}{-}O{-}H \qquad O{=}\overset{\displaystyle O}{\overset{\|}{Cl}}{-}O{-}H \qquad O{=}\overset{\displaystyle O}{\overset{\|}{\underset{\|}{\underset{\displaystyle O}{Cl}}}}{-}O{-}H .
$$

[52] Ramsay, "Electron as an Element," p. 781.
[53] Ibid., pp. 781–82.

Further, since oxygen's charge was never electropositive but always -2 in compounds[54] and hydrogen always carried a charge of $+1$ in acids, the electron theory of valence predicted the chlorine atom would have a positive charge in each of the oxyacids. It had, together with hydrogen, donated electrons to the oxygen atoms, giving it electrovalences of $+1$, $+3$, $+5$, and $+7$ in the series $HClO$, $HClO_2$, $HClO_3$, and $HClO_4$.

Rewriting Ramsay's equation for the reaction of sodium with chlorine and showing the chlorine atom with its seven valence electrons gave

$$ENa + ClE_7 = NaEClE_7 .$$

Upon solution, the solid salt yielded sodium ion Na and chloride ions $EClE_7$, clearly revealing that the chloride ion now had eight valence electrons, which Ramsay claimed was the maximum number present in any ion. His electron formula for the chloride ion was a structural representation of Abegg's rule of eight, and his electron formulas were, therefore, an early attempt to indicate with the symbol **E** the number of electrons lying in the outermost region of an atom or ion.

Ramsay used his electron formulas to represent the dissolving of other salts, bases, and acids whose aqueous solutions were electrolytes. Electrolysis showed that these solutions obviously contained oppositely charged ions, but it was still an open question whether the ions were present in the compound before solution, as Ramsay and Thomson believed, or resulted from the solution process, as Arrhenius claimed. Though neither interpretation proved to be entirely correct, Ramsay's formulas below illustrated distinctly the existence of the ions in a solution, particularly in solutions of inorganic salts:

$$E_3Fe + 2ClE_7 \rightarrow EFe(EClE_7)_2 \quad \xrightarrow{H_2O} \quad EFe \text{ (aq.)} + 2EClE_7 \text{ (aq.)} \tag{3.1}$$

or

$$Fe + 2Cl \rightarrow FeCl_2 \quad \xrightarrow{H_2O} \quad Fe^{+2} + 2Cl^-$$

Iron atom	Chlorine atom	Ferrous chloride		Ferrous ion	Chloride ion

$$E_3Fe + 3ClE_7 \rightarrow Fe(EClE_7)_3 \quad \xrightarrow{H_2O} \quad Fe \text{ (aq.)} + 3EClE_7 \text{ (aq.)} \tag{3.2}$$

[54] In peroxides each oxygen atom has a charge of -1.

or

$$Fe + 3Cl \rightarrow FeCl_3 \quad \xrightarrow{H_2O} \quad Fe^{+3} + 3Cl^-$$

| Iron | Chlorine | Ferric | | Ferric | Chloride |
| atom | atom | chloride | | ion | ion |

His formula for ferrous chloride (equation 3.1) showed the iron atom, E_3Fe, donating two electrons, one to each chlorine atom, ClE_7, forming ferrous and chloride ions in the undissolved solid. The solution process merely permitted a separation of the existing ions. In the second equation, the interpretation was the same, except that the iron atom now donated three electrons, one to each of the three chlorine atoms producing ferric and chloride ions that also separated on solution.

The Electron Bond Applied to Nonelectrolytes

Ramsay's proposal that the electron served as the bond of union clearly suggested that whenever bond formation occurred between two atoms, one of the atoms was the electron donor and the other, the electron acceptor. In other words, one atom was active and the other passive—the dominating idea of nearly all electron valence theories in the first two decades of the twentieth century. Indeed, this idea influenced Ramsay's thinking when he applied his theory to the formation of gaseous molecules such as hydrogen chloride, HCl, hydrogen, H_2, and chlorine, Cl_2.

According to the usual rules of valence, both hydrogen and chlorine were monovalent and possessed one bond of affinity represented as H— and Cl—. "Now when they unite," Ramsay asked, "are there two bonds or one? Should we write H—Cl with one bond or H— —Cl with two?" Because each chemical bond was equivalent to a single electron, Ramsay argued that the symbol Cl— was incorrect for an atom of chlorine. Chlorine had "strictly speaking, no bond, that is, no electron, but merely possesses the power of receiving one from the hydrogen." Hydrogen chloride, therefore, had the formula H—Cl or electronically $HEClE_7$, with the bonding electron coming from the hydrogen atom. Similarly, the hydrogen molecule was HEHE and not HEEH, for though the two atoms were identical, one of the hydrogen atoms had to be the active atom or electron donor and the other, the electron acceptor or passive atom.[55]

[55] Ramsay, "Electron as an Element," p. 782.

Assigning an electron formula to the chlorine molecule was more difficult. Because each chlorine atom possessed seven valence electrons, one
or perhaps several electrons donated entirely by either chlorine atom or
partly by both atoms might bind the two atoms together. Allowing chlorine's valence to vary as it did in its oxygenated compounds, $+1$, $+3$, $+5$,
and $+7$, Ramsay considered the following to be possible electron formulas
for the chlorine molecule:

$$E_6ClECIE_7, \quad E_4ClE_3ClE_7, \quad E_2ClE_5ClE_7, \quad \text{and} \quad ClE_7ClE_7.$$

But he believed it best to select the simplest formula, $E_6ClECIE_7$, ClECl,
or Cl—Cl if we ignore the "latent" electrons, arguing that "a structural
formula shows by bonds those electrons which we deem it serviceable to
represent." [56]

Ammonia, $H_3E_3NE_5$, and methane, $H_4E_4CE_4$, were other examples of
Ramsay's electron formulas. In each case the bonding electrons, three in
ammonia and four in methane, came from the hydrogen atoms; each compound had a central atom surrounded by an outer group of eight electrons,
which clearly illustrated Abegg's rule of eight.

In assigning electron formulas either to chemical compounds or to the
elementary diatomic gases, Ramsay clearly had used an identical representation for the chemical bond, regardless of whether the substance dissociated into ions upon solution. Compare his formulas for sodium chloride,
$NaClE_7$, or ferric chloride, $Fe(ClE_7)_3$, which dissociated, with those
for chlorine, $E_6ClECIE_7$, or methane, $H_4E_4CE_4$, molecules, which did
not. Yet nowhere in his discussion on valence in *Modern Chemistry* or in
his presidential address did Ramsay explain why he employed an identical
mechanism of bond formation for these two distinctly different classes of
substances. [57]

Indeed, in his *Essays, Biographical and Chemical* (1908) Ramsay
spoke of the ions present in all substances, stating, in other words, that his
electron bond of union was in all cases an electrostatic bond, though once
again his discussion hardly explained why only certain substances released
their ions upon solution. In the essay "What Is Electricity?" he argued that
in the formation of sodium chloride the transfer of an electron from sodium
to chlorine took place at the moment of combination and that sodium and

[56] Ibid., p. 783.
[57] Ibid., pp. 781–82; Ramsay, *Theoretical Chemistry*, p. 161.

chloride ions were present in the solid salt. To support his argument, Ramsay pointed out that when the salt solution gradually evaporated to dryness leaving the solid salt, "the electron does not leave the chlorion and attach itself to the sodion; if that happened the result would be metallic sodium and chlorine gas; and they are certainly not formed." The solid salt, when melted, conducted an electric current, and this, he said, was due to the separation of the sodium and chloride ions, which in the solid had no freedom of motion. In fact, Ramsay continued, "if this conception [of the chemical bond] be extended, all chemical combinations should be regarded as the transference of electrons from one set of elements to the next." [58] He added:

It may be conjectured that in the case for instance of such a compound as sugar, which dissolves in water as such, the atoms of carbon, hydrogen, and oxygen of which it consists, have interchanged electrons, otherwise chemical combination would not exist; but that the ions do not part from each other, even when opportunity is given by dissolving the sugar in water. Although facilities for motion in many cases lead to separation of ions, it does not follow that when facilities are present separation will always take place.[59]

In maintaining that the bond of union between a pair of atoms was always an electrostatic bond, Ramsay, like Thomson and Abegg, left unexplained why certain compounds dissociated and others did not. In addition, Ramsay's symbol for the chemical bond, **E**, failed to indicate which atom was the electron donor and which the acceptor, unless, of course, he included all the valence electrons in each atom.

On the other hand, Ramsay's formulas were the only ones in use at that time that showed both the valence electrons involved in bonding and any remaining valence electrons in an atom. But as Pattison Muir pointed out in *A History of Chemical Theories and Laws*, "Until chemists are convinced by facts that many reactions can be expressed in terms of 'the electron as an element' view more simply, more forcibly, and more suggestively than by the use of the language they have already learned, they will not trouble to acquire the new tongue." [60] Muir seemed to speak for all chemists, for apparently not one of them adopted Ramsay's language, with the possible exception of G. N. Lewis, who in 1916 used the symbol **E** to

[58] Ramsay, *Essays*, pp. 200–201.
[59] Ibid., p. 221.
[60] Muir, *Chemical Theories and Laws*, p. 546.

represent not the chemical bond but the number of valence electrons in a molecule.

Actually Ramsay recognized that his symbolism failed to indicate the directional character of the electron bond and that his formulas were "cumbrous." For these reasons he also represented an electron bond by an arrow. Sodium chloride, $NaECIE_7$, had the alternate formula $Na \rightarrow Cl^{vii}$, in which the arrowhead pointed toward the atom that received the electron, and the Roman numeral indicated "latent bonds or unused electrons." In later publications this latter method completely replaced the symbol E, which previously had represented both the electron bond and the latent electrons.[61]

Ramsay's Theory of Coordination Compounds

The electron, according to Ramsay, was a material body whose transfer from one atom to another resulted in oppositely charged ions that attracted each other. Chemists and physicists had used this theory of chemical union, the electron theory of valence, to account for the formation of compounds such as $NaCl$, NH_3, and H_2SO_4. An even greater test of the theory, Ramsay thought, would be its ability to account for the structures of the so-called molecular compounds, particularly the long-standing problem of ammonium chloride, and for Werner's more recent structures of the coordination compounds.

Upon applying the theory of electron transfer to molecular and coordination compounds, Ramsay found that in many instances the application led to a violation of Abegg's rule of eight—some of the combined atoms had more than eight valence electrons surrounding them. To solve this difficulty, he adopted an idea that Thomson had introduced in 1907, namely, that an atom undergoing bond formation could simultaneously gain and lose an electron, a process represented symbolically with a pair of arrows pointing in opposite directions $\rightleftarrows$.

Thomson employed this symbolism to indicate a double bond in his electron structures of unsaturated hydrocarbons, and at the time it suggested to him a new kind of isomerism later called *electron isomerism* or *electromerism*:

[61] William Ramsay, "Elements and Electrons," *J. Chem. Soc.* 95 (1909), 625; idem, "Compounds of Electrons," *Rice Institute Pamphlet* 1 (July 1915), 410–24.

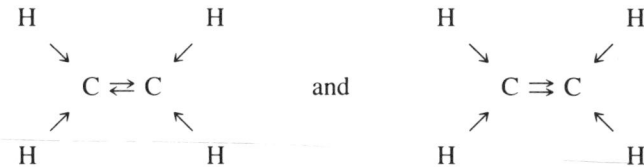

Ramsay never considered the existence of electromers or whether the pair of arrows constituted a double bond. He needed the simultaneous gain and loss of an electron only to prevent an atom from acquiring more than eight valence electrons. His hypothesis was in fact an electron interpretation of Arrhenius's suggestion that an atom possessed one or more pairs of latent valences that were always of opposite sign. The simultaneous loss and gain of an electron, in other words, corresponded to the positive and negative charges of Arrhenius's latent valences.

In arriving at his electron structure of ammonium chloride, Ramsay assumed that the nitrogen atom had five valence electrons and that it could receive only three more electrons from three atoms of hydrogen, as it did in ammonia:

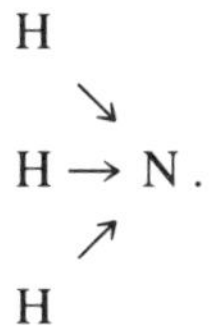

$$
\begin{array}{c}
H \\
\searrow \\
H \rightarrow N\,. \\
\nearrow \\
H
\end{array}
$$

Nitrogen now had eight valence electrons, the maximum number permitted. It was, therefore, impossible to add another hydrogen atom, giving NH_4, unless the nitrogen atom donated one of its five valence electrons to the chlorine atom that already had seven electrons but still had a single vacancy to fill before becoming saturated.[62] His structure was:

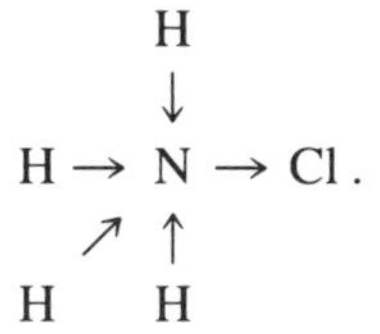

$$
\begin{array}{c}
H \\
\downarrow \\
H \rightarrow N \rightarrow Cl\,. \\
\nearrow \uparrow \\
H \quad H
\end{array}
$$

Thus, in forming ammonium chloride from ammonia and hydrogen chloride, Ramsay believed that the nitrogen atom in ammonia gained one

[62] Ramsay, "Electron as an Element," p. 785.

electron from the hydrogen atom of hydrogen chloride, HCl, and lost one to chlorine. Upon solution the electron linking the nitrogen and chlorine atoms remained with chlorine giving Cl^- and NH_4^+.[63]

As examples of the electron structures of coordination compounds, Ramsay selected the following cobaltammine nitrites:

$[Co(NH_3)_3(NO_2)_3]$, $[Co(NH_3)_4(NO_2)_2]NO_2$,

$[Co(NH_3)_5(NO_2)](NO_2)_2$,   and $[Co(NH_3)_6](NO_2)_3$.

These compounds dissociated in solution to give zero, two, three, and four ions, respectively, and this behavior, he argued, resulted directly from their electron structures.

Just as the nitrogen atom in NH_4Cl takes one electron from the hydrogen of the HCl and gives one up to the chlorine, so it appears reasonable to suppose that in these cobaltammines each nitrogen atom of the three ammonia groups takes from the cobalt atom one electron, whilst it gives one at the same time. The formula of the triammino-nitrite would therefore be:

$$
\begin{array}{c}
NH_3 \\
\updownarrow \\
H_3N \leftrightarrows Co \equiv (NO_2)_3 \\
\Updownarrow \\
NH_3 \\
[Co(NH_3)_3(NO_2)_3]
\end{array}
\qquad (1)
$$

If another molecule of ammonia be added, then the cobalt atom gives to the nitrogen of the ammonia an electron, but does not receive one in return. The nitrogen atom of that ammonia group is then "overloaded," for it had received four electrons in addition to the normal five, making nine in all; now it appears that no element can be associated with more than eight in all. Hence that nitrogen atom must lose an electron. This it imparts to one of the (NO_2)-groups which parts company with the cobalt atom, [as a nitrite ion, NO_2^-] and as a complex ammonium nitrite is now present, it is ionizable on solution in water.

$$
\begin{array}{c}
NH_3 \rightarrow NO_2 \\
\uparrow \\
H_3N \leftrightarrows Co == (NO_2)_2 \\
\nearrow\swarrow \qquad \nwarrow\searrow \\
H_3N \qquad NH_3 \\
[Co(NH_3)_4(NO_2)_2]^+ + NO_2^-
\end{array}
\qquad (2)
$$

[63] Looking at this process from another point of view, one could argue that the ammonium group, NH_4, carried a positive charge because the nitrogen atom's loss of one of its original five valence electrons was not completely balanced by the addition of the electron from the

For the remaining formulas,

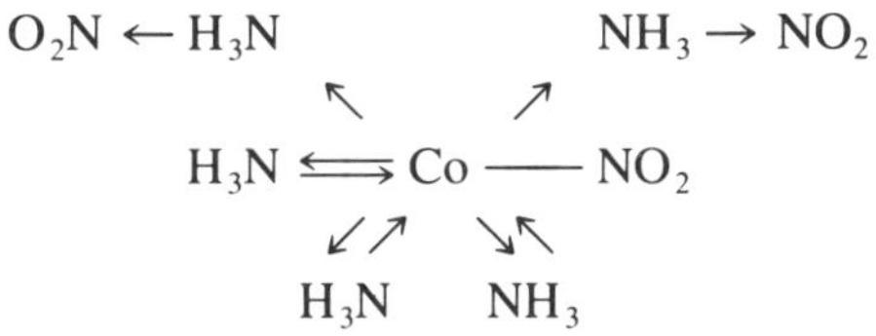

$$O_2N \leftarrow H_3N \qquad\qquad NH_3 \rightarrow NO_2$$

$$H_3N \Longleftrightarrow Co \underline{\quad\quad} NO_2 \qquad\qquad (3)$$

$$H_3N \quad NH_3$$

$$[Co(NH_3)_5(NO_2)]^{++} + 2NO_2^-$$

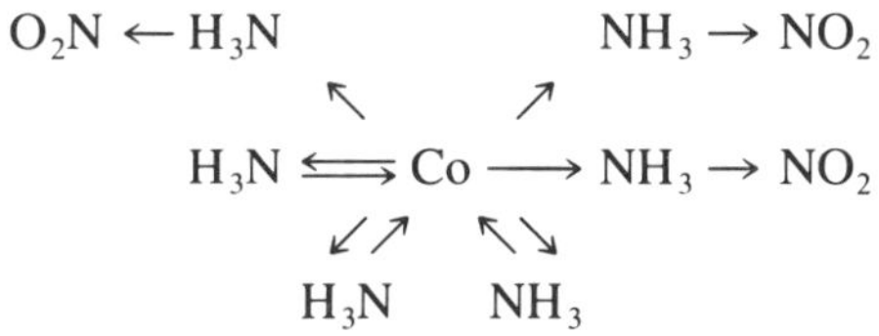

$$O_2N \leftarrow H_3N \qquad\qquad NH_3 \rightarrow NO_2$$

$$H_3N \Longleftrightarrow Co \longrightarrow NH_3 \rightarrow NO_2 \qquad\qquad (4)$$

$$H_3N \quad NH_3$$

$$[Co(NH_3)_6]^{+++} + 3NO_2^-$$

the successive addition of a fourth, a fifth, and a sixth ammonia molecule resulted in the stepwise liberation of one, two, and three nitrite groups as nitrite ions.[64]

In each of Ramsay's formulas, cobalt had a total valence of six, represented by three different symbols for the chemical bonds. The meaning of the single and double arrows we have already seen; the dash was only a temporary symbol that Ramsay employed to indicate that he had not yet established the direction of electron transfer between the cobalt atom and the nitrite groups. Ramsay believed, however, that the failure of $[Co(NH_3)_3(NO_2)_3]$ to dissociate in solution influenced the direction of electron transfer, and he tried to establish the direction by analogy with a simpler cobalt salt.

Selecting cobalt nitrate, $Co(NO_3)_3$, a cobalt salt that dissociated (he could not choose cobalt nitrite because it was unknown), Ramsay said dissociation occurred because cobalt had given up an electron to each of the three nitrate groups, $3NO_3^-$, becoming "cobaltion." In the ionizable cobaltammine $[Co(NH_3)_6](NO_2)_3$, Ramsay argued that once again the cobalt atom no longer had three electrons at its disposal, for it had also given

---

fourth hydrogen atom. This electron was not a free electron but belonged to a neutral hydrogen atom, and it failed, therefore, to neutralize completely the positive charge on the nitrogen atom and on the NH_4 group as a whole.

[64] Ramsay, "Electron as an Element," pp. 785–86.

them up, in this case to the three NH_3—NO_2 groups. It followed, therefore, that in the "nonionizable compound the cobalt does not, as in its ordinary salts part with three electrons, but that it receives them from the nitro groups." [65]

Ramsay thus arrived at the somewhat paradoxical conclusion that in the dissociation of a simple cobalt salt like nitrate (or the nitrite) the cobalt atom gave up an electron to each of the three nitrate groups. But upon the addition of ammonia molecules to the cobalt salt as in the triammine nitrite $[Co(NH_3)_3(NO_2)_3]$, the cobalt atom now received three electrons from the three nitrite groups. In trying to explain the different behavior of the cobalt atom in the two compounds, Ramsay suggested that the presence of another group in a molecule, such as ammonia, might influence an atom's affinity for electrons. His statement, however, was hardly an explanation of the bonding in coordination compounds; it revealed rather an aspect of the electron theory of valence that clearly required further attention.

John Newton Friend's Electron Theory of Valence

In the same volume of the *Journal of the Chemical Society* that contained Ramsay's valence theory, John Newton Friend (1881–1966) published his version of the electron theory of valence.[66] Friend, who had recently completed graduate studies in chemistry at Würzburg and later became a well-known textbook author and instructor at the University of Birmingham, introduced a theory requiring three kinds of valence: (1) free negative valence, (2) free positive valence, and (3) residual or latent valences. The first two were the usual negative and positive valences of atoms, while the residual or latent valences were those that an atom used in pairs of equal and opposite sign. All atoms, including the rare gases, possessed residual or latent valences. Fluorine, chlorine, and oxygen were the only atoms that had free negative but no free positive valences. Hydrogen and the metals had free positive valences though they lacked free negative valences, and the remaining atoms, excluding the rare gases, were amphoteric, or capable of using both free positive and free negative valences.

Friend, therefore, found it difficult to accept Abegg's rule of eight. He

[65] Ibid., pp. 786–87.
[66] John Newton Friend, "Valency," *J. Chem. Soc.* 93 (1908), 262–64.

acknowledged that though it was very remarkable that the sum of the free positive and negative valences of many atoms often totaled eight, this assumption obviously did not hold in all cases, most notably the rare gases, hydrogen, and the alkali metals. Even certain amphoteric elements had a total free valence less than eight. Bromine's valence sum was only 4: -1 in HBr, and $+3$ in BrF$_3$; the valence of nitrogen was $+3$ in the compound NOF and -3 in ammonia, NH$_3$, giving a total free valence of 6.[67]

In his 1908 publication Friend had not used (at least not explicitly) the electron to account for these valence changes. An electron interpretation, due in part to Ramsay's influence, first appeared in his text *The Theory of Valency*, published in 1909.[68] Free positive valence resulted from the loss of an electron or electrons by an atom, and free negative valence from the gain of one or more electrons. Residual or latent valences, which an atom used only in pairs of equal and opposite sign and which resembled Leopold Spiegel's neutral affinities and Arrhenius's electrical double valences, were now the result of an atom's simultaneously gaining and losing an electron in bond formation.

In binary compounds, such as the oxides and halides, chemical combination took place because the atoms acquired opposite charges, the result of electron transfer from one atom to the other. The electrostatic bond in these compounds Friend attributed to the union of oppositely charged free valences. To account for the bond in the diatomic hydrogen molecule and in the metallic hydrides, in each case of which the atoms had only free positive valences, Friend maintained that these atoms used their latent valences. The hydrogen molecule thus had the formula

$$\blacksquare + H\pm \quad \mp H + \blacksquare \,,$$

which showed the latent pairs connecting the two hydrogen atoms as well as the thicker lines representing the free positive valence of each atom.[69] When written electronically Friend's formula was

$$\cdot H \rightleftharpoons H \cdot \,,$$

[67] Ibid. Abegg believed eight was the maximum possible valence sum.

[68] Friend, *Theory of Valency*. For one of Friend's later publications on valence see "Electrochemical Conceptions of Valency." See also comments on it by Smith, "Friend's Theory of Valency."

[69] Friend, "Valency," p. 263.

where the arrows $\rightleftarrows \rightrightarrows$ now indicated the simultaneous gain and loss of an electron by each hydrogen atom (which was equivalent to the oppositely charged pair of latent valences) and the dot, the hydrogen atom's free positive valence. For the metallic hydrides, the electron formula was essentially the same, the only difference being the pair of dashes as an alternate way of representing the latent valences,

$$\cdot K{=}H\cdot \, .$$

Because the atoms of oxygen, fluorine, and chlorine possessed only free negative valences, Friend believed that only with latent valences could he account for the diatomic molecules of these atoms:

$$:O \rightleftarrows \rightrightarrows O: \qquad \cdot F \rightleftarrows \rightrightarrows F\cdot \qquad \cdot Cl \rightleftarrows \rightrightarrows Cl\cdot$$

Of course, utilization of latent valences immediately suggested the possibility that many atoms might form a greater number of bonds than the number usually assigned to them. Hydrogen and fluorine had one free valence and were monovalent. But by using their latent pair in combination with their free valence, these atoms could conceivably be divalent, though Friend acknowledged they were never known to function as such.[70] The variable valence of chlorine, on the other hand, was well established, and Friend's suggestion of its variable valence presented nothing essentially new.

For atoms of the rare gases (Group 0), the latent valences were so weak that their molecules were monatomic even at atmospheric temperatures. But possibly, Friend wrote, if we determined their densities at temperatures near their boiling points, we might observe some indication of association to diatomic molecules.[71] Friend concluded, however, that in most cases, the latent valences of an atom were very weak and did not in general produce any kind of chemical union. The nitrogen molecule's remarkable inertness was, therefore, probably due to the three free positive valences of one nitrogen atom completely saturating the three free negative valences of the other and did not result from the use of latent valences:

$$\pm N \,\substack{+\\+\\+} \,\substack{-\\-\\-}\, N \pm$$

Indeed, Friend argued that nitrogen was the only gaseous molecule in

[70] Friend, *Theory of Valency*, pp. 152, 153.
[71] Friend, "Valency," p. 264.

which such combination was possible, for it alone had an equal number of free positive and negative valences.[72]

In putting forward his valence theory, Friend, like Ramsay, did not develop in any detail an electron model of the atom. He suggested only that electron addition and removal took place at definite points on a spherical atom. The nitrogen atom when trivalent, according to Friend, had its three valences in one plane because all attempts to prepare optically active derivatives of the type

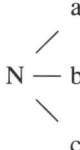

had failed. The three electrons lay at equal distances from one another on the circumference of a great circle of the sphere.

When nitrogen used its latent valences and behaved as a pentavalent atom, rearrangement took place, however. Four electrons constituting the four negative valences (three free valences and one latent valence) arranged themselves on the circumference of the great circle at equal distances from one another, while the fifth valence, a positive latent valence, resulted from the escape of an electron from either one of the sphere's two poles.[73] The following diagrams will make Friend's model clear:

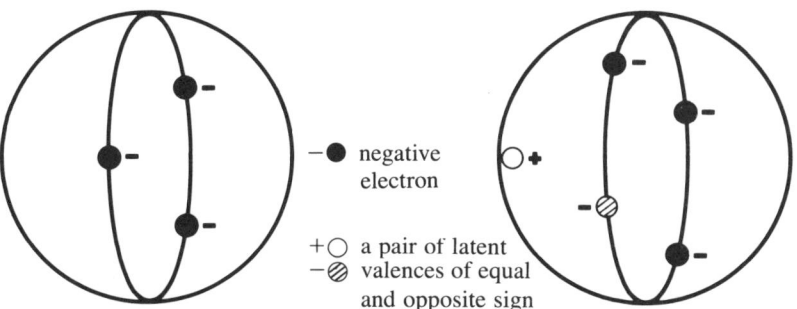

−● negative electron

+○ a pair of latent
−⊘ valences of equal and opposite sign

Arrangement of valence electrons   Arrangement of valence electrons
   in trivalent nitrogen (NH₃)   in pentavalent nitrogen (NH₄Cl)

[72] Friend, *Theory of Valency*, p. 157.
[73] Ibid., p. 167.

Conclusion

In the first decade of the twentieth century chemists, like physicists, assumed the chemical bond in all compounds to be electrostatic or polar. But neither group could explain why some compounds dissociated into oppositely charged ions and others did not, especially when all compounds presumably consisted of ions. The view that perhaps those compounds not undergoing dissociation, chiefly the organic compounds, contained no ions, had not yet become prevalent.

The idea of residual or latent valences continued to persist and received an electronic interpretation in the theories of Abegg, Ramsay, and Friend. These chemists tried to solve the long-standing problem of the ammonium chloride structure with their various interpretations of residual valence and were the first to provide electron structures for complex ions and for Werner's coordination compounds.

An important innovation appearing at this time was Abegg's rule of eight. Abegg assumed that all atoms had a maximum of eight electrons available for bonding. His rule was central to Ramsay's electron theory, J. J. Thomson adopted it, and later it appeared as the foundation of Kossel's electron theory of valence. Kossel believed every atom reacted in order to achieve an electron structure identical to that of the rare gas nearest it in the periodic table, that is, to attain a structure containing eight outermost electrons.

Lewis also attached considerable importance to the presence of eight outermost electrons in an atom, as we shall see upon discussing his theory of the cubic atom. But the rule of eight worked best in describing the formation of polar compounds. Something more was needed to account for the union of nonpolar compounds. This was the rule of two, the theory of the shared electron pair, which Lewis proposed in 1916, more than ten years after Abegg had introduced his rule of eight.

4. J. J. Thomson's Electrostatic Theory of Valence Comes to America

Introduction

In proposing that the chemists' valence bond was a unit Faraday tube connecting charged atoms in a molecule, J. J. Thomson, in the period 1904–7, had shown how to translate the organic chemists' well-established structural formulas into an electrical theory of valence. Applied to the structures of simple saturated hydrocarbons and their substitution products, his proposal led to the following electron formulas:

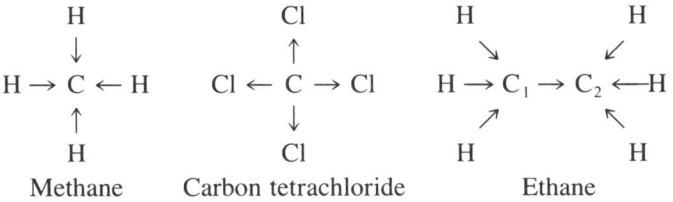

| | | |
| Methane | Carbon tetrachloride | Ethane |

In each case the arrow pointed in the direction of electron transfer, indicating that the carbon atom in methane had a charge of negative four and of positive four in carbon tetrachloride. For the two carbon atoms in ethane, the *net* charge on C_1 was negative two and negative four on C_2.

Richard Abegg, in his 1904 publication on valence, took a similar position regarding the charge distribution on the carbon atoms in hydrocarbons. In zinc dimethyl, $Zn(\overset{+}{C}\overset{-}{H}_3)_2$, the methyl group was apparently negative, but in methyl chloride, $\overset{+}{C}H_3\overset{-}{Cl}$, it seemed to have a positive charge. Hence, Abegg found it highly likely that in ethane the two methyl groups were oppositely charged and held together by an electrostatic attraction: $H_3\overset{+}{C}.\overset{-}{C}H_3$.

The ideas of Thomson and Abegg constituted the first application of the electrostatic theory of valence to the structures of organic molecules. But it was only a beginning. American chemists, among them K. George Falk

(1880–1953) and his collaborators John M. Nelson (1876–1965) and Hal T. Beans (1876–1960) at Columbia University, Harry S. Fry and Lauder W. Jones at the University of Cincinnati, Julius Stieglitz at the University of Chicago, and William A. Noyes at the University of Illinois undertook a systematic application of the theory to the numerous compounds of organic chemistry as well as extending it to the compounds of inorganic chemistry. Their electrostatic theory, being at that time the most widely held electrical theory of valence, simply became the electron theory of valence. Its central idea was that the chemical bond between two atoms resulted from the complete transfer of an electron from one of the atoms to the other. We should compare it with the valence theory G. N. Lewis put forward a few years later, which assumed the bond to result instead from electron sharing, not electron transfer between two atoms undergoing combination.

Falk and Nelson's Application of the Electron Theory of Valence to Organic Chemistry

Falk and Nelson's development of the electron theory of valence appeared in a series of articles published from 1909 to 1915 in the *Journal of the American Chemical Society*.[1] In agreement with Thomson, they saw the theory as only a modification of the structural formulas then employed, a simple addition to them: "an arrow being used instead of a dash to denote a bond, the direction of the arrow showing the direction in which the corpuscle [electron] is transferred." Following Abegg, they assumed that the relative positions of the elements in the periodic table ordinarily determined the direction of electron transfer between combining atoms. In each horizontal row the elements of greater atomic weight were usually negative with respect to those of lesser atomic weight, and in the vertical columns, an increase in atomic weight led to increasing positive character. Thus, chlorine was more negative than iodine and iodine chloride normally had the formula I → Cl, though Falk and Nelson thought quite possibly a less stable form, I ← Cl, existed.[2]

[1] K. George Falk and John M. Nelson's first paper, "The Electron Conception of Valency in Organic Chemistry," appeared in the *School of Mines Quarterly* 30 (April 1909), 179–98. They also presented its contents in their next paper, "The Electron Conception of Valence," *J.A.C.S.* 32 (1910), 1637–54.

[2] Falk and Nelson, "The Electron Conception of Valency," p. 181; idem, "The Electron Conception of Valence," p. 1639.

For the saturated hydrocarbons—carbon-hydrogen compounds containing only single bonds—Falk and Nelson acknowledged that introducing directive valences would be of limited value since chemists had established their structures fairly completely and satisfactorily without directive valences. It was also difficult to assign directive valences to saturated hydrocarbons and their derivatives because of the carbon atom's ability either to gain or to lose electrons as indicated by the stability of both methane and carbon tetrachloride. The general inertness of saturated hydrocarbons did not make it easier to arrive at any definite conclusions regarding their structures and relative stabilities.[3]

Moses Gomberg's studies on the triphenylmethyl radical offered some support for the presence of directive valences in saturated hydrocarbons. Falk and Nelson pointed out that if one assumed triphenylmethyl to be hexaphenylethane, $(C_6H_5)_3C$—$C(C_6H_5)_3$, an explanation for the electrical conductivity of triphenylmethyl in liquid sulfur dioxide, which, as Gomberg (1866–1947) had shown, followed upon postulating the existence of the "pseudo-ions" $(C_6H_5)_3C^+$ and $(C_6H_5)_3C^-$. Indeed, a similar relation might exist between the methyl groups in ethane, though Falk and Nelson never made any attempt to test experimentally their analogy's validity.[4]

As a second argument to support directive valences, Falk and Nelson turned to the well-known fact that the saturated dicarboxylic acids differed in properties depending on whether they contained an even or an odd number of carbon atoms. The acids containing an even number of carbon atoms had higher melting points and lower solubilities than the odd-numbered acids immediately preceding and succeeding them but varied regularly when compared with other acids having an even number of carbon atoms. Of course a similar regularity held for the acids with an odd number of carbon atoms. A graph of melting point, or solubility, as a function of the number of carbon atoms gave a saw-toothed pattern if it included both kinds of acids, but produced a smooth curve when it contained only acids with either an even or an odd number of carbon atoms.

The dicarboxylic acids, therefore, fell into two well-defined groups, and directive valences, Falk and Nelson maintained, accounted for these groups. Those acids with an even number of carbon atoms had asymmetri-

[3] Falk and Nelson, "The Electron Conception of Valence," p. 1639.

[4] Ibid., p. 1640; Moses Gomberg, "Über das Triphenylmethyl," *Berichte* 35 (1902), 2397–408. See also Paul Walden, "Über abnorme Elektrolyte," *Zeit. phys. Chemie* 43 (1903), 451.

cal directive valences, while acids having an odd number of carbon atoms had symmetrically directed valences. In each group the acids were strictly homologous only with those belonging to that group.

To illustrate their proposal, Falk and Nelson assigned the following formulas to selected members of the two groups, leaving as unknown in all but the simplest acids most of the directions of the valence bond connecting adjacent carbon atoms:

$$CO_2H \leftarrow CH_2 \rightarrow CO_2H \qquad CO_2HCH_2 \rightarrow CH_2CO_2H$$
$$\text{Malonic acid} \qquad\qquad \text{Succinic acid}$$

$$CO_2HCH_2 \leftarrow CH_2 \rightarrow CH_2CO_2H \quad CO_2HCH_2CH_2 \rightarrow CH_2CH_2CO_2H \,.$$
$$\text{Glutaric acid} \qquad\qquad\qquad \text{Adipic acid}$$

Falk and Nelson also assigned directive valences to unsaturated hydrocarbons containing a double bond, $R_2C{=}CR_2$, which they believed was the equivalent of two single bonds. The single bond resulted from the transfer of an electron from one atom to another, though the electron did not separate entirely from the first atom, "there remaining some connection between the two which may be imagined as lines of forces holding the two atoms in chemical combination."[5] Thus, in those hydrocarbons in which the valences of the double bond were in opposite directions, the same kind of physical connection existed between the pair of bonded carbon atoms as when the valences were in the same direction.

The oppositely directed valences did not give electrically neutral carbon atoms, for Falk and Nelson assumed that in all cases the electrons transferred were at specific locations on the atoms, producing positive and negative points or regions on each atom's surface. They made no further assumptions regarding the spatial distributions of the electrons in an atom except for the one case in which the central carbon atom had four unlike atoms or groups attached to it and the optical activity of such compounds demonstrated that the direction of the four valences did not lie in one plane.[6]

Falk and Nelson's structures, therefore, took into account the possible existence of electron isomers. But they believed, that if two or more electron isomers of a given compound were theoretically possible, one of them would always be more stable than the others. If only one isomer was

[5] Falk and Nelson, "The Electron Conception of Valency," p. 181.
[6] Falk and Nelson, "The Electron Conception of Valence," p. 1641.

known, it was because the others were too unstable to exist, going instead over to the stable form. A study of a compound's chemical reactions, they said, would decide which formula or formulas to assign to the possible isomers.[7]

As an example of how chemists might establish the directions of the valences in unsaturated hydrocarbons, Falk and Nelson selected the addition reactions of simple binary compounds to propylene, $CH_3CH=CH_2$, whose three possible electron structures are shown below:

$$CH_3CH \rightleftarrows CH_2 \qquad CH_3CH \Rrightarrow CH_2 \qquad CH_3CH \Lleftarrow CH_2 .$$

They argued as follows. On treating with hydrogen iodide, $H \rightarrow I$, if either of the last two structures was correct, all of the iodine would go to the more positive carbon atom and the hydrogen to the other carbon atom of the double bond, forming only one product. But, if the first structure was correct, then two products would form due to the hydrogen and iodine dividing themselves between the two carbon atoms, each of which had a positive and a negative region on its surface and therefore attracted both the positive hydrogen and the negative iodine of hydrogen iodide.

According to Falk and Nelson, Arthur Michael's study of hydrocarbon addition reactions,[8] in each case of which he got a mixture of products, definitely settled the argument in favor of the first formula, $CH_3CH \rightleftarrows CH_2$:

$$CH_3CH=CH_2 + HI \quad \rightarrow CH_3CHICH_3 \text{ (principally)}$$
$$\text{and } CH_3CH_2CH_2I \text{ (very little)};$$
$$CH_3CH=CH_2 + BrCl \rightarrow CH_3CHBrCH_2Cl \text{ (5 parts)}$$
$$\text{and } CH_3CHClCH_2Br \text{ (7 parts)};$$
$$CH_3CH=CH_2 + ICl \quad \rightarrow CH_3CHICH_2Cl \text{ (1 part)}$$
$$\text{and } CH_3CHClCH_2I \text{ (4 parts)};$$
$$CH_3CH=CH_2 + ClOH \rightarrow CH_3CHOHCH_2Cl \text{ (principally)}$$
$$\text{and } CH_3CHClCH_2OH \text{ (very little)}.$$

Falk and Nelson thus represented the addition of hydrogen iodide to propylene:

$$CH_3CH \rightleftarrows CH_2 + H \rightarrow I = CH_3CH_2 \rightarrow CH_2I$$
$$\text{and}$$
$$CH_3CH \rightleftarrows CH_2 + I \leftarrow H = CH_3CHI \rightarrow CH_3 .$$

[7] Ibid., p. 1640.

[8] Arthur Michael, "Über einige Gesetz und deren Anwendung in der organischen Chemie," *J. prakt. Chemie* 60 (1899), 286–384, 409–86.

But they could not explain why in all of these reactions the addition products did not form in equal proportion as their structure predicted. Two factors, they said, probably affected the course of the propylene–hydrogen iodide reaction: (1) the influence of the attached groups—methyl on one carbon, hydrogen on the other—and (2) the polarity differences between the hydrogen and iodine—the smaller the difference the more nearly equal the amounts of products.[9]

Despite the inconclusiveness of this suggestion as well as the paucity of their experimental evidence, Falk and Nelson maintained that in general the form $RC \rightleftarrows C\acute{R}$ best represented the most stable valence distributions in a double bond. The less stable isomers, if they existed at all, had structures of the type $RC \rightrightarrows C\acute{R}$ or $RC \leftrightarrows C\acute{R}$. Indeed, Falk and Nelson believed that the maleic-fumaric acid isomerism and that of their derivatives, usually thought to result from a difference in spatial configuration, was due to directive valences. The more stable fumaroid type had the general formula $CR\acute{R} \leftrightharpoons CR\acute{R}$; the malenoid type, $CR\acute{R} \rightrightarrows CR\acute{R}$. Similarly, the configuration $RC \rightleftharpoons C\acute{R}$ represented best the stable form of triple-bonded hydrocarbons.[10]

Falk and Nelson's Extension of Directive Valences to Other Elements in the Periodic Table

Though in their 1909 and 1910 publications Falk and Nelson concerned themselves chiefly with applying directive valences to hydrocarbons, in these and in later publications they investigated the structures of many other organic and inorganic compounds.[11] In a paper on oxygen compounds read in 1912 before the Eighth International Congress of Applied Chemistry in New York, they argued that oxygen in its single-bonded combinations would always have the electron transfer directed toward it because of its highly negative character.[12] This made the electron structures of these compounds relatively easy to establish:

---

[9] Falk and Nelson, "The Electron Conception of Valence," p. 1642.

[10] Ibid., p. 1643.

[11] Ibid., pp. 1645–49, 1652–53; John M. Nelson, Hal T. Beans, and K. George Falk, "The Electron Conception of Valence: IV. The Classification of Chemical Reactions," *J.A.C.S.* 35 (1913), 1810–21; K. George Falk and John M. Nelson, "The Electron Conception of Valence: VI. Inorganic Compounds," *J.A.C.S.* 37 (1915), 274–86.

[12] Nelson and Falk, "The Electron Conception of Valence: III. Oxygen Compounds,"

$$H \rightarrow O \leftarrow H \qquad H_4 \lessgtr N \rightarrow O \leftarrow H \qquad H_3 \lessgtr C \rightarrow O \leftarrow H.$$

Water   Ammonium hydroxide   Methyl alcohol

On the other hand, in oxygen's double-bonded compounds, or indeed whenever a double bond occurred between any pair of unlike atoms, Falk and Nelson believed that no experimental evidence available established conclusively the valence directions. The relative positions of the atoms in the periodic table fixed the direction of one valence—the arrow going from the less negative atom to the more negative—but this criterion, they said, did not necessarily hold for the second valence. Its direction could be the same or different from the first. Two isomers were possible, and Falk and Nelson's choice of which isomer represented the more stable structure was purely arbitrary: "the more stable one may for the present be assigned the structure in which both valences possess the same direction, since this appears to be the most reasonable view in the union of two dissimilar atoms." [13]

Thus, for the carbonyl group, $C{=}O$, three structures were theoretically possible:

$$C \Rrightarrow O \qquad\qquad C \rightleftarrows O \qquad\qquad C \Lleftarrow O.$$

$$\text{I} \qquad\qquad\qquad \text{II} \qquad\qquad\qquad \text{III}$$

They rejected structure III outright because the oxygen atom's greater negative character compared with that of carbon made this isomer too unstable ever to isolate. Of the two remaining structures, Falk and Nelson believed II to be intermediate in stability and hence capable of isolation and I to represent the most stable isomer. To the benzophenone molecule, which contained a carbonyl group and of which two crystalline modifications actually existed, Falk and Nelson assigned the structures

$$(C_6H_5)_2C \Rrightarrow O \qquad \text{and} \qquad (C_6H_5)_2C \rightleftarrows O.$$

The first, the more stable isomer, melted at 48.5°C; the second or more labile form, at 25°C.

To illustrate the application of directive valences to triple-bonded

---

Original Communications to the Eighth International Congress of Applied Chemistry 6 (September 1912), 213.

[13] Falk and Nelson, "The Electron Conception of Valence," p. 1648; Nelson and Falk, "Valence: Oxygen Compounds," p. 213. They said nothing about valence directions on the oxygen molecule in the latter paper or in the others on valence.

structures, Falk and Nelson selected the nitrogen molecule. Assuming a charge of negative three on the nitrogen atom in ammonia and accepting Ramsay's hypothesis that in ammonium compounds the nitrogen atom simultaneously gained and lost a fourth electron, they tried to show that the triple bond had the following directive valences: N≣N. Their argument rested on the fact that substitution of the three hydrogen atoms of an ammonium salt by a single nitrogen atom converted the latter into a diazonium salt:

$$R \rightarrow N \equiv N \qquad\qquad R \rightarrow N \equiv H_3$$
$$\downarrow \qquad\qquad\qquad\qquad \downarrow$$
$$X \qquad\qquad\qquad\qquad X$$

Diazonium salt   Ammonium salt

According to Falk and Nelson, the tendency of a diazonium salt to liberate nitrogen gas was entirely analogous to an ammonium salt's liberating ammonia. Hence, if the formula for ammonia was N≣H$_3$, it seemed logical to conclude that the nitrogen molecule had the structure N≣N.

Falk and Nelson also pointed out that the liberation of nitrogen upon heating ammonium nitrite, NH_4NO_2, though not establishing the direction of the nitrogen molecule's three valences, did show that the nitrogen atoms had opposite charges. The nitrogen atom attached to the electropositive hydrogen atoms was negative; the other, bonded to electronegative oxygen, was positive. When combined in the nitrogen molecule there was no reason to doubt that the two atoms had not retained their positive and negative charges.[14]

Falk and Nelson's arguments held only for the nitrogen molecule. Their general conclusion was that any atom, depending on its reaction conditions, could have a positive or a negative valence. This was of course not a new idea. William A. Noyes and Julius Stieglitz in 1901 and Richard Abegg and J. J. Thomson in 1904 had proposed the same idea.[15]

Falk and Nelson's extensive use of directive valences also provided additional illustrations of Abegg's rule of eight—that atoms upon combining with other atoms could gain or lose a maximum of eight electrons. In the structures of methane, CH_4, and carbon tetrachloride, CCl_4, the change in

[14] Falk and Nelson, "The Electron Conception of Valence," p. 1648.

[15] William A. Noyes and Albert C. Lyon, "The Reaction between Chlorine and Ammonia," *J.A.C.S.* 23 (1901), 460–63; Julius Stieglitz, "On Positive and Negative Halogen Ions," *J.A.C.S.* 23 (1901), 797–99.

carbon's valence, from -4 in the former to $+4$ in the latter, already suggested that the carbon atom had the capacity to surround itself with eight outermost electrons. It was equally clear that the nitrogen atom in Group V of the periodic table gained three electrons in ammonia, $N{\leftrightarrows}H_3$, while in nitric acid,

$$O \leftleftarrows N \rightarrow O \leftarrow H ,$$
$$\Downarrow$$
$$O$$

it lost five. The difference between its two extreme valences was once again eight. In Groups VI and VII, the valence of several of the atoms varied in a similar way:

$$\nwarrow \quad \nearrow \qquad\qquad\qquad\qquad\qquad \nwarrow \quad \nearrow$$
$$\leftarrow S \rightarrow {}^{+6} \quad \text{and} \quad \rightarrow S \leftarrow {}^{-2} \qquad\qquad \leftarrow I \rightarrow {}^{+7} \quad \text{and} \quad I \leftarrow {}^{-1}.$$
$$\swarrow \quad \searrow \qquad\qquad\qquad\qquad\qquad \swarrow \downarrow \searrow$$

The Application of Directive Valences to the Structures of Organic Acids

In the second of the series of papers published on the electron theory of valence, Falk used directive valences to account for the experimentally measured values of the ionization constants (K) of organic acids. Wilhelm Ostwald had shown in the 1880s that substituted electropositive and electronegative atoms or radicals acted on the acid's carboxyl group, CO_2H, affecting the relative strength of the acid and hence its K value.[16] Falk went a step further. He suggested in 1911 that the substituent's influence resulted from its ability to receive or donate an electron when it formed a bond with one of the acid's carbon atoms, indeed that its influence was greatest when it formed a bond with the alpha (α) carbon atom, that is, the carbon atom adjacent to the carboxyl group.

The direction of the bond or valence established the electron's location—it was always at the arrowhead—and according to Falk, the electron's location determined the magnitude of the ionization constant. Thus, an electropositive substituent, one that gave up an electron, upon forming a bond with the α-carbon atom increased the concentration of negative

---

[16]K. George Falk, "The Electron Conception of Valence: II. The Organic Acids," *J.A.C.S.* 33 (1911), 1140–52. See discussion on Ostwald in chapter 1.

charge around the carboxyl group, making it more difficult to remove the positively charged hydrogen atom of the carboxyl group, COO ← H. Because fewer charged hydrogen atoms or ions were set free, the ionization constant of the substituted acid decreased in value compared with that of the previously unsubstituted acid. Conversely, an electronegative substituent, one that received an electron, decreased the negative charge in the region of the carboxyl group and consequently increased the ionization constant's value.

Falk's examination of ionization constants enabled him to divide the organic acids into four fairly well defined classes, which depended on the valence directions on the α-carbon atom of each acid.[17] The differences in ionization constants were great enough to assign directive valences to each class, and for acids with ionization constants lying between these values he thought that knowledge of their composition was also necessary to establish the acid's class:

Class (type of group attached to α-carbon atom)	$K \times 10^5$	Examples
I ⇶ C.CO₂H 3 electropositive	< 0.01	butyric, isobutyric, n-heptylic, benzoic
II ⇶ C.CO₂H 2 electropositive 1 electronegative	0.1–0.4	iodoacetic, chloroacetic, α-bromobutyric, α-chlorobutyric, fumaric
III ⇌ C.CO₂H 1 electropositive 2 electronegative	> 2	dichloroacetic, maleic, α-α-dibromopropionic
IV ⇶ C.CO₂H 3 electronegative	value of K is too high	trichloroacetic

In his examples Falk included the isomeric pair, maleic and fumaric acids. These unsaturated acids were cis-trans isomers and had the structures

[17]Falk, "Valence: Organic Acids," p. 1141.

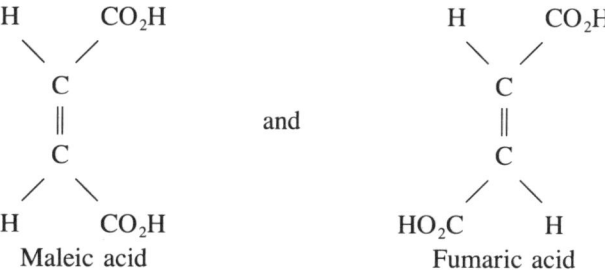

Falk included them because in his interpretation of directive valences a double bond was the equivalent of two single bonds and had the same effect on the ionization constant as two single bonds, taking into account, of course, the valence directions. Thus, maleic, fumaric, and other unsaturated acids should fit as easily into the classification as the saturated acids.[18] Indeed, Falk pointed out that the first ionization constant of fumaric acid was comparable to the value of chloroacetic acid (Class II). With the usual assumption that the direction of electron transfer in the carbon-hydrogen bond was from hydrogen to carbon, $H \rightarrow C$, the valences of the carbon-carbon double bond had to be in opposite directions,

$$\overset{\overset{\text{H}}{\downarrow}}{\text{H}_2\text{OC.C}} \rightleftarrows \overset{\overset{\text{H}}{\downarrow}}{\text{C.CO}_2\text{H}},$$

in order to place this acid in Class II.

Maleic acid fell in Class III because its first ionization constant (that for the carboxyl group on the left in the formula, according to Falk) was of the same order as that of dichloroacetic acid. The valences of its carbon-carbon double bond were in the same direction, $C \Rrightarrow C$, and when combined with the carbon-hydrogen valence, it gave the structure of a Class III acid:

$$\overset{\overset{\text{H}}{\downarrow}}{\text{HO}_2\text{C.C}} \Rrightarrow \overset{\overset{\text{H}}{\downarrow}}{\text{C.CO}_2\text{H}}.$$

Falk discussed the electron structure of the aromatic acid, benzoic acid, C_6H_5COOH. Due to the small value of its ionization constant, he

[18] Ibid., p. 1149.

placed this acid in the first or weakest class of acids. Benzoic acid had the partial electron formula

$$
\begin{array}{c}
CO_2H \\
| \\
H \rightarrow C \rightarrow C \leftrightharpoons C \leftarrow H \\
\| \qquad\quad | \\
H \rightarrow C - C = C \leftarrow H, \\
\uparrow \\
H
\end{array}
$$

where the dashes indicated unknown directive valences.[19]

In classifying the organic acids according to the directive valences on the α-carbon atom, Falk took into account only three of carbon's four valences. The direction of the fourth valence, connecting the carboxyl group with the α-carbon atom, remained unspecified simply because Falk knew of no experimental evidence he could use to establish its direction. Harry Shipley Fry at the University of Cincinnati quickly criticized Falk for this omission. He called Falk's classification incomplete, and in 1912, a year after Falk's paper appeared, Fry proposed eight different classes of organic acids, taking into consideration the fourth valence of the α-carbon atom.[20]

The real significance of Falk's attempt to relate directive valence to acid strength was not whether chemists divided the organic acids into four or into eight classes. Rather, with the electron and directive valences, Falk explained why an increase in the number of electronegative substituents on the α-carbon atom resulted in a corresponding increase in the acid's ionization constant: the electronegative substituents removed electrons from the α-carbon atom, leaving it positively charged and hence capable of repelling more strongly the positively charged hydrogen atom of the carboxyl group.

Falk's hypothesis resembled the successful explanation of acid strength that G. N. Lewis first put forward in 1916. Falk required the complete transfer of the single electron constituting the chemical bond. But Lewis, assuming that the chemical bond was a pair of electrons shared between two atoms, believed that an electronegative atom or radical attached to the α-carbon atom caused only a shift and not a complete transfer of the elec-

[19] Ibid., p. 1150.
[20] Harry S. Fry, "A Critical Survey of Some Recent Applications of the Electron Conception of Valence," *J.A.C.S.* 34 (1912), 664–73.

tron pair or pairs away from the carboxyl group and its ionizable hydrogen. The partial loss of bonding electrons from the carboxyl group facilitated the removal of its hydrogen atom and gave a higher value to the ionization constant of the substituted acid.

Fry's Theory of Positive and Negative Valences

Harry Shipley Fry first began to elaborate his ideas on the electron theory of valence in a paper read before the Cincinnati section of the American Chemical Society in January 1908. In this paper and again in a 1911 publication, Fry introduced the terms "electromer" and "electronic tautomerism," denoting a new kind of isomer (the electronic isomer) and a new kind of tautomerism involving electromers in dynamic equilibrium, respectively. He also described an electron formula for benzene that included an explanation of the *ortho-*, *para-*, or *meta-*directing ability of substituents on the benzene ring.[21] Fry published on the electron theory of valence over the next thirteen years, bringing together in 1921 all his ideas

[21] Harry S. Fry, "An Hypothesis Relative to the Constitution of the Benzene Nucleus: An Application of the Corpuscular Atomic Conception of Positive and Negative Valencies to the Constituent Atoms of Benzene" (paper delivered at a meeting of the Cincinnati section of the American Chemical Society, Cincinnati, Ohio, January 15, 1908). The proceedings were not published. The ideas Fry presented in this paper appeared in subsequent papers. See Harry S. Fry, "Die Konstitution des Benzols vom Standpunkte des Korpuskular-atomistischen Begriffs der positiven und negativen Wertigkeit," *Zeit. phys. Chemie* 76 (1911), 387.

Adolf von Baeyer apparently discovered tautomerism during his work with indigo derivatives in 1882. See Baeyer, "Über das Isatin," *Berichte* 15 (1882), 2093–102; Baeyer, "Über die Verbindungen der Indigogruppe," *Berichte* 16 (1883), 2188–204. Peter Conrad Laar (1853–1929), Privatdozent at the Technische Hochschule, Hanover, suggested the name *tautomerism* in "Über die Möglichkeit mehrerer Strukturformeln für dieselbe chemische Verbindung," *Berichte* 18 (1885), 648–57. See also Laar, "Über die Hypothese der wechselnden Bindung," *Berichte* 19 (1886), 730–41. Paul Heinrich Jacobson (1859–1923) at the University of Berlin introduced the term *desmotrophy* in "Zur-Kenntnis der orthoamidinten aromatischen Mercaptane," *Berichte* 20 (1887), 1895–903. See also Jacobson, "Zur Kenntnis der orthoamidinten aromatischen Mercaptane: III," *Berichte* 21 (1888), 2624–31. Ludwig Knorr (1858–1921) at the University of Jena demonstrated that tautomerism involved an intramolecular change and wrote "Studien über Tautomerie," *Annalen der Chemie* 293 (1897), 70–120; 303 (1898), 133–49; 306 (1899), 332–93. Two isomeric forms actually exist. In acetoacetic ester, $CH_3.CO.CH_2.COOC_2H_5$, for example, the hydrogen atom occupies two different positions in the molecule, enabling the molecule to exist in either the keto form, —CO—CH$_2$—, or the enol form, —C(OH)=CH—. See also William H. Perkin, "Tautomerism," presidential address, *J. Chem. Soc.* 105 (1914), 1176–89; Thomas Lowry, "Dynamic Isomerism" *B.A.A.S. Report* 74 (1904), 193–224; Alexander Findlay, *A Hundred Years of Chemistry*, pp. 121–24.

in a single monograph, *The Electronic Conception of Valence and the Constitution of Benzene.*[22]

Following J. J. Thomson's example, Fry maintained that the electron theory of valence was a further development of the organic chemist's theory. He called it a formulative hypothesis, "a conception of positive and negative valences applied to structural formulas," which was not primarily dependent on the existence of electrons or on the fundamental nature of chemical affinity. Indeed, the structural theory of organic chemistry, according to Fry, was itself only a formulative hypothesis, for its established utility did not depend on the existence of atoms or the fundamental nature of valence or chemical affinity.[23]

Fry called attention to a statement on the real existence of atoms Kekulé made in 1867 to support the use of formulative hypotheses in chemistry:

The question whether atoms exist or not has but little significance from a chemical point of view; its discussion belongs rather to metaphysics. In chemistry we have only to decide whether the assumption of atoms is an hypothesis adapted to the explanation of chemical phenomena. More especially have we to consider whether a further development of the atomic hypothesis promises to advance the knowledge of the mechanism of chemical reactions.[24]

Paraphrasing Kekulé, he urged chemists to adopt the same attitude toward the electron theory of valence:

In chemistry we have to decide whether the electronic conception of positive and negative valences is an hypothesis adapted to the explanation of chemical (and physical) phenomena. More especially have we to determine whether the further development of this hypothesis of positive and negative valences promises to advance our knowledge of the mechanism of chemical reactions. In view of the fact that electronic formulas, in many instances, have proven to be more precise and more significant than the ordinary structural formulas in the explanation of chemical phenomena and the mechanisms of reactions, the hypothesis of positive and negative valences may possibly become a necessary adjunct to the structure theory.[25]

Fry's contemporaries Falk and Nelson justified their application of the electron to valence theory and structural formulas in much the same way.

---

[22] Harry S. Fry, *The Electronic Conception of Valence and the Constitution of Benzene.*

[23] Harry S. Fry, "The Electronic Conception of Positive and Negative Valences," *J.A.C.S.* 37 (1915), 2370, 2372.

[24] Ibid., p. 2372, quoted by Fry. Kekulé's statement is from "On the Existence of Chemical Atoms," *American Journal of Science* 44 (1867), 270.

[25] Ibid., pp. 2372–73.

They found this approach of value in explaining the facts more simply than some explanations in current use and in a few cases accounting for previously unexplained facts, such as the physical properties of the organic dibasic acids. Nor did they consider it a weakness that some of the electrical structures predicted or suggested by their valence theory were unknown. On the contrary, Falk and Nelson found their theory better than one that could not account for known compounds. The nonexistence of predicted compounds, they argued, could very well be due to the limited range of conditions under which chemists had prepared most presently known substances.[26]

Because Falk and Nelson and, of course, Fry looked upon the electron theory of valence only as a formulative hypothesis, not one of them ever discussed in any detail the electronic constitution of the different atoms to which they applied their theory of valence. In his early papers Fry examined the structures of the elementary diatomic gases, accepting the view commonly held at the time that the two atoms constituting each gas, though chemically identical, had to have opposite charges in order for the molecule to exist. Indeed, such an assumption had already led Thomson in 1907 to foresee the possibility of what Fry called electronic isomers, some of which might be only slightly stable and hence incapable of isolation.[27]

Thus, the reaction between two elementary gases, ordinarily represented by the equation

$$X_2 + Y_2 \rightleftarrows 2XY,$$

when written using Fry's electronic notation was

$$
\begin{aligned}
X_2 &= X^+ - X^- \rightleftarrows X^+ \;+\; X^- \\
Y_2 &= Y^- - Y^+ \rightleftarrows Y^- \;+\; Y^+ \\
&\qquad\qquad\qquad \Updownarrow \qquad\;\; \Updownarrow \\
&\qquad\qquad X^+ - Y^- \;\; X^- - Y^+.
\end{aligned}
$$

The dynamic instability of the electromers, $X^+ - Y^-$ and $X^- - Y^+$, served as the basis for Fry's second new conception, that of "electronic tautomerism."[28]

[26] Falk and Nelson, "The Electron Conception of Valence," pp. 1637, 1638.

[27] J. J. Thomson, *The Corpuscular Theory of Matter*, p. 132; Fry, "Die Konstitution des Benzols," p. 387.

[28] Harry S. Fry, "Positive and Negative Hydrogen, the Electronic Formula of Benzene, and the Nascent State," *J.A.C.S.* 36 (1914), 263; idem, *The Electronic Conception of Valence*, p. 11; idem, "Die Konstitution des Benzols," p. 387.

Since the late nineteenth century chemists had used tautomerism to indicate an equilibrium mixture of only structural isomers, such as the well-known keto-enol modification:

$$
\begin{array}{ccc}
\text{O} & & \text{OH} \\
\| & & | \\
CH_3\text{—}C\text{—}CH_2 & \text{and} & CH_3\text{—}C\text{=}CH\text{—} .
\end{array}
$$

But electronic tautomerism, according to Fry, was "an equilibrium mixture of two (or more) electromers in the sense that one electromer may be assumed to revert to another electromer through the transposition of valence electrons,"[29] causing a reversal in polarity of certain atoms or radicals in the respective electromers:

$$
\overset{+}{X} - \overset{-}{Y} \rightleftarrows \overset{+}{X} + \overset{-}{Y} \rightleftarrows \overset{+}{X} + \ominus + Y \rightleftarrows X + Y \rightleftarrows \overset{-}{X} + \ominus + \overset{+}{Y} \rightleftarrows \overset{-}{X} +
$$

$$
\overset{+}{Y} \rightleftarrows \overset{-}{X} - \overset{+}{Y} .
$$

In the transition from electromer $\overset{+}{X} - \overset{-}{Y}$ to $\overset{-}{X} - \overset{+}{Y}$ or vice versa, $\overset{+}{X} - \overset{-}{Y} \rightleftarrows \overset{-}{X} - \overset{+}{Y}$, when either Y lost or X gained an electron it became a neutral atom or radical. The presence of these neutral substances, X and Y, in the electronic tautomeric equilibrium, Fry pointed out, provided new evidence supporting the existence and behavior of free radicals, most notably triphenylmethyl and its derivatives. Indeed, Fry maintained that chemists could understand free radicals only by employing his electron formulas and theory of positive and negative valences, for ordinary structural formulas afforded very little help in this matter.[30]

Structural theory, he said, had not accounted for the electrical conductivity of triphenylmethyl in liquid sulfur dioxide,[31] whereas his electronic formulas showed that the conductivity, and general chemical properties, of this and other free radicals depended on their developing positive or negative valences through electron loss or gain. Fry compared the conductivity of the triphenylmethyl solution with that of sodium in liquid ammonia. Just as each sodium atom by losing an electron formed a positive sodium ion in ammonia solution, so the triphenylmethyl radical formed a positive ion in liquid sulfur dioxide solution:

[29] Fry, "Positive and Negative Hydrogen," p. 264; idem, *The Electronic Conception of Valence*, p. 12.

[30] Fry, *The Electronic Conception of Valence*, pp. 146–67.

[31] Moses Gomberg and Lee H. Cone, "Über Triphenylmethyl," *Berichte* 37 (1904), 2033–51; Paul Walden, "Über abnorme Elektrolyte," *Zeit. phys. Chemie* 43 (1903), 443.

$$(C_6H_5)_3C \rightarrow \ominus + (C_6H_5)_3C^+ \,.$$

The sulfur atom having a valence of four in the sulfur dioxide molecule still had a pair of unused latent valences, and Fry, following the suggestion of Friend, Ramsay, and others, believed they were oppositely charged, $\pm SO_2$. The sulfur atom's positive valence, therefore, attracted the electron liberated by the triphenylmethyl radical, forming a negative ion,

$$O_2S\pm + \ominus \rightarrow \overset{-}{S}O_2 \,,$$

and the presence of these ions, $(C_6H_5)_3C^+$ and $\overset{-}{S}O_2$, explained the conductivity of the triphenylmethyl solution.

Fry's equations describing the electrical conductivity of sodium in liquid ammonia, which Hamilton P. Cady (1874–1943) in 1897 and later Edward C. Franklin (1862–1937) and Charles A. Kraus (1875–1967)[32] had discovered, were similar:

$$Na \rightarrow \overset{+}{Na} + \ominus$$
$$H_3N\pm + \ominus \rightarrow \overset{-}{N}H_3 \,.$$

But Fry claimed that in addition to the ions $\overset{+}{Na}$ and $\overset{-}{N}H_3$ the solution also contained some undissociated molecules of a compound that he called sodium ammonia, $NaNH_3$. Indeed, the existence of such a compound, though exceptional at that time because of the nitrogen atom's tetravalence, was essential, for it not only aided Fry's explanation of the sodium-ammonia reaction but, by analogy, the reactions of free radicals such as triphenylmethyl in sulfur dioxide. Sodium ammonia, $NaNH_3$, behaved as a free radical, he said, because it contained tetravalent nitrogen; thus triphenylmethyl, $(C_6H_5)_3C$, with a trivalent carbon atom was also a free radical. The reactivity in both cases was due to the abnormal valences of the nitrogen and carbon atoms and their tendencies to revert to a normal valence by electron gain or loss.

The Electron Clarifies the Meaning of Oxidation and Reduction

Throughout the nineteenth century chemists generally defined oxidation as the gain of oxygen in a chemical reaction and reduction as the gain

---

[32] Hamilton P. Cady, "The Electrolysis and Electrolytic Conductivity of Certain Substances Dissolved in Liquid Ammonia," *J. Phys. Chem.* 1 (1897), 707–13; Edward C. Franklin and Charles A. Kraus, "The Electrical Conductivity of Liquid Ammonia Solutions," *Amer. Chem. Journal* 23 (1900), 306.

of hydrogen. By 1901, however, Ostwald in the first edition of his *Wissenschaftlichen Grundlagen der analytischen Chemie* recognized that oxidation-reduction included reactions other than those involving oxygen and hydrogen and dealt with a definite change in an atom's electrical state. Albert B. Prescott (1832–1905) and Otis C. Johnson (1839–1912) in the fifth edition of their *Qualitative Chemical Analysis* and Julius Stieglitz in *The Elements of Qualitative Analysis* also gave electrical accounts of oxidation-reduction reactions.[33] Similar ideas appeared in the writings of Richard Abegg, J. J. Thomson, and William A. Noyes.

The electrical interpretation of oxidation, the loss of electrons by an atom, and of reduction, the gain of electrons, thus were processes of electron transfer easily illustrated with the electron formulas of Fry and Falk and Nelson. An understanding of the electron's role in oxidation-reduction reactions became much more widespread after these authors published their papers. Indeed, Fry proposed in 1911 the adoption of oxidation-reduction reactions as one of three methods to use in determining the electrical charge carried by an atom in a molecule.[34] The other two, hydrolysis and ionization-electrolysis, Abegg had already suggested in his 1904 publication on valence and need no repeating here.[35] Fry believed that, since the conversion of a structural formula into an electron formula was not an arbitrary procedure, carefully studying these processes provided the means of correctly assigning a positive or negative charge to an atom and hence establishing the chemical bond's direction.

To illustrate how oxidation-reduction reactions established an atom's charge or valence, Fry chose the slow combustion of methane at temperatures below its ignition point. The reaction went through the successive oxidations:

---

[33] Wilhelm Ostwald, *Wissenschaftlichen Grundlagen der analytischen Chemie*; Albert B. Prescott and Otis C. Johnson, *Qualitative Chemical Analysis*, pp. 238–42; Julius Stieglitz, *The Elements of Qualitative Analysis with Special Consideration of the Application of the Laws of Equilibrium and of the Modern Theories of Solution*, pp. 251–98.

[34] Fry, "Die Konstitution des Benzols," p. 405; idem, "A Critical Survey," p. 669; John M. Nelson, Hal T. Beans, and K. George Falk, "The Electron Conception of Valence: IV. The Classification of Chemical Reactions," *J.A.C.S.* 35 (1913), 1810–21; Fry, "Die Konstitution des Benzols," p. 405; idem, "A Critical Survey," p. 669; Fry, *The Electronic Conception of Valence*, pp. 15–19.

[35] Richard Abegg, "Die Valenz und das periodische System," *Zeit. anorg. Chemie* 39 (1904), 338–44. Feodor Selivanov ("Beitrag zur Kenntnis der gemischten Anhydride der unterchlorigen Säure und analoger Säuren," *Berichte* 25 [1892], 3617–23) and Julius Stieglitz

$$CH_4 \rightarrow CH_3OH \rightarrow HCHO \rightarrow$$

Methane Methyl Formaldehyde
alcohol

$$HCOOH \rightarrow HOCOOH \qquad \text{or} \qquad CO_2 .$$

Formic   Carbonic   Carbon
acid   acid   dioxide

When written electronically and assuming that the hydrogen atoms were positive and the oxygen atoms negative and divalent, Fry's formulas showed clearly the changing charge on the carbon atom as it went through the successive oxidations.[36] Because the gain of electrons or the development of negative valences corresponded to reduction, and the loss of electrons or formation of positive valences corresponded to oxidation, his formulas represented the carbon atom's successive oxidation states:

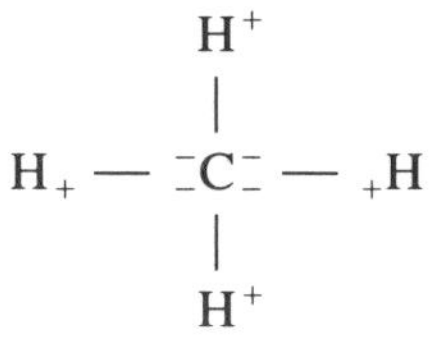

I

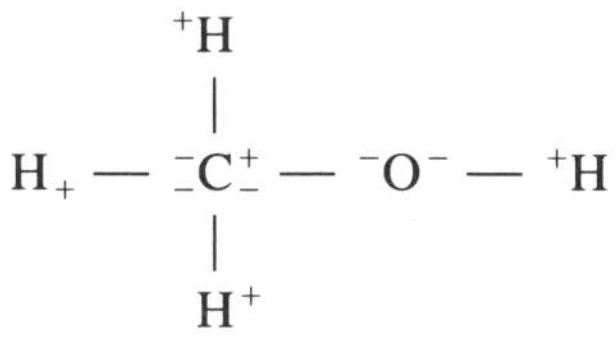

II

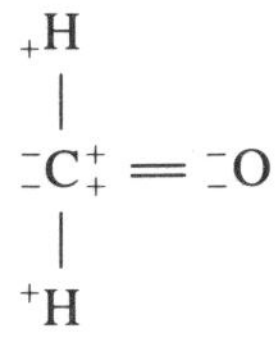

III

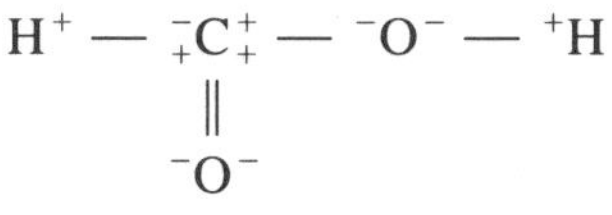

IV

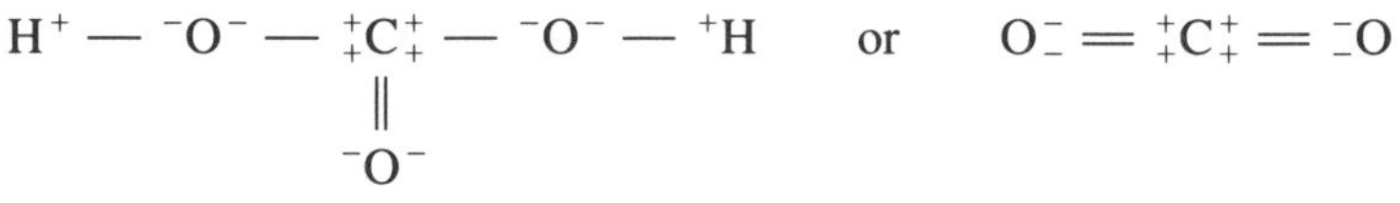

V

("On the Beckmann Rearrangement: I. Chlorimidoesters," *Amer. Chem. Journal* 18 [1896], 756) used hydrolysis for determining the electric charge on certain atoms.

[36] Fry, "Die Konstitution des Benzols," p. 405; idem, "A Critical Survey," p. 669.

or

$$-\ ^-_-C^-_- -\qquad -\ ^-_-C^+_- -\qquad -\ ^-_-C^+_+ -\qquad -\ ^-_+C^+_+ -\qquad -\ ^+_+C^+_+ -$$

I   II   III   IV   V

On the basis of these oxidation reactions, Fry proposed in 1911 a new valence rule: if an atom's empirically determined valence was n, then that atom existed in $(n + 1)$ electrical states, that is, it had $n + 1$ different electrovalences.[37] Carbon with a valence n equal to four had, therefore, $n + 1$ or five different electrovalences; chlorine, $n = 1$, and oxygen, $n = 2$, had, respectively, two and three electrovalences: Cl^- and Cl^+; $^-O^-$, $^-O^+$, and $^+O^+$.

Fry maintained that his rule of $n + 1$ electrovalences was more general and more comprehensive than any other. It should replace such earlier hypotheses as Abegg's theory of normal and contravalences, the first theory to relate an atom's positive and negative valences, and Friend's free positive, free negative, and residual or latent valences. Abegg's normal and contravalences were, of course, compatible with Fry's ideas regarding electromers and electronic tautomerism; Fry in fact introduced the term "electronic amphoterism" to describe an atom's capacity to gain or lose electrons and thereby acquire negative and positive electrovalences. Alternatively, Fry rejected Friend's contention that the metals and hydrogen had only positive valence, and that chlorine, fluorine, and oxygen had only negative valence. Abegg had already pointed out hydrogen's apparent negative valence in the metallic hydrides that Henri Moissan investigated, and in 1901 both William A. Noyes's study of the ammonia-chlorine gas reaction and Stieglitz's suggested mechanism for the reaction between water and chlorine gas appeared to indicate the presence of positive chlorine.[38]

In his examination of the electron theory of valence as it existed in this period, 1901–12, Fry also criticized two statements Falk and Nelson made in their publications. Falk, following Abegg, had written in 1911 that "the

[37] Fry, "Die Konstitution des Benzols," p. 388; idem, "A Critical Survey," p. 669.

[38] Henri Moissan, "Étude de la combinaison de l'acide carbonique et de l'hydrure de potassium," *Comptes rendus* 136 (1903), 723–27; idem, "Sur la non-conductibilité électrique des hydrures métalliques," *Comptes rendus* 136 (1903), 591–92; Noyes and Lyon, "The Reaction between Chlorine and Ammonia," pp. 460–63; Stieglitz, "On Positive and Negative Halogen Ions," pp. 797–99.

arrangement of the elements in the periodic system serves in general to indicate the electrical relations of the elements to each other in the production of a bond." [39] The electropositive metallic elements were on the left side of the periodic table; the electronegative nonmetals, on the right.

This statement, according to Fry, contradicted a fundamental tenet of the electron theory of valence, namely, the ability of any atom to have a positive and a negative electrovalence. How did chlorine's location in Group VII of the periodic table show that this atom had a positive and a negative electrovalence? Because of hydrogen's usual positive valence when combined with a nonmetal, chlorine was negative in hydrogen chloride, H^+—^-Cl; yet in combination with the more electronegative oxygen atom in the oxyacid hypochlorous acid, it was positive H^+—$^-O^-$—^+Cl. An atom's location in the periodic table could not establish its valence directions, for they depended solely on the polarity of the atoms or radicals combined with that atom. Fry believed that his valence rule—an atom of valence n had $(n + 1)$ electrovalences—illustrated each atom's electrical behavior much more completely than its location in the periodic table. Indeed, he maintained that his $(n + 1)$ rule was a necessary extension of the general arrangement found in the periodic table. [40]

Of course Fry's criticism that Falk and Nelson relied only on the periodic table to establish an atom's electrovalence was totally unwarranted. Using directive valences, Falk and Nelson arrived at essentially the same electron theory of valence as Fry, including the different possible electron structures that Fry called electromers. Fry was correct, however, in pointing out the periodic table's inability to represent an atom's electrovalences. The periodic table only indicated each atom's most common valence. It could not show Fry's electrovalences because they were really the atom's oxidation states. The electron's discovery occurred in 1897, nearly thirty years after Mendeleev introduced the periodic table in 1869, and only then did chemists begin to use the electron in valence theory. They established relations between an atom's oxidation states and its place in the periodic table, but they did not completely understand the relation until the wave-mechanical theory of atoms and molecules appeared in the mid-1920s.

Fry directed a second criticism at Falk's 1911 paper on organic acids. [41]

[39] Falk, "Valence: Organic Acids," p. 1141. Falk and Nelson said essentially this in their 1910 paper ("Electron Conception of Valence," p. 1639). This idea also appeared throughout their 1909 paper ("Electron Conception of Valency") but was not explicitly stated.

[40] Fry, "A Critical Survey," pp. 668, 669.

[41] Falk, "Valence: Organic Acids," p. 1141.

In this paper Falk divided the acids into four classes that he said depended on the directive valences of the α-carbon atom:

$$\equiv\!\!\!\!\!/\,C.CO_2H \qquad \rightleftharpoons\!\!\!\!\!=C.CO_2H \qquad \rightleftharpoons\!\!\!=C.CO_2H \qquad \equiv\!\!\!=C.CO_2H .$$

But as Fry pointed out, Falk had considered only three of the α-carbon atom's valences, omitting in his discussion the fourth valence, which, Fry claimed, was the most important because it bound the α-carbon atom to the carboxyl group and thus influenced the carboxyl group's behavior.[42] Instead of Falk's four classes of organic acids, Fry demanded eight:

$$\equiv\!\!\!\!\!/\,C\rightarrow CO_2H \qquad \rightleftharpoons\!\!=C\rightarrow CO_2H \qquad \rightleftharpoons\!=C\rightarrow CO_2H \qquad \equiv\!=C\rightarrow CO_2H$$

$$\equiv\!\!\!\!\!/\,C\leftarrow CO_2H \qquad \rightleftharpoons\!\!=C\leftarrow CO_2H \qquad \rightleftharpoons\!=C\leftarrow CO_2H \qquad \equiv\!=C\leftarrow CO_2H .$$

Fry did not attempt to show experimentally the existence of the eight classes by measuring ionization constants as Falk had done. Rather, he wanted to demonstrate that not only logically was it necessary to consider the α-carbon atom's fourth valence but that experimentally there was also direct evidence indicating a difference in polarity of the carboxyl group. It was either the electron acceptor or donor in forming the bond with the α-carbon atom. But before considering Fry's evidence, we must first examine his electron structure of the benzene molecule, for on it Fry rested his argument. Indeed, Fry considered his benzene structure a major contribution to the electron theory of valence.

Fry's Electron Structure of Benzene

Henry Armstrong (1848–1937) once wrote: "The determination of the 'structure' of this hydrocarbon [benzene] has given rise to a large amount of paper warfare." Armstrong was referring to the great number of benzene formulas that chemists had proposed since Kekulé's original structure in 1865, among them the diagonal formula of Adolf Claus (1866), the prismatic formula of Albert Ladenburg (1869), his own centric formula (1887) and that of Adolf von Baeyer (1888), and the dynamic formulas of John N. Collie (1897).[43] Despite the variety of benzene structures, none was very

---

[42] Fry, "A Critical Survey," p. 670.

[43] Henry Armstrong, "Valency," *Encyclopaedia Britannica*, p. 848; August Kekulé, "Untersuchungen über aromatische Verbindungen," *Annalen der Chemie* 137 (1866), 129–77; Adolf Claus, *Theoretische Betrachtungen und deren Anwendung zur Systematik*

satisfactory. Benzene's structure remained an unsolved problem among the otherwise very successful empirical structures of organic chemistry.

Alfred W. Stewart (1880–1947) summarized the merits of benzene's various structures in his *Stereochemistry* of 1907 and reached the conclusion that any successful structure had to be a dynamic structure.[44] Only Collie's structure, which depicted the benzene molecule in a state of continual vibration and included all the other proposed structures as transient phases of its motion, met Stewart's requirement.

Collie's model contained a six-membered ring of carbon atoms, each carbon atom having its four valences distributed tetrahedrally. Two of the valences connected each carbon atom to its two neighboring carbon atoms. A third valence went to a hydrogen atom. The fourth valence was sometimes directed inward as in the centric formula, sometimes joined to the fourth valence of an adjacent carbon atom forming a double bond as in Kekulé's formula, or assumed still some other spatial direction. According to Collie (1859–1942), each tetrahedron could rotate about its own center and about the center of gravity of the whole system:

First phase   Kekulé's formula   Centric formula

Kekule's formula   Last phase

Collie's structure, Stewart wrote, was "in complete accord both with the Kekulé and the centric formula, showing that they are mutually convertible into one another. It also shows how the supposed double linkages

der organischen Chemie; Albert Ladenburg, "Bemerkungen zur aromatischen Theorie," *Berichte* 2 (1869), 140–42; Henry Armstrong, "An Explanation of the Laws Which Govern Substitution in the Case of Benzenoid Compounds," *J. Chem. Soc.* 51 (1887), 258–68, 583–90; Adolf von Baeyer, "Über die Konstitution des Benzols," *Annalen der Chemie* 245 (1888), 103–90; John N. Collie, "A Space Formula for Benzene," *J. Chem. Soc.* 71 (1897), 1013–23.

[44] Alfred W. Stewart, *Stereochemistry*, pp. 530–31.

of the Kekulé formula shift between the carbon atoms. . . . But it differs from both in that . . . in two out of five configurations there are two distinct sets of hydrogen atoms. . . ." The Collie formulas are "in perfect accord with the formulae of Ladenburg, Claus, or Dewar, or that of Baly, Edwards, and Stewart. This adaptability is not possessed in any degree by previous space formulae, and it is this which makes the Collie formula superior to the others." [45]

Believing that benzene's structure provided a critical test for the electron theory of valence, Fry accepted Stewart's judgment that chemists should grant "preeminence" to Collie's space formula. But in making the conversion from Collie's formula to an electron formula, Fry found it necessary to consider only one of Collie's several spatial projections, namely, the planar centric arrangement. He had shown from his study of the methane to carbon dioxide oxidation that the carbon atom had five electrovalences:

$$
\begin{array}{ccccc}
\overset{+}{\underset{+}{+\,C\,+}} & \overset{+}{\underset{+}{-\,C\,+}} & \overset{+}{\underset{-}{-\,C\,+}} & \overset{+}{\underset{-}{-\,C\,-}} & \overset{-}{\underset{-}{-\,C\,-}} \\[1.5em]
\text{I} & \text{II} & \text{III} & \text{IV} & \text{V}
\end{array}
$$

If the benzene ring contained these five electrovalences arranged in symmetrical combinations of the pairs I and V, II and IV, or III and III, Fry argued, six centric electronic formulas or electromers were possible.[46] The combination of II and IV, for example, gave two electromers:

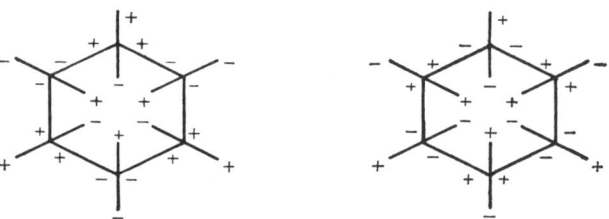

[45] Ibid., p. 531.

[46] Fry, "Die Konstitution des Benzols," pp. 388–89; idem, "A Critical Survey," p. 669. These were the only combinations that gave alternating positive and negative charges on both the carbon and the hydrogen atoms.

J. J. Thomson.

Richard Abegg. From *Berichte* 46 (1913).

Alfred Mayer. From *Biographical Memoirs*, vol. 8 (Washington, D.C.: National Academy Press, 1916).

Johannes Stark. From N. H. Heathcote, *Nobel Prize Winners in Physics* (New York: Henry Schuman, 1953).

Lauder Jones. From *National Cyclopedia of American Biography*, vol. 51, © 1969, James T. White & Co.

K. George Falk. From *National Cyclopedia of American Biography*, vol. 42, © 1958, James T. White & Co.

Julius Stieglitz. From *Biographical Memoirs*, vol. 21 (Washington, D.C.: National Academy Press, 1941).

Harry S. Fry. *National Cyclopedia of American Biography*, vol. 38, © 1953, James T. White & Co.

William C. Bray. From *Biographical Memoirs*, vol. 26 (Washington, D.C.: National Academy Press, 1951).

J. U. Nef. From *Biographical Memoirs*, vol. 34 (Washington, D.C.: National Academy Press, 1960).

Albert A. Noyes. From *Biographical Memoirs*, vol. 31 (Washington, D.C.: National Academy Press, 1958).

Howard J. Lucas. From *Biographical Memoirs*, vol. 43 (Washington, D.C.: National Academy Press, 1973).

Irving Langmuir. From *Biographical Memoirs*, vol. 45 (Washington, D.C.: National Academy Press, 1974).

Gilbert N. Lewis. Courtesy, University of California, Berkeley.

Wolfgang Pauli. Courtesy, Nobel Institute.

William A. Noyes. From *Biographical Memoirs*, vol. 27 (Washington, D.C.: National Academy Press, 1952).

The remaining combinations gave two more similar pairs (four electromers),[47] and, according to Fry, the formula below, omitting the double bonds, represented all six of benzene's possible electronic structures.

Thus, while the Kekulé and centric formulas failed to provide any structural basis for substitution in the *ortho*, *para*, or *meta* positions on the benzene ring, Fry pointed out that his electron formula clearly revealed a relation. If a given hydrogen atom or substituent were negative, then those hydrogen atoms or substituents *ortho* and *para* to it were positive; those *meta* to it were negative. On the other hand, if a given hydrogen atom or substituent were positive, then the hydrogen atoms or substituents *ortho* and *para* to it were negative; those *meta* to it, positive.[48] Fry's structure

[47] Fry, "Die Konstitution des Benzols," pp. 389–90. For the same discussion, see also idem, "Interpretations of Some Stereochemical Problems in Terms of the Electronic Conception of Positive and Negative Valences: Part II. Halogen Substitution in the Benzene Nucleus and in the Side Chain," *J.A.C.S.* 36 (1914), 1035–46, and idem, *The Electronic Conception of Valence*, pp. 45–51.

[48] Collie's space formula had also indicated the existence in benzene of two distinct groups of hydrogen atoms—a 1,3,5 and a 2,4,6 group—and it enabled him to present a stereochemical explanation of the well-known Crum Brown and John Gibson substitution rule. See Alexander Crum Brown and John Gibson, "A Rule for Determining Whether a Given Benzene Monoderivative Shall Give a Meta-di-derivative or a Mixture of Ortho- and Para Di-derivatives," *J. Chem. Soc.* 61 (1892), 367–69. This was a purely empirical rule used to determine whether a monosubstituted benzene derivative yielded chiefly a mixture of *ortho* and *para* derivatives or chiefly the *meta* derivative. For example, according to this rule, C_6H_5Cl was a derivative of H.Cl not directly oxidizable to HO.Cl. Thus C_6H_5Cl would yield *ortho*-chloronitrobenzene and *para*-chloronitrobenzene on nitration. On the other hand, $C_6H_5NO_2$ was as a derivative of $H.NO_2$ directly oxidizable to $HO.NO_2$. Therefore $C_6H_5NO_2$ gave *meta*-chloronitrobenzene on chlorination. Fry's electron formula of benzene permitted an entirely new interpretation of the Crum Brown–Gibson rule (Fry, "Stereochemical Problems: Halogen Substitution in Benzene," p. 251; idem, "Die Konstitution des Benzols," p. 391).

predicted that when substituents were of the same sign or polarity they always occupied *meta* positions on the benzene ring, but when the substituents were of opposite sign or polarity, they went only to *ortho* or *para* positions. This behavior was in agreement with the established directing ability of different substituents and with the Crum Brown–Gibson substitution rule.[49]

Fry believed that the well-known rearrangements of the substituted nitrogen halides supported his electron structure of benzene. Acetanilide, on treatment with hypochlorous acid, yielded phenylacetyl nitrogen chloride, which readily rearranged to *para*-chloroacetanilide. Additional treatment of the latter with hypochlorous acid gave *para*-chlorophenylacetyl nitrogen chloride and 2,4-dichloroacetanilide. Repeating the process resulted in the final product, 2,4,6-trichlorophenylacetyl nitrogen chloride, which did not undergo further rearrangement.[50] Fry thus had to show why the halogen atom (X), which wandered from the nitrogen, always occupied a position on benzene either *ortho* or *para* to the amino group, provided these positions were unoccupied:

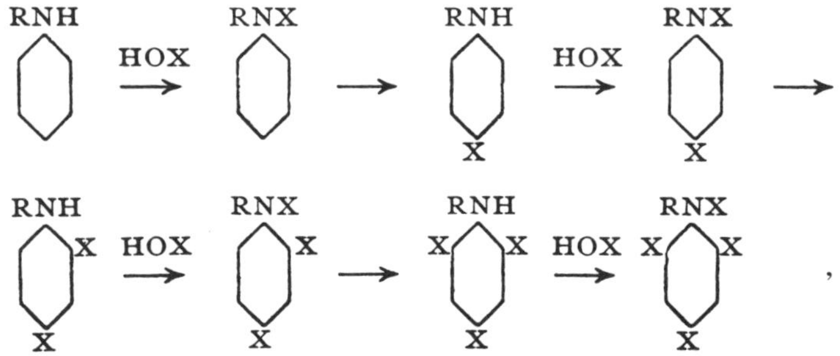

where X = Cl, Br; R = acetyl, formyl, benzoyl.

Arguing that in ammonia, NH_3, the nitrogen atom was negative and the three hydrogen atoms, positive, Fry pointed out that, since aniline, $C_6H_5NH_2$, was a derivative of ammonia, the group —NH_2 was negative and, accordingly, it occupied a negative hydrogen's position in the benzene ring. Aniline's electron formula and its reaction with hypochlorous acid, H^+—$^-O^-$—^+Cl, were, therefore:

[49] Fry, "A Critical Survey," p. 667.
[50] Fry, "Die Konstitution des Benzols," p. 396; idem, "A Critical Survey," p. 667.

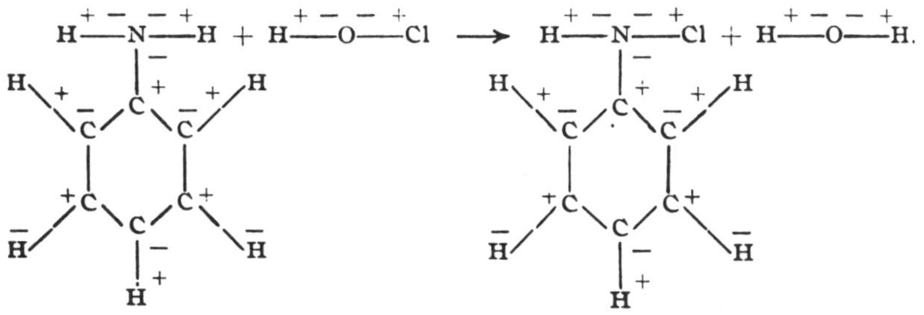

$$\text{H}\overset{+}{-}\overset{-}{\text{N}}\overset{-}{-}\overset{+}{\text{H}} + \text{H}\overset{+}{-}\overset{-}{-}\overset{-}{\text{O}}\overset{+}{-}\text{Cl} \longrightarrow \text{H}\overset{+}{-}\overset{-}{\text{N}}\overset{-}{-}\overset{+}{\text{Cl}} + \text{H}\overset{+}{-}\overset{-}{\text{O}}\overset{-}{-}\overset{+}{\text{H}}.$$

Fry then claimed that only positive halogen atoms could exchange positions with the positive hydrogens of the benzene ring. The positive halogen atom of phenylacetyl (or a *para*-substituted phenylacetyl) nitrogen chloride had to go to a position *ortho* or *para* to the negative —NH$_2$ or RNH— group. When these positions were occupied, as in a 2,4,6-trichloroacyl nitrogen chloride, rearrangement was both actually and theoretically impossible. This behavior, Fry maintained, proved the identical polarity of benzene's 2, 4, and 6 positions, since a positive chlorine atom occupied each of them, and consequently it proved his electron structure of benzene.[51]

Fry also showed how his structure accounted for the products formed on the nitration and chlorination of benzene.[52] In the direct nitration of benzene the positive nitro group replaced one of benzene's three positive hydrogen atoms:

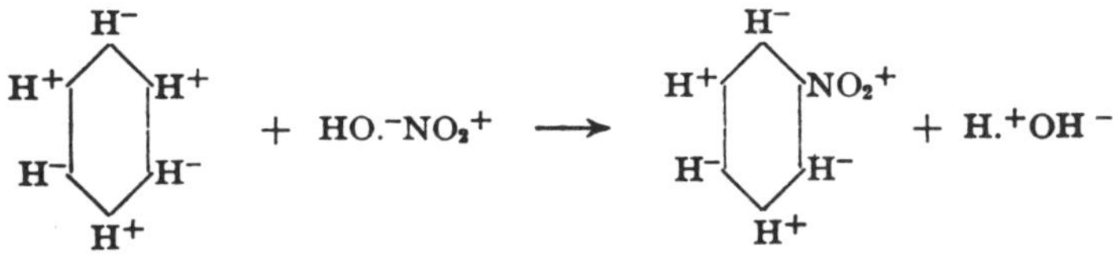

Treatment of nitrobenzene with chlorine produced only *meta*-chloronitrobenzene, a result that followed, Fry said, because both the nitro group and the substituted chlorine atom were positive and occupied positions *meta* to each other. Hence, in the reaction with chlorine, Cl$^+$.$^-$Cl, the positive

[51] Fry, "Die Konstitution des Benzols," p. 396; idem, "A Critical Survey," p. 668.
[52] Fry, "Die Konstitution des Benzols," pp. 393–95.

chlorine atom replaced a positive hydrogen atom that then combined with the negative chlorine atom forming hydrogen chloride, $H^+ . {}^-Cl$:

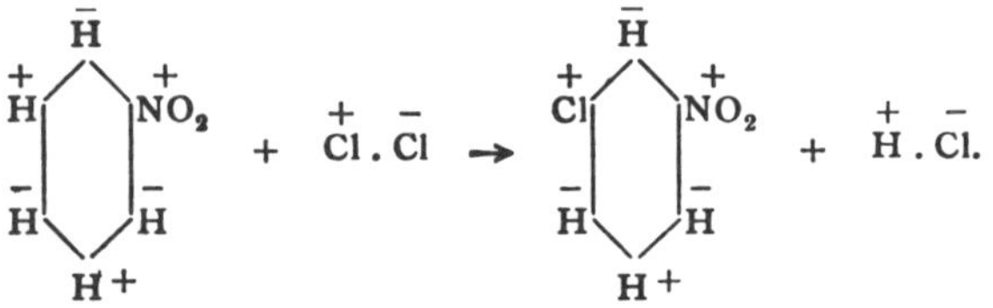

$$\overset{-}{H} \quad \overset{+}{H}\bighexagon\overset{+}{NO_2} \quad + \quad \overset{+}{Cl} . \overset{-}{Cl} \quad \rightarrow \quad \overset{-}{H} \quad \overset{+}{Cl}\bighexagon\overset{+}{NO_2} \quad + \quad \overset{+}{H} . \overset{-}{Cl}.$$

On the other hand, the nitration of chlorobenzene gave *ortho*-chloronitrobenzene and *para*-chloronitrobenzene. The nitro group and the chlorine atom were, therefore, of opposite sign: the nitro group was once again positive; the chlorine atom in the original chlorobenzene, negative. The equation below shows the positive nitro group taking positions *ortho* and *para* to the negative chlorine atom:

$$\overset{-}{Cl} \quad \bighexagon \quad + \quad HO.NO_2 \quad \longrightarrow \quad \overset{-}{Cl} \quad \bighexagon NO_2 \quad \text{or} \quad \overset{-}{Cl} \quad \bighexagon \quad + \quad H.OH$$

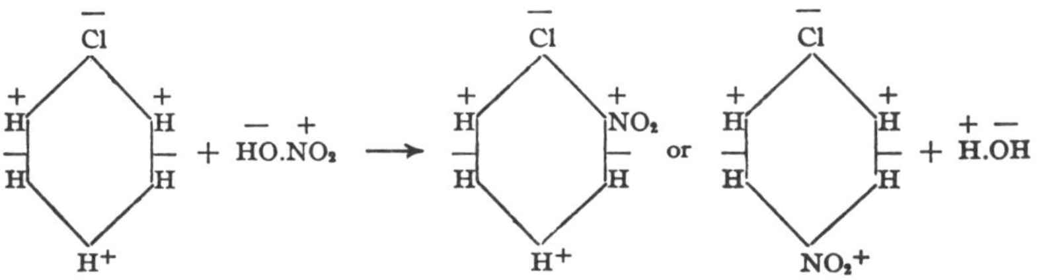

We return now to Fry's criticism of K. George Falk's classification of organic acids, recalling that Fry found it incomplete because Falk omitted the direction of the valence that bound the α-carbon atom to the carboxyl group. This fourth valence, Fry insisted, was of prime importance because it determined the polarity of the group attached to the carboxyl group and, of course, influenced the carboxyl group's behavior. He could demonstrate its importance, Fry said, by examining the reactions of any acid containing two carboxyl groups attached to different carbon atoms.[53]

Phthalic acid met Fry's requirement because it had two adjacent carboxyl groups attached to a benzene ring, and as Baeyer discovered in 1892 they behaved differently. Certain phthalic acids, the dihydrophthalic acids, when heated lost only one molecule of carbon dioxide. Baeyer had attributed the loss of carbon dioxide to a "shock" (*Erschütterung*) felt by the

[53] Fry, "A Critical Survey," pp. 670–72.

α-carbon atom. But Julius Brühl in Heidelberg said that was really no explanation, because it left unanswered why the shock removed only one molecule of carbon dioxide and not the expected two. Brühl (1850–1911) believed that the behavior of the phthalic acids resulted from their different stabilities; his arguments were essentially independent of any structural theory. But as Julius Cohen (1859–1935) at Leeds University correctly noted, Brühl was also begging the question by avoiding any reference to the influence of structure on stability.[54]

Recognizing that none of these interpretations accounted satisfactorily for the phthalic acids' behavior, Fry pointed out that according to his electron structure, the two carboxyl groups were of opposite polarity. The fourth valences of the two carbon atoms binding the carboxyl groups were in opposite directions and, therefore, the carboxyl groups should react differently:

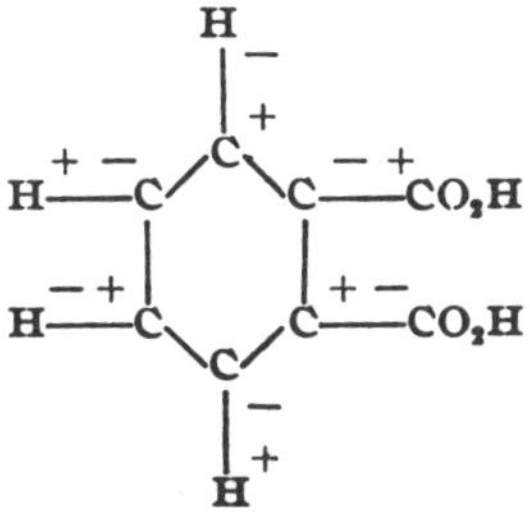

How did polar valences account for this difference, that is, how did the opposite polarities of the carboxyl groups cause only one of them to lose carbon dioxide? To answer this question, Fry turned to the electron structures of two simple organic acids, formic and carbonic:

$$H^+ \!\!-\!\! {}^-CO_2H \qquad\qquad HO^- \!\!-\!\! {}^+CO_2H .$$

Formic acid   Carbonic acid

His structures showed (1) that the carboxyl groups in these acids were, respectively, negative and positive and (2) that in formic acid three of car-

[54] Baeyer, "Über die Konstitution des Benzols," p. 178; Julius W. Brühl, "Neue Beiträge zur Frage nach der Konstitution des Benzols," J. prakt. Chemie 49 (1894), 229; Julius B. Cohen, Organic Chemistry, p. 461.

bon's four valences were positive, while in carbonic acid the carbon atom had four positive valences.

$$H^+ - {}^-_+C^+_+ - {}^-O^- - {}^+H \quad H^+ - {}^-O^- - {}^+_+C^+_+ - {}^-O^- - {}^+H.$$
$$\|\qquad\qquad\qquad\qquad\quad\|$$
$${}^-O^- \qquad\qquad\qquad\qquad {}^-O^-$$

Thus, both structurally and electronically, Fry argued, carbonic acid could lose carbon dioxide according to the electronic equation:

$$H^+ - {}^-O^- - {}^+_+C^+_+ - {}^-O^- - {}^+H \rightarrow$$
$$\|$$
$${}^-O^-$$

$$H^+ - {}^-O^- - {}^+H \ + \ O^-_- = {}^+_+C^+_+ = {}^-_-O.$$

In formic acid, however, the carbon atom's fourth valence was negative and the spontaneous loss of carbon dioxide could not occur because all of the carbon's valences had to be positive.

Fry's electron formulas also indicated that an atom attained its maximum degree of oxidation when all of its valences were positive. The carbon atom in formic acid, having one negative and three positive valences, had not yet attained this state. A positive carboxyl group, therefore, corresponded to the oxidation state of the carbon atom in carbonic acid (or carbon dioxide) and could lose carbon dioxide. Alternatively, a negative carboxyl group corresponded to the oxidation state of the carbon atom in formic acid, from which the loss of carbon dioxide was electronically impossible.

Thus while phthalic acid had both a positive and a negative carboxyl group, only the positive acid group lost carbon dioxide. And this difference, Fry concluded,

constitutes the explanation of the fact that one molecule instead of two molecules of carbon dioxide is removed from the several phthalic acids noted above, and furthermore, is evidence that the direction of the valence which binds a carboxyl radicle to a compound has a definite influence upon the properties of the radicle and the compound, and therefore, is not to be disregarded in any applications of the electron conception of directive valences.[55]

[55] Fry, "A Critical Survey," p. 673.

Conclusion

The publications of Falk, Nelson, and Fry in the period 1909–15 represented a serious effort to account for the structures and reactions of molecules, especially those of organic chemistry, with an electrostatic or polar theory of valence. But these authors went to the extreme in assuming only polar electron bonds and in interpreting all reactions electrostatically. Their use of electron bonds rather than the organic chemist's nonelectrical bond, the traditional dash, indicated the direction chemistry had to take if chemists were ever to explain electrically double bond addition, the directing ability of substituents on the benzene ring, and the influence of substituents on organic acid strength. What they really needed was a less extreme theory to account for the well-established polarity of numerous organic molecules. An incomplete electron transfer or a sharing of electrons rather than complete electron transfer was the key.

On the other hand, the electrostatic theory of valence led to a clear statement of oxidation and reduction. Chemists would have had far more difficulty understanding this conception without it. Indeed, once the meaning of oxidation-reduction was clear, Fry used it to establish the positive and negative valences of an atom in a molecule, actually the atom's oxidation number, and to illustrate the electron structures of his electromers. These structures in some ways resembled the later resonance forms of organic chemistry.

Falk, Nelson, and Fry first applied the electron theory of valence to organic molecules because these were the only molecules with well-established structures. Fry, in fact, believed he had solved the long-standing controversy regarding the benzene structure when he published his electron model of this molecule. With it he accounted for benzene's stability, its reactivity, and most important, the directing ability of the benzene ring substituents.

5. The Search for Electromers

Introduction

The ability of the same atom to acquire a positive or a negative charge was the fundamental principle that inspired the search for electromers. It was central to every electron theory of valence in the early twentieth century. Abegg's theory of normal and contravalences required every atom to have a positive and a negative charge. J. J. Thomson included oppositely charged carbon atoms in his electron formulas for simple hydrocarbons such as ethylene.

A few years after Abegg's and Thomson's publications, in the years 1908–11, Fry presented his arguments for electromers and the need for chemists to isolate them. He pointed out that since chemists already believed each of the elementary diatomic gases (H_2, O_2, Cl_2) contained atoms with opposite and equal charges that accounted for its electrostatic bond, then surely these atoms could have the same positive or negative charge when they combined with other atoms:

$$H_2 = H^+ - H^-$$
$$Cl_2 = Cl^- - Cl^+$$

$$2HCl = H^+ - Cl^- + H^- - Cl^+ .$$

The hypothetical forms H^+—Cl^- and H^-—Cl^+ were the two resulting electromers, though, according to Fry, they were seldom present in equal proportion because of greater stability of one of the forms, generally the one in which the atoms displayed their usual valences. Fry also claimed that electromers resulted from an intramolecular oxidation-reduction reaction taking place between the atoms or groups of atoms (radicals) in a compound.[1]

[1] Harry S. Fry, "Die Konstitution des Benzols vom Standpunkte des Korpuskular-

Even before the electron theory of valence had developed sufficiently to include electromers, the persistence of the electrochemical theories of Davy and Berzelius throughout the nineteenth century led to continuing speculations on electrical atoms. William A. Noyes in 1901 had already suggested the possibility of what Fry later called electromers; other electrochemical speculations appeared that same year in the papers of Arthur Lapworth (1872–1941) in Manchester and Julius Stieglitz at the University of Chicago. Lapworth described the bromine-phenol reaction with the ionic scheme

$$Br^+ + Br^- + H^+ + C_6H_5O^- \rightarrow C_6H_5OBr + HBr,$$

concluding that "it is to electrolytic dissociation, often doubtless in extremely minute amount, that the majority of changes in organic compounds may be most probably assigned." Stieglitz wrote that for a number of years in lectures before the student and professional body at Chicago he had used an ionic mechanism and the hypothesis of positive halogens to account for the behavior of hypochlorous, hypobromous, and hypoiodous acid.[2]

Noyes and Stieglitz, joined by Stieglitz's former pupil Lauder W. Jones, continued their efforts to establish mechanisms for chemical reactions using only a polar or electrostatic theory of valence. They correctly recognized the need for polarity in accounting for many organic reactions but failed for the most part to distinguish between their organic polarity and the more pronounced polarity of inorganic compounds. Their hypothesis of electromers proved false because it, too, depended on an electrostatic theory of valence.

William A. Noyes and the Search for Positive Chlorine and the Electromers of Nitrogen Trichloride

Noyes accepted the electrolytic dissociation of substances not usually considered electrolytes, and it enabled him in 1901 to devise an ionic

---

atomistischen Begriffs der positiven und negativen Wertigkeit," *Zeit. phys. Chemie* 76 (1911), 387.

[2] William A. Noyes and Albert C. Lyon, "The Reaction between Chlorine and Ammonia," *J.A.C.S.* 23 (1901), 460–63; Arthur Lapworth, "The Form of Change in Organic Compounds and the Function of the α-Meta-Orientating Groups," *J. Chem. Soc.* 79 (1910), 1266; Julius Stieglitz, "On Positive and Negative Halogen Ions," *J.A.C.S.* 23 (1910), 797–99.

mechanism that accounted for the unexpected results he obtained in studying the reaction between ammonia and chlorine. His proposed mechanism was unique at that time, because it assumed that when the reactants separated into positive and negative atoms, the same atom was sometimes positive and sometimes negative.[3] Indeed, Noyes suggested for the first time the existence of two different electrical structures for the same molecule. His search for them lasted for twenty years, and his failure to establish their separate existence, particularly the electrical structures of nitrogen trichloride, resulted in Noyes's eventually abandoning the polar theory of valence and accepting G. N. Lewis's theory of the shared electron pair bond.[4]

The reaction between chlorine and ammonia, which Noyes investigated with Albert C. Lyon, an undergraduate student at Rose Polytechnic Institute, was a well-known demonstration originally introduced by August von Hofmann to illustrate the composition of ammonia.[5] In this reaction, when one added a concentrated solution of ammonia to a tube filled with chlorine gas, the chlorine combined with an equal volume of the hydrogen from ammonia, liberating nitrogen. Since ammonia consisted of one volume of nitrogen in combination with three volumes of hydrogen, the addition of water at the end of the experiment left the tube one-third full of nitrogen gas, thus confirming the formula of ammonia, NH_3. The equation for the reaction was, therefore,

$$3Cl_2(g) + 2NH_3(aq) \rightarrow N_2(g) + 6HCl(aq).$$
$$\text{3 vol.} \qquad\qquad\qquad \text{1 vol.}$$

When Noyes and Lyon used a dilute solution of ammonia (0.5% by weight), they found that instead of the expected one-third reduction, only a one-sixth reduction in volume actually occurred, and a new product, nitrogen trichloride, resulted. On the basis of this reduction in volume (6 volumes of chlorine to 1 volume of nitrogen) and the knowledge that nitrogen

[3] Noyes and Lyon, "The Reaction between Chlorine and Ammonia," pp. 460–63. See also William A. Noyes, "Present Problems of Organic Chemistry," *Science* 20 (1904), 490–501.

[4] William A. Noyes, "An Attempt to Prepare Nitro-Nitrogen Trichloride: II. The Conduct of Mixtures of Nitrogen and Chlorine in a Flaming Arc," *J.A.C.S.* 43 (1921), 1774–82 (especially p. 1781); idem, "The Relation of Shared Electrons to Potential and Absolute Polar Valences," *Chemical Reviews* 5 (1928), 552–55.

[5] Noyes and Lyon, "The Reaction between Chlorine and Ammonia," pp. 460–63.

and nitrogen trichloride formed in equimolar quantities, they gave as the equation for this reaction

$$6Cl_2(g) + 3NH_3(aq) \rightarrow NCl_3(aq) + N_2(g) + 9HCl(aq).$$

6 vol.   1 vol.

Noyes and Lyon's study also showed that on titrating nitrogen trichloride with an arsenious acid solution, one molecule of the trichloride (three chlorine atoms) was equivalent in oxidizing power to three molecules (six atoms) of free chlorine.[6] Nitrogen trichloride's anomalous behavior led them to speculate on its electrical constitution to explain its oxidizing power. Indeed, they suggested that the positive chlorine atoms in the chlorine molecules retained their positive charge in nitrogen trichloride and that the trichloride existed in two electrical forms.

Noyes and Lyon's graphical representation of the reaction appears below:

	H^+	Cl^-	Cl^+	
N^{---}	H^+	Cl^-	Cl^+	N^{---}
	H^+	Cl^-	Cl^+	
N^{+++}	H^-	Cl^+	Cl^-	H^+
	H^-	Cl^+	Cl^-	H^+
	H^-	Cl^+	Cl^-	H^+

When the oppositely charged ions present in the intermediate state were appropriately paired off, the charges on the reactants were

$$3NH_3 \rightarrow 2(N^{---} + 3H^+) + (N^{+++} + 3H^-),$$
$$6Cl_2 \rightarrow 6(Cl^+ + Cl^-),$$

and on the products

$$NCl_3 \rightarrow (N^{---} + 3Cl^+),$$
$$N_2 \rightarrow (N^{---} + N^{+++}),$$
$$9HCl \rightarrow 6(H^+ + Cl^-) + 3(H^- + Cl^+).$$

[6] Willibald Hentschel, "Über Chlorstickstoff," *Berichte* 30 (1897), 1434–37; Noyes and Lyon, "The Reaction between Chlorine and Ammonia," p. 462. In 1918, Noyes gave the following account of the reaction between nitrogen and chlorine: "In terms of the electron theory this is easily explained by assuming that in nitrogen trichloride the nitrogen atom has gained three electrons, becoming negative, while each chlorine atom has lost an electron be-

Ammonia and hydrogen chloride thus had ionized in two ways giving two electrical structures for each molecule:

$$(N^{+++} + 3H^-) \leftrightarrows NH_3 \rightleftarrows (N^{---} + 3H^+)$$
$$(H^- + Cl^+) \leftrightarrows HCl \rightleftarrows (H^+ + Cl^-)   .$$

I   II

In concluding their study of the chlorine-ammonia reaction, Noyes and Lyon added that the formation of nitrogen and nitrogen trichloride were not independent reactions. They assumed that the six chlorine molecules reacted simultaneously with the three ammonia molecules. The reaction, in other words, was a complicated ninth-order ionic reaction: "If we suppose what seems not inherently improbable, that all reactions involving the decomposition of molecules are preceded by an ionization of the parts of those molecules, it would follow that elementary molecules, as well may ionize into positive and negative parts." [7]

Noyes and Lyon admitted putting forward their ionic hypothesis with some hesitation, at the same time inviting other chemists to comment on it. They received a brief reply from Arthur A. Noyes (1866–1936) at the Massachusetts Institute of Technology. He criticized their suggestion that ammonia (and presumably hydrogen chloride) had two ionic structures, claiming that if such structures existed, ammonia would spontaneously decompose into nitrogen and hydrogen. Responding to this criticism in a later publication, William Noyes argued that his ionic structures separated into free ions only when the two gases reacted with each other and not when they were in the uncombined state. [8] Thus no spontaneous decomposition of ammonia occurred.

---

coming positive. In order to convert the chlorine of the trichloride to the form in ordinary chlorides each chlorine atom must gain *two* electrons. In free chlorine, on the other hand, one chlorine atom has lost an electron, while the other has gained one, hence the two atoms of free chlorine require only two electrons to convert both atoms to the negative form" ("The Electron Theory," *Journal of the Franklin Institute* 185 [January 1918], 70).

[7] Noyes and Lyon, "The Reaction between Chlorine and Ammonia," pp. 462, 463. In a later article (William A. Noyes and Arthur B. Haw, "The Reaction between Chlorine and Ammonia: II," *J.A.C.S.* 42 [1920], 2167–73), Noyes and Haw wrote that William C. Bray and Carr Thomas Dowell ("Experiments with Nitrogen Trichloride," *J.A.C.S.* 39 [1917], 907) had pointed out that the complicated ninth-order reaction that Noyes and Lyon had proposed was highly improbable. Bray and Dowell demonstrated that the nitrogen and nitrogen trichloride formed in the experiment resulted from independent reactions.

[8] Arthur A. Noyes, "Die Reaktion zwischen Chlor und Ammoniak," *Zeit. phys. Chemie* 41 (1902), 378; William A. Noyes, "Present Problems of Organic Chemistry," p. 498.

Noyes and Lyon's hypothesis, especially their suggestion of positive chlorine, brought a favorable reply from Julius Stieglitz, however. Stieglitz believed that the quantitative experiments of Alexander A. Jakowkin (1860–1936) in Saint Petersburg showed "that the existence of *positive chlorine ions*, in aqueous solutions both of chlorine and of hypochlorous acid, can no longer be considered as a mere hypothesis, but must be accepted as proved experimentally."[9] Jakowkin, he pointed out, had already demonstrated the reversibility and ionic nature of the chlorine-water reaction:

$$Cl_2 + H_2O \rightarrow (HOCl + H^+ + Cl^-)aq .$$

But no one recognized that, since the reaction was ionic with respect to hydrochloric acid and water (which according to Stieglitz was universally recognized), it was also ionic with respect to hypochlorous acid and chlorine. Hypochlorous acid dissociated very slightly into negative hydroxyl and positive chlorine ions; chlorine, partially into positive and negative chlorine ions. The cause of the hydrolysis and of the reversibility, he insisted, "lies, not more in the minimal, but actual, ionization of water, than in the formation of these ions of hypochlorous acid and of chlorine."[10] In Stieglitz's opinion, the ionization of hypochlorous acid and of chlorine was the obvious and most important result of Jakowkin's experiments.

Stieglitz thus concluded that hypochlorous acid ionized in two ways—as a weak acid and as a still weaker base:

$$HOCl \rightleftarrows H^+ + OCl^- \qquad \text{(Acidic ionization)}$$
$$HOCl \rightleftarrows HO^- + Cl^+ \qquad \text{(Basic ionization)} .$$

He maintained that no matter how small hypochlorous acid's basic ionization constant might be, compared with the well-established but slight ionization of water, it would demonstrate the existence of positive chlorine ions and was a fact that chemists could no longer neglect.[11]

[9] Alexander A. Jakowkin, "Über die Hydrolyse des Chlors," *Zeit. phys. Chemie* 29 (1899), 613–57; Stieglitz, "On Positive and Negative Halogen Ions," p. 798.

[10] Alexander A. Jakowkin, "Über die Dissociation des Chlorhydrates in wässeriger Lösung bei 0°," *Berichte* 30 (1897), 518–21; Stieglitz, "On Positive and Negative Halogen Ions," p. 798.

[11] Stieglitz, "On Positive and Negative Halogen Ions," p. 799. For later developments regarding the study of positive chlorine in hypochlorous acid, see William A. Noyes and Thomas A. Wilson, "The Ionization Constant of Hypochlorous Acid: Evidence for Amphoteric Ionization," *J.A.C.S.* 44 (1922), 1630–37; and William A. Noyes, "The Electronic Interpretation of Oxidation and Reduction," *J.A.C.S.* 51 (1929), 2391–96.

Two years after Noyes and Lyon published their paper on the chlorine-ammonia reaction, J. J. Thomson visited Yale University to deliver the Silliman Lectures. In these lectures, which appeared in 1904 as *Electricity and Matter*, Thomson introduced his electron theory of electrostatic attraction.[12] Though Noyes was already using electrostatic attraction to account for the union of atoms, his 1904 address to the International Congress of Arts and Science in St. Louis was the first time he included a discussion of the electron's role in chemical union. Noyes referred to the "ingenious experiments of J. J. Thompson [*sic*]" that had shown that electrons were capable of an independent existence. "If we accept the theory of electrons," he continued, "it is evident that the electrons must be present in the molecule of an electrolyte no matter in what manner it is formed. It is but a step further to the conclusion that the electrons are involved in every combination or separation of atoms and, indeed, may be the chief factor in chemical combination."[13]

In the ensuing years, Noyes continued to publish articles on the electron theory of valence. His presidential address to the Illinois Academy of Science in 1912 had the title "The Electron Theory," and in other papers he interpreted with the electron theory the reactions of inorganic and organic compounds, ionization phenomena, and the connection between electrical conductivity and electron loss in metals.[14]

During this same period, Noyes renewed his study of the chlorine-ammonia reaction, searching for evidence to support the ionic mechanism he had cautiously proposed to account for nitrogen trichloride's unexpected formation. The mechanism Noyes finally adopted required the presence of positively charged chlorine atoms and included two different electrical structures for nitrogen trichloride, one of which contained positively charged chlorine atoms. In Fry's terminology, Noyes claimed there were two electromers of nitrogen trichloride. A positively charged chlorine atom seemed to be a fairly novel idea in 1901, but by 1913 Noyes knew that it had a rather long history. In a paper published that year he cited

[12] Thomson's 1904 paper, "On the Structure of the Atom," *Phil. Mag.* 39 (1904), 237–65, contains the same ideas.

[13] W. Noyes, "Present Problems of Organic Chemistry," p. 468.

[14] William A. Noyes, "The Electron Theory," *Proceedings, Illinois Academy of Science* 5 (1912), 20–28; idem, "Molecular Rearrangements," *J.A.C.S.* 31 (1909), 1368–74; idem, "A Possible Explanation of Some Phenomena of Ionization by the Electron Theory," *J.A.C.S.* 34 (1912), 663–64.

these earlier references to positive chlorine to support his claim that one of nitrogen trichloride's electromers had the structure $N^{---} Cl_3^{+++}$.[15]

Noyes described how Paul Schutzenberger (1827–97) in Paris as early as 1861 had prepared compounds of the type R.C.(OCl):O and had believed that chlorine in these compounds reacted like an electropositive metal in salts.[16] He pointed out that Feodor Selivanov (1859–?), at one time Mendeleev's assistant in Saint Petersburg, in 1892 observed that, on hydrolyzing the chloroamines RNHCl, $RNCl_2$, and R_2NCl, hydrogen replaced the chlorine. Since these compounds reacted with hydrogen iodide, liberating two atoms of free iodine (I_2) for each atom of combined chlorine,

$$R_2NCl + 2HI = R_2NH + HCl + I_2,$$

Selivanov concluded that the chlorine atom's peculiar behavior occurred because even in combination it existed as "hypochlorous chlorine" and behaved as if it had a positive charge. Selivanov remarked that the chlorine atoms in nitrogen trichloride showed the same strange behavior.[17]

Noyes also referred to the publications of Stieglitz and Paul Walden, which supported the presence of positively charged chlorine and the other halogens bromine and iodine. Indeed, in a publication, "On Positive and Negative Halogen Ions," which appeared shortly after Noyes and Lyon's 1901 paper, Stieglitz wrote: "The interesting paper of Professor Noyes and Mr. Lyons [sic] had made it necessary to publish these lines rather prematurely; but their invitation for discussion encourages me to do so before the completion of my own experiments."[18] Positive halogens had thus be-

[15] William A. Noyes, "An Attempt to Prepare Nitro-Nitrogen Trichloride, an Electromer of Ammono-Nitrogen Trichloride," *J.A.C.S.* 35 (1913), 767–75.

[16] Paul Schützenberger, "Über die Substitution electronegativer Körper an die Stelle der Metalle in Sauerstoffsalzen," *Annalen der Chemie* 120 (1861), 113–18; idem, "On the Substitution of Electro-Negative Bodies (Chlorine, Bromine, Iodine, Cyanogen, Sulphur, etc.) for the Metals in Oxygenised Salts: Production of a New Class of Salts in Which the Electro-Negative Bodies Replace the Basic Hydrogen," *Chemical News* 3 (13 April 1861), 225–26.

[17] Feodor Selivanov, "Beitrag zur Kenntnis der gemischten Anhydride der unterchlorigen Säure und analoger Säuren," *Berichte* 25 (1892), 3617–23. Both Schützenberger and Selivanov seem to have recognized the atomicity of electricity and that electricity reacted with the weights of elements in unit quantities according to the laws of definite and of multiple proportions.

[18] Julius Stieglitz, "On the Beckmann Rearrangement: I. Chlorimidoesters," *Amer. Chem. Journal* 18 (1896), 751–61; idem, "On Positive and Negative Halogen Ions," pp. 797–99; Paul Walden, "Über abnorme Elektrolyte," *Zeit. phys. Chemie* 43 (1903), 385–

come a part of the electron theory of valence starting from Abegg's and Thomson's earliest publications to the latest developments of Fry, Falk, Nelson, and others.

Noyes's conclusive proof for the first of nitrogen trichloride's two electromers followed from Abegg's rule of hydrolysis. This rule (which Abegg published after Noyes's 1901 paper) said that in hydrolysis the positive and negative components of a compound combined, respectively, with the water molecule's negative hydroxyl group and its positive hydrogen atom. Noyes demonstrated in his 1913 paper that nitrogen trichloride hydrolyzed reversibly, giving ammonia and hypochlorous acid, and thus argued that these products could have resulted only if the reaction proceeded accordingly:

$$N^{---}Cl_3^{+++} + 3H^+O^{--}H^+ = N^{---}H_3^{+++} + 3Cl^+O^{--}H^+ .$$

His equation showed clearly the presence of positive chlorine in the first nitrogen trichloride electromer.

Noyes, of course, recognized that the nitrogen trichloride hydrolysis did not give the products normally obtained from nonmetallic chlorides, for in those chlorides, chlorine was always negative and the other nonmetal, positive. Phosphorus trichloride when hydrolyzed gave, as expected, phosphorous acid, $P(OH)_3$, and hydrogen chloride, HCl, indicating a positive phosphorus atom and three negative chlorine atoms: $P^{+++}Cl_3^{---}$.

Since nitrogen and phosphorus belonged to the same periodic group, Noyes expected their chlorides to hydrolyze the same way. When they did not, he introduced his second nitrogen trichloride electromer, which contained positive nitrogen and negative chlorine atoms and would hydrolyze as phosphorus trichloride and other nonmetallic chlorides:

$$N^{+++}Cl_3^{---} + 3H^+O^{--}H^+ =$$
$$H^+O^{--}N^{+++}O^{--} + H^+O^{--}H^+ + 3H^+Cl^- .$$

Noyes called the second electromer, $N^{+++}Cl_3^{---}$, nitro-nitrogen trichloride, because it supposedly produced nitrous acid on hydrolysis. He named the first, $N^{---}Cl_3^{+++}$, which gave ammonia on hydrolysis, ammono-nitrogen trichloride.[19]

464; Paul Walden, "Über einige anorganische Lösungs- und Ionisierungsmittel," *Zeit. anorg. Chemie* 25 (1900), 209–26.

[19] W. Noyes, "An Attempt to Prepare Nitro-Nitrogen Trichloride," p. 769.

To prepare the unknown electromer, nitro-nitrogen trichloride, Noyes reacted nitrosyl chloride, NOCl, and phosphorus pentachloride.[20] Nitrosyl chloride was an obvious choice because upon hydrolysis it gave nitrous acid, HONO, which contained a positive nitrogen atom and, according to Noyes, proved the presence of positive nitrogen in nitrosyl chloride. Thus, after chlorinating nitrosyl chloride with phosphorus pentachloride the nitrogen atom retained its positive charge, yielding nitro-nitrogen trichloride according to the following reactions:

$$NaNO_2 + PCl_5 \xrightarrow{high\ temperature} NOCl + NaCl + POCl_3$$
$$NOCl + PCl_5 \xrightarrow{high\ temperature} NCl_3 + POCl_3 .$$

By establishing the separate existence of nitro-nitrogen trichloride and ammono-nitrogen trichloride, Noyes hoped to show that an atom retained its electrical charge in its compounds and that its charge was not the result of the compound's ionizing on solution. Their existence clearly would support the hypothesis that electrical forces held atoms in combination. Noyes's reaction scheme was largely a failure, however. He obtained free nitrogen and chlorine as products but no nitro-nitrogen trichloride. Indeed, his failure to prepare nitrogen trichloride's two electromers was the chief reason why Noyes abandoned a purely electrostatic theory of valence and in 1921 accepted G. N. Lewis's shared electron pair bond.[21]

Lauder W. Jones: The Synthesis and Structures of Electromers

Along with the increasing use of polar valences to describe the structure and behavior of compounds not usually considered electrolytes went a corresponding recognition of hydrolysis as the principal means of identifying the charges on the atoms or radicals of a compound. Noyes used hydrolysis, with Abegg's rule that in the water molecule the hydrogen atom

[20] Ibid., pp. 774–75.

[21] Ibid.; W. Noyes, "An Attempt to Prepare Nitro-Nitrogen Trichloride: II," pp. 1774–82. See also idem, "The Relation of Shared Electrons to Potential and Absolute Polar Valences," *Chemical Reviews* 5 (1928), 552–55. For Noyes's later publications on the problem of nitrogen trichloride, see W. Noyes and Haw, "The Reaction between Chlorine and Ammonia: II"; William A. Noyes, "The Reaction between Chlorine and Ammonia: III. Probable Formation of Trichloro-Ammonium Chloride," *J.A.C.S.* 42 (1920), 2173–79; W. Noyes, "An Attempt to Prepare Nitro-Nitrogen Trichloride: II," pp. 1774–82; W. Noyes, "The Interaction between Nitrogen Trichloride and Nitric Oxide: Reactions of Compounds with Odd Electrons," *J.A.C.S.* 50 (1928), 2902–10.

was positive and the hydroxyl group, negative to show that ammono-nitrogen trichloride's electron structure was $N^{---}Cl_3^{+++}$. He had hoped that hydrolysis would also demonstrate the existence of nitrogen trichloride's second electromer nitro-nitrogen trichloride, $N^{+++}Cl_3^{---}$.

Lauder W. Jones (1869–1960), Stieglitz's former pupil at the University of Chicago and now with Fry at the University of Cincinnati, was another who recognized the value of hydrolysis in assigning electrical charges to atoms and radicals, particularly those of organic compounds. In a 1912 paper Jones argued that John Ulric Nef's studies in 1899[22] on chlorocyanogen (cyanogen chloride) and iodocyanogen (cyanogen iodide) hydrolysis led at once to the charges on the products and hence the charges originally on the reactants:

$$Cl^{-+} CN + H^{+-} OH = H^{+-} Cl + HO^{-+} CN \qquad \text{(Cyanic acid)}$$
$$I^{+-} CN + H^{+-} OH = HO^{-+} I + H^{+-} CN \qquad \text{(Hydrocyanic acid)}.$$

He also pointed out that though hydrolysis, depending on reaction conditions, often showed the same atom or radical, with either a positive or a negative charge, an electron interpretation of oxidation-reduction easily accounted for the variation. In the well-known benzenesulfonic acid hydrolysis, hydrolysis in acid medium gave benzene and sulfuric acid, but in basic medium phenol and sulfurous acid. Oxidation-reduction related the pairs—benzene and phenol, sulfurous and sulfuric acid—though like Fry, Jones believed that the relation resulted from an intramolecular oxidation-reduction of benzenesulfonic acid.[23] This gave two electromers in equilibrium:

$$C_6H_5^{-+} SO_3H \rightleftarrows C_6H_5^{+-} SO_3H .$$

Electromer I   Electromer II

In acid medium, in which the concentration of hydrogen ion (H^+) was high, electromer I predominated, and the products were, therefore, benzene and sulfuric acid:

$$C_6H_5^{-+} H = C_6H_6 ,$$
$$HSO_3^{+-} OH = H_2SO_4 .$$

---

[22] John U. Nef, "Über das Phenylacetylen, seine Salze und seine Halogensubstitutionsprodukte," *Annalen der Chemie* 308 (1899), 320; idem, "Über das Verhalten der tri- und tetrahalogen substituierten Methane," *Annalen der Chemie* 308 (1899), 329–33.

[23] Lauder W. Jones, "The Beckmann Rearrangement of Hydroxamic Acids," *Amer. Chem. Journal* 48 (1912), 26.

But, in basic medium, with a high concentration of hydroxyl ion (OH^-), the reaction required electromer II to yield phenol and sulfurous acid:

$$C_6H_5{}^{+-} OH = C_6H_5OH \, ,$$
$$HSO_3{}^{-+} H = H_2SO_3 \quad .$$

Practically the same explanation of this hydrolysis appeared the following year in a paper by William C. Bray (1879–1946) and Gerald E. K. Branch (1886–1954), who were at the University of California, Berkeley.[24] Unlike Jones, they did not refer to Fry's theory of electromers in accounting for the observed reactions but claimed instead that benzenesulfonic acid was an equilibrium mixture of two electrically tautomeric forms:

$$C_6H_5 \leftarrow S \rightarrow O \leftarrow H \rightleftarrows C_6H_5 \rightarrow S \rightarrow O \leftarrow H .$$

I   II

In the two tautomers, the sulfur atom's polar number (now called the oxidation number) changed from +6 in I to +4 in II, and that of phenyl, C_6H_5, went from −1 in I to +1 in II. The polar number of two adjacent atoms or groups, sulfur and phenyl, had changed, and for this reason Bray and Branch argued that tautomers of this kind were inseparable. Alternatively, in the more familiar keto-enol tautomerism, though an atomic rearrangement occurred, each atom's polar number remained the same, making it possible to isolate the two equilibrium forms.

Thus in maintaining that tautomers were only sometimes separable, Bray and Branch believed that tautomeric equilibrium differed from Fry's electronic isomerism, in which the electromers were always separable. Yet, their tautomers were for the most part identical with Jones's electromers. Indeed, in 1914 Fry raised precisely this point, insisting that the experimental facts, interpreted from either viewpoint, led to the conclusion that benzenesulfonic acid had two electronic structures, $C_6H_5{}^+.SO_3H^-$ and $C_6H_5{}^-.SO_3H^+$.[25] They were the electromers that hydrolyzed according to the following ionic mechanism:

[24] William C. Bray and Gerald E. K. Branch, "Valence and Tautomerism," *J.A.C.S.* 35 (1913), 1440–47.

[25] Harry S. Fry, "Positive and Negative Hydrogen, the Electronic Formula of Benzene, and the Nascent State," *J.A.C.S.* 36 (1914), 265.

$$C_6H_5^+.SO_3H^- + H^+.OH^- \rightarrow C_6H_5^+.OH^- + H^+.SO_3H^-$$
$$C_6H_5^-.SO_3H^+ + H^+.OH^- \rightarrow C_6H_5^-.H^+ + HO^-.SO_3H^+$$

The major challenge that Jones, like Fry and Noyes, faced in his search for electromers was to prove that he had actually prepared and isolated different electronic structures. This presented far greater experimental difficulties than the separation of structural tautomers and led Jones to compare the present status of electromers to that of an earlier time, when chemists first recognized the isomerism of organic compounds. Faraday had written in 1825 that once chemists began to look for isomers they would probably "multiply on us." Jones, writing in 1914, believed that chemists were "in much the same position at the present time with respect to the new isomers which depend upon the distribution of electrons and their exchange; and it is not improbable that they will 'multiply on us' in the future." [26]

To establish the existence of electromers, Jones had to demonstrate an atom's different electrical distributions in its compounds. Fry's $(n + 1)$ valence rule, n being the atom's usual valence, was available, Fry having already used it to illustrate the five possible ways of distributing electronically the four bonds around a carbon atom.

Applied to nitrogen, Fry's rule showed that trivalent nitrogen had four different charge distributions:

$$
\begin{array}{cccc}
- & + & + & + \\
N\ - & N\ - & N\ + & N\ + \\
- & - & - & +
\end{array}
$$

Each positive and negative sign indicated, respectively, the loss and gain of one electron by the nitrogen atom in forming its bonds. For pentavalent nitrogen, six different distributions existed, ranging from N^{-----} to N^{+++++}. The problem at hand, therefore, was to demonstrate that the nitrogen atom actually had these charge distributions.

In 1913 and 1914, Jones studied the reactions of a large number of nitrogen-containing compounds, including amines, nitrites, and nitrile oxides, but to follow his arguments we need to consider only a few specific cases, beginning with the simplest. In nitrous and nitric acid, he assigned to nitrogen the electrovalences N^{+++} and N^{+++++}, in agreement with the fact that oxygen was always negative when combined with another ele-

[26] Michael Faraday, "On New Compounds of Carbon and Hydrogen, and on Certain Other Products Obtained during the Decomposition of Oil by Heat," *Phil. Trans.* 115 (1825), 440–66; Lauder W. Jones, "Electromers and Stereomers with Positive and Negative Hydroxyl," *J.A.C.S.* 36 (1914), 1290.

ment. In ammonia, nitrogen had the electrovalence N^{---} because hydrogen was usually positive in combination with nonmetals, but in ammonium compounds, the ammonia-like nitrogen also had a pair of oppositely charged valences, the latent valences of Arrhenius, Friend, and others, giving it the distribution N^{----+}.[27]

Jones used hydrolysis to determine the direction of the valence bonds between carbon and nitrogen. Amines, though hydrolyzing with difficulty, gave ammonia and an alcohol or phenol, and this suggested the formula

$$R.C_{+}\!-N^{-+H}_{-+H},$$
$$H_2$$

which hydrolyzed accordingly:

$$R.C_{+}\!-N^{-+H}_{-+H} + H_{+}\!-OH \rightleftarrows R.C_{+}\!-OH + N^{-+H}_{-+H}.$$
$$H_2 \qquad\qquad\qquad H_2$$

The other possible electronic structure,

$$R.C_{-+}N^{-+H}_{-+H},$$
$$H_2$$

would give products never obtained on hydrolysis—a hydrocarbon and hydroxylamine,

$$HO_{-+}N^{-+H}_{-+H}.$$

In the same way Jones assigned electron structures to nitriles, nitroparaffins, and aldimides.

From his study Jones concluded further that in compounds with a carbon-nitrogen bond the carbon atom did not oxidize the nitrogen. This result was important because it suggested an electronic mechanism for the Beckmann rearrangement, a system containing a carbon-nitrogen bond. He believed that the determining factor in the rearrangement was the tendency of the carbon-nitrogen bond to undergo an intramolecular oxidation leaving the carbon atom completely oxidized, C^{+4}, and the nitrogen atom completely reduced, N^{-3}.[28] He illustrated the rearrangement

$$\begin{array}{c} R \\ \pm \end{array}$$
$$O^{-+}_{-+}C_{+}\!-N \rightarrow (O^{-+}_{-+}C^{-+}_{-+}N_{-+}R) \rightarrow O^{-+}_{-+}C^{+-}_{+-}N_{-+}R.$$

[27]Lauder W. Jones, "Applications of the Electronic Conception of Valence," *Amer. Chem. Journal* 50 (1913), 416.

[28]Ibid., p. 441; idem, "Electromers and Stereomers," p. 1287.

In fact, Jones thought that some kind of intramolecular oxidation-reduction was responsible for all rearrangement reactions, including the Curtius and Hofmann rearrangements.

Jones's continued search for electromers resulted in his examining hydroxylamine's behavior, in particular its rather complicated decomposition into ammonia, nitrogen, and some nitrous oxide, N_2O, and nitrous acid, HNO_2. Jakob Meisenheimer (1876–1934) in Berlin was also studying the reactions of hydroxylamine, and in 1913 he succeeded in preparing two isomeric derivatives that he represented by the general formula $(R)_3N(OCH_3)(OH)$. Meisenheimer synthesized the two structures identically except for the order of introducing the hydroxyl and alkoxyl groups, yet they appeared fundamentally different. When heated in aqueous solution, one decomposed to give a tertiary amine and an aldehyde, while the other yielded a tertiary amine oxide and an alcohol.[29]

Meisenheimer could not explain the difference in the decompositions, remarking only that the bond between the nitrogen atom and the alkoxyl group was not the same in the two isomers. In agreement with the widely accepted view that nitrogen had a maximum valence of five, he believed that in one isomer nitrogen's fourth valence held the alkoxyl group but in the other a nonequivalent, fifth valence held it. Meisenheimer's results convinced Jones that the theory of electromers furnished the key to the unexplained isomerism and to his own studies on hydroxylamine's decomposition. Indeed, he claimed that Meisenheimer's hydroxylamine derivatives represented the first known isolation of electromers.

To follow Jones's argument,[30] we begin with his interpretation of hydroxylamine's decomposition—an interpretation that of course was highly satisfying, for it required hydroxylamine to exist as an equilibrium mixture of four electromers:

$$\text{(I) } (H^+)_2\text{}^-N\text{-}{}^+O\text{-}{}^+H \qquad \text{(Ia) } (H^+)_3\text{}^-N\text{}^+_-\text{}^+O$$

$$\Updownarrow \qquad\qquad\qquad\qquad \Updownarrow$$

$$\text{(II) } (H^+)_2\text{}^-N\text{+}\text{-}O\text{-}{}^+H \qquad \text{(IIa) } (H^+)_3\text{}^-N\text{}^+_+\text{}^-O$$

The decomposition of electromer (Ia) accounted for the appearance of ammonia, one of the reaction's main products. This structure also contained an oxygen atom with one positive bond, and as a result of the decomposi-

[29] Jakob Meisenheimer, "Über die Ungleichartigkeit der fünf Valenzen des Stickstoffs," *Annalen der Chemie* 397 (1913), 273–84.
[30] Jones, "Electromers and Stereomers," pp. 1271–80.

tion it liberated an unusual species of oxygen that Jones called active oxygen:

$$(H^{+})_3{}^{-}_{-}N^{+-}_{-+}O \rightarrow (H^{+})_3{}^{-}_{-}N^{+}_{-} + {}^{-}_{+}O \,.$$

Active oxygen

Electromer (II) decomposed to give an active imide,

$$(H^{+})_2{}^{-}_{-}N_{+}\text{-}O_{-+}H \rightarrow H_{+}\text{-}N^{-}_{+} + H_{+}\text{-}O_{-+}H \,,$$

Active imide

which then combined with active oxygen from electromer (Ia), forming hyponitrous acid,

$$H_{+}\text{-}N^{+}_{-} + O^{+-}_{-+}N^{-}_{-}(H^{+})_3 \rightarrow H_{+}\text{-}N^{-+}_{+-}O + (H^{+})_3{}^{-}_{-}N^{+}_{-} \,.$$

This acid decomposed yielding the other observed products, nitrous oxide, nitrous acid, and water.

Jones did not try to account electronically for nitrogen's formation. He suggested only that hydroxylamine's decomposition was slow enough to allow the resulting hyponitrous acid to react with the hydroxylamine, giving an intermediate product that immediately decomposed into nitrogen and water.

The unexplained behavior of Meisenheimer's isomers followed directly from Jones's assumption that hydroxylamine was an equilibrium mixture of four electromers. In electromer (II) the hydroxyl group had its customary negative charge, but in (I) its charge was positive. Thus in the two alkoxyl isomers of hydroxylamine, the alkoxyl (or hydroxyl) group was positive in one and negative in the other.[31] This, Jones believed, accounted for the difference in their decomposition reactions:

$$(CH_3{}^{+})_3{}^{-}_{-}N^{-}_{+}\begin{smallmatrix} +O-+CH_3 & (4) \\ -O-+H & (5) \end{smallmatrix} \rightarrow (CH_3{}^{+})_3{}^{-}_{-}N^{-}_{+}\begin{smallmatrix} +H \\ -O-+H \end{smallmatrix} + H_2C^{-+}_{+-}O$$

Isomer A

and

$$(CH_3{}^{+})_3{}^{-}_{-}N^{-}_{+}\begin{smallmatrix} +O-+H & (4) \\ -O-+CH_3 & (5) \end{smallmatrix} \rightarrow (CH_3{}^{+})_3{}^{-}_{-}N^{-+}_{+-}O + CH_{3+}\text{-}O_{-+}H$$

Isomer B

[31] Ibid., p. 1287. Jones's assumption of positive hydroxyl and his interpretation of these decompositions received support from Julius Stieglitz and Paul N. Leech, "The Molecular Rearrangement of Triarylmethylhydroxyl Amines and the Beckmann Rearrangement of Ket-

Jones also pointed out that in Meisenheimer's isomers, the two groups OH and OCH_3, or the two oxygen atoms upon which the electromerism depended, were not connected to each other but to a third atom, nitrogen:

$$CH_3O + -N + -OH \quad \text{and} \quad HO + -N + -OCH_3 .$$

Such a structural arrangement, he thought, was undoubtedly responsible for the relative stability of these electromers compared with the stability of those having atoms of different polarity directly bonded as in $A + -B$ and $A - +B$.

Jones, like Stieglitz, believed, therefore, that electromers of hypochlorous acid, $H + -O - +Cl$ and $H + -O + -Cl$, had to exist. To account for the well-known fact that the chlorine of hypochlorous acid replaced positive hydrogen in organic compounds clearly required a structure that reacted accordingly,

$$H + -O - +Cl + H + -CH_3 \rightarrow Cl + -CH_3 + H + -O - +H ,$$

though Jones claimed that methyl chloride subsequently underwent an intramolecular oxidation to $H_3C + -Cl$. On the other hand, the dissociation of hypochlorous acid in light demanded the structure $H + -O + -Cl$.[32] The result was

$$2H + -O + -Cl \rightarrow 2H + -Cl + O_+^{-+}O .$$

Jones, of course, acknowledged that chemists had not succeeded beyond any doubt in separating these or any other electromers, but he maintained that the problem of isolating electromers was analogous to that which existed for structural tautomers. Chemists had not yet separated tautomers of hydrocyanic acid and other compounds in which a wandering atom moved from the first to the second of two atoms bonded to each other: $HCN \rightleftarrows CNH$. The majority of successful separations—as Jones believed would be the case with electromers—involved compounds in which the

---

oximes," *J.A.C.S.* 36 (1914), 288. Later Arthur Michael attacked Stieglitz's work, in "On the Nonexistence of Valence and Electronic Isomerism in Hydroxylammonium Derivatives," *J.A.C.S.* 42 (1920), 1232–45.

[32] Jones, "Electromers and Stereomers," p. 1282.

tautomeric change was between two atoms not directly bonded but separated by a third atom as in the familiar keto-enol tautomerism:

$$O = C - CH_2 \rightleftarrows HO - C = CH.$$

The Contributions of Julius Stieglitz

We have mentioned Julius Stieglitz (1867–1937) as belonging to that group of pioneers who applied the electron theory of valence to organic chemistry. Stieglitz studied under Victor Meyer at Göttingen and for a few months with the organic chemist John Ulric Nef (1862–1915) at Clark University before moving permanently to the newly founded University of Chicago in 1892. There he again served under Nef, who came to Chicago to become head of its chemistry department, and upon Nef's death in 1915 Stieglitz succeeded him as department head, remaining at Chicago until his own death in 1937.

While at Chicago, Stieglitz carried out his twenty-year study of molecular rearrangements, and like his former pupil, Lauder W. Jones, he used an electronic mechanism to account for these rearrangements. In Stieglitz's words, he hoped "to shed some light, from a purely chemical side, on the forces holding atoms in place in molecular structures."[33] Chemists could correlate electronically the study of all organic reactions, he said, once they recognized that any atom's or radical's reactivity depended on its electronic state, or, to use Abegg's term, on its polar character. Indeed, by 1915 Stieglitz showed that the alkyl halide reactions were consistent with the polar formula $\overset{+}{R}\overset{-}{X}$, where X = Cl, Br, or I, and that other negative groups, OH, OR, or NH_2, could replace the compound's negative halogen atom. He found the behavior of compounds containing positive halogens entirely analogous.[34]

As a consequence of his work, Stieglitz emerged as one of the leading defenders of polarity in un-ionized compounds. Arthur E. Remick (b. 1899) in his book *Electronic Interpretations of Organic Chemistry* re-

[33]Julius Stieglitz, "Molecular Rearrangements of Triphenylmethane Derivatives," *Proc. Nat. Acad. Sci.* 1 (1915), 196.

[34]Ibid., p. 197; Julius Stieglitz and Peter P. Peterson, "Über Stereoisomere Chlorimido-Ketone," *Berichte* 43 (1910), 782; Stieglitz and Leech, "The Molecular Rearrangement of Triarylmethylhydroxyl Amines," pp. 272–301.

marked that Stieglitz so convincingly marshaled his evidence "that he established the validity of the concept of polarity upon purely chemical evidence before the advent of Debye's evidence based on measurements of dielectric constants." [35]

In his electron theory of molecular rearrangement, Stieglitz's central argument was that rearrangement occurred whenever a molecule contained an oxygen, nitrogen, halogen, or other atom or radical in the positive state rather than in its usual negative state. The tendency of these atoms or radicals to attain their more stable negative state, he said, was the driving force behind the reaction, for this disturbed the valence electrons of the neighboring atoms in the compound, making rearrangement inevitable. [36]

Thus, according to Stieglitz, the acyl halogen amines, since they underwent the Hofmann rearrangement, contained a positive halogen atom—a positive chlorine atom, [37] as illustrated in the following equation:

$$(C_6H_5)_3C.NH(Cl) + NaOH \rightarrow NaCl + HOH + (C_6H_5)_2:NC_6H_5,$$

or electronically

$$(C_6H_5^+)_2{=}C\overset{+}{\,}{}^-N\overset{-}{\,}{}^+H \xrightarrow{\;-(H^+Cl^-)\;} (C_6H_5^+)_2{=}C\overset{+}{\,}{}^-N \rightarrow$$
$$\overset{|}{(C_6H_5^+)} \quad Cl^+ \qquad\qquad \overset{|}{(C_6H_5^+)}$$

$$(C_6H_5^+)_2{=}C_+^{+-}N_-^- \longrightarrow (C_6H_5^+)_2{=}C_+^{+}{}_-^-N^-({}^+C_6H_5).$$
$$(C_6H_5^+)$$

In the initial step the positive chlorine atom took two electrons from the nitrogen atom. This oxidizing action resulted in the release of H^+ and Cl^-, which reacted with sodium hydroxide to form sodium chloride and water. The unsaturated univalent nitrogen atom then removed two electrons from the carbon atom adjacent to it, and the resultant positive valence on the

[35] Arthur E. Remick, *Electronic Interpretations of Organic Chemistry*, p. 23. For Peter Debye's work, see *Polar Molecules*.

[36] Stieglitz and Leech, "The Molecular Rearrangement of Triarylmethylhydroxyl Amines," p. 275; Julius Stieglitz and Bert A. Stagner, "Molecular Rearrangements of β-Triphenylmethyl–β-Methylhydroxylamines and the Theory of Molecular Rearrangements," *J.A.C.S.* 38 (1916), 2047; Jones, "Applications of the Electronic Conception of Valence," p. 441.

[37] Chlorine also had a positive charge because in ammonia or amines the nitrogen atom was always negative.

carbon atom caused the release of one of the positive phenyl groups, which moved to the negative nitrogen atom to complete the rearrangement.[38]

The acyl halogen amines, Stieglitz pointed out, also exhibited other properties characteristic of positive chlorine, namely, (1) they were prepared from hypochlorous acid, (2) they regenerated that acid on hydrolysis, (3) the chlorine atom retained its oxidizing power, and (4) all attempts to replace the chlorine by negative groups (OH, NH_2) proved futile. Conversely, in nonrearranging compounds such as the alkyl chlorides $R_3C.Cl$ and acyl chlorides $R.OC.Cl$, which were prepared from hydrogen chloride and regenerated hydrochloric acid on hydrolysis, Stieglitz claimed chlorine was undoubtedly present as negative chlorine.[39] The chloroamines thus had the structure R_2N-+Cl; the alkyl and acyl chlorides, the structures R_3C+-Cl and $R.OC+-Cl$, respectively.

Stieglitz insisted that only a polar theory of valence could show why chlorine's properties differed in the two classes of compounds. He believed that both the chloroamines and the alkyl and acyl chlorides might be minutely ionizable, but they were certainly not electrolytes in the ordinary sense of the term. There was no need for strongly polarized compounds always to produce electrolytes, he said, and no necessity to assume any kind of chemical union other than polar chemical bonds: "We thus find that strongly polarized combinations of atoms need not produce electrolytes. This being the case in any instance open to demonstration, one is led inevitably to the inquiry, why it should not be true for a great many polarized combinations and whether the assumption of non-polar unions is at all necessary."[40]

Several years earlier, in 1908, Arthur A. Noyes at the Massachusetts Institute of Technology had suggested the existence of two distinct kinds of chemical union.[41] Noyes had been studying the conductivity and ionization of salt solutions and of strong and weak acidic and basic solutions. As expected, the weak acids and bases behaved according to the law of mass action, whereas the salts and the strong acids and bases did not. These solutions fell naturally into two classes, and their obvious differences in be-

---

[38] Stieglitz, "Molecular Rearrangements of Triphenylmethane Derivatives," p. 200.

[39] Ibid., pp. 197–98.

[40] Ibid., p. 199.

[41] Arthur A. Noyes, "The Conductivity and Ionization of Salts, Acids, and Bases in Aqueous Solutions at High Temperatures," *J.A.C.S.* 30 (1908), 351–52.

havior led Noyes to suggest two different kinds of molecules, which he called *electrical molecules* and *chemical molecules*. The names, he said, corresponded to the type of force that gave rise to each kind of molecule.

In electrical molecules, only a loose union existed among the ions making up the structure, and the ions still retained their electrical charges. Salts and most inorganic acids and bases were of this type. But when the ions combined in a more intimate way, forming molecules whose constitutents completely lost their charges and all other original characteristics, Noyes called them chemical molecules. Most organic acids belonged to this class. Noyes also assumed that neither class consisted exclusively of electrical or chemical molecules. In the strong acid, hydrochloric acid, the electrical form H^+Cl^- predominated, with only a small amount of the chemical form HCl present. On the other hand, for a weak acid or any slightly ionized compound, a reversal occurred in the proportions of the two forms.

William Bray, Gerald Branch, and G. N. Lewis adopted Noyes's suggestion of two different kinds of union in their 1913 publications on valence. Lewis and Bray had been Noyes's associates at the Massachusetts Institute of Technology at the time Noyes introduced his classification and were quite familiar with it. By 1914 J. J. Thomson also broadened his polar bond theory to include nonpolar bonds, which he said resulted from the electrical attraction between the electrons of one atom and the "positive core" of another. There was no electron transfer between the two atoms as in a polar bond, thus permitting each atom to remain, as a whole, electrically neutral.[42]

Stieglitz, nevertheless, continued to argue against the necessity of nonpolar bonds. Indeed, of all those—Fry, Jones, and Falk and Nelson— who were developing a purely polar electron theory of valence, he alone seemed to grasp the consequence of denying outright nonpolar bonds. In the intramolecular oxidation-reduction of the electromeric pair X^+—^-Y $\rightleftarrows X^-$—^+Y, a transfer of two electrons took place but, as Stieglitz pointed out, in single steps resulting in an intermediate nonpolar state XY, or X^+—$^-Y \rightleftarrows XY \rightleftarrows X^-$—$^+Y$. Denial of the nonpolar union, he said, also denied the existence of the electromers in equilibrium:

[42] Bray and Branch, "Valence and Tautomerism," pp. 1440–47; G. N. Lewis, "Valence and Tautomerism," *J.A.C.S.* 35 (1913), 1448–55; J. J. Thomson, "The Forces between Atoms and Chemical Affinity," *Phil. Mag.* 27 (1914), 757–89. Chapter 6 contains a fuller discussion of Thomson's work.

If we use the ordinary conceptions of electronic valence . . . this should lead . . . to an intermediate stage, where neither X nor Y would be charged with excess of positive or negative electricity; in other words, the usual valence force between the two atoms would be lacking and, consequently, *dissociation into X and Y should take place*. The only alternative to this conclusion . . . is that in the intermediate product XY, *nonpolar valences*, of the type suggested by Bray and Branch, G. N. Lewis, Thomson . . . unite X and Y. Indeed the evident absence of such dissociation in indicated cases of electromerism would . . . be excellent chemical evidence in favor of the existence of some form of nonpolar valence.

Yet, despite this obvious shortcoming Stieglitz believed that the chemical evidence clearly favored a polar theory of valence: "The well-known specific and generally one-sided actions of absorption, metathesis, etc., which are obviously directed by the nature of the specific *polar* charges on atoms both in inorganic and organic compounds would [still] find their explanation in the fact that the *chemically active molecules are those whose* reacting atoms have polar charges."[43]

In later publications, particularly his 1922 paper "The Electron Theory of Valence as Applied to Organic Compounds," Stieglitz discussed further the advantages of a polar theory of valence, citing numerous examples of reactions that clearly indicated a separation of compounds into positive and negative components.[44] This theory, he maintained, was still "unqualifiedly superior" in interpreting molecular rearrangements, oxidation-reduction, and addition reactions when compared with any theory of nonpolar valences, including the shared electron pair theory that G. N. Lewis had introduced in 1916 and that by 1922 had become the dominating electron theory of valence.

Stieglitz also called attention in his paper to the confusion that continued to surround the meaning of polarity. Chemists applied it not only to the extreme polar compounds such as sodium chloride but to the less polar organic molecules—the alcohols, hydroxylamines, carboxylic acids, and other compounds. The two polarities, he said, were "of a quite different order," the cause of the difference being a problem of "atomistics" (of atomic theory), and he had no desire to propose any theory to account for the difference:

[43] Stieglitz and Stagner, "Molecular Rearrangements of β-Triphenylmethyl–β-Methyl-hydroxylamines," p. 2053.

[44] Julius Stieglitz, "The Electron Theory of Valence as Applied to Organic Compounds," *J.A.C.S.* 44 (1922), 1293–1313.

That one and the same word is used to express the two kinds of polarity [and the] attempt to escape from the resulting confusion of thought by emphasizing the non-polar character of organic compounds when they really show distinctly polar behavior has added confusion to confusion. Polarity exists wherever there is a difference in charge, positive and negative, between two atoms and it is so valuable a conception in the treatment of organic compounds that it must be clearly insisted upon.[45]

Indeed, in a long letter written to Fry a few years later, Stieglitz complained that because chemists failed to differentiate between the two types of polarity, G. N. Lewis misread his 1922 publication. Stieglitz used positive and negative signs to indicate "organic" polarity, which the physical chemists, including Lewis, mistakenly believed to represent ordinary ionic polarity. In his letter, Stieglitz also claimed that he stated explicitly at the outset of his 1922 paper his acceptance of what he called the "Lewis-Bohr doublet theory" of the chemical bond (the shared electron pair) but at the same time insisted that polarity resulted from the proximity of the doublets to certain atoms.[46]

Actually, Stieglitz's alleged acceptance of the Lewis shared electron pair was the very noncommittal statement: "of the various theories of non-polar valence brought forward the author would incline to the type of non-polar valence first presented by Bohr and to the somewhat similar views of G. N. Lewis." Stieglitz's claim of acceptance is even harder to justify because the primary purpose of his paper was to show the superiority of polar valences over nonpolar. The only sign of Stieglitz's willingness to compromise on the two valence theories, if any, does not appear until the concluding section of his paper:

The polarity need not be of the extreme character shown by common salts but may very well be of the character proposed by Bohr, Lewis, and Kossel, where the transfer of electrons from atom to atom is not as complete as in the case of common electrolytes. Polarity exists, nevertheless, and the application of the theory of polar valence represents a decided advance in the interpretation of reactions of organic compounds.[47]

Stieglitz appeared to give equal credit to Lewis, Bohr, and Kossel for using the doublet theory to account for the polarity of molecules. But, in

[45] Ibid., pp. 1296, 1311.

[46] Stieglitz's letter is reproduced in Harry S. Fry, "A Pragmatic System of Notation for Electronic Valence Conceptions in Chemical Formulas," *Chemical Reviews* 5 (1928), 566–67.

[47] Stieglitz, "The Electron Theory of Valence," pp. 1294, 1313.

fact, Lewis pointed out in 1916 that in a molecule the more electronegative atom would attract more strongly the electron pair constituting the bond between two atoms. This produced a positive-negative polarity similar to that resulting from a complete electron transfer as in the polar or electrostatic theory of chemical bonding. Indeed, William A. Noyes first called attention to this similarity in 1923 when he published a short paper showing how to reconcile Stieglitz's view and, in general, the theory of polar valences with the Lewis theory. If chemists simply assumed that the two electrons forming the bond remained together when the atoms separated upon reaction, Noyes wrote, the atom that retained the electron pair acquired a negative charge while the atom that separated without the electrons was, therefore, positive.[48] The two theories became one.

Conclusion

Noyes, Jones, Fry, and Stieglitz tried to place the hypothesis of electromers on a solid experimental foundation with their extensive studies on molecular structure and reaction mechanisms. They relied on hydrolysis, the known electropositive and electronegative character of elements such as hydrogen and oxygen, and an electronic interpretation of oxidation and reduction (which lent itself very easily to the polar theory) in their effort to establish each atom's positive and negative charges. They firmly believed that eventually they would isolate the elusive electromers and thus contribute significantly to the polar theory's correctness. But their search for electromers ended in failure and instead cast considerable doubt on the theory's validity, for electromers were one of the polar theory's cornerstones.

The polar theory also failed to distinguish between the polarity present in organic compounds and the obviously different polarity characteristic of inorganic compounds. Of its adherents, only Stieglitz apparently recognized the confusion surrounding the dual meaning of polarity and showed some willingness to accept a nonpolar valence theory to complement the polar theory.

[48] G. N. Lewis, "The Atom and the Molecule," *J.A.C.S.* 38 (1916), 762–85; William A. Noyes, "A Possible Reconciliation of the Octet and Positive-Negative Theories of Chemical Combination," *J.A.C.S.* 45 (1923), 2959–61. See also William A. Noyes, "Electronic Theories," *Chemical Reviews* 17 (1935), 17.

6. The Decline of the Electrostatic or Polar Theory of Valence

Introduction

The electron theory of polar valences assumed that the chemical bond was always an electrostatic attraction between a pair of oppositely charged atoms that resulted from the complete transfer of an electron from one of the atoms to the other. It was therefore a unitary theory of valence. A compound's ability to dissociate when placed in solution supposedly offered direct experimental evidence for the existence of charged atoms in the compound. For compounds not liberating charged atoms upon solution, particularly organic compounds, the evidence was indirect, based usually on the results of hydrolysis reactions. Regardless of statements to the contrary by John M. Nelson, Hal T. Beans, and K. George Falk and by Julius Stieglitz[1] in the second decade of the twentieth century, chemists had not firmly established the presence of electrically charged atoms in organic compounds.

Those chemists who adopted the unitary polar theory had a firm conviction that the same fundamental law of electrostatic attraction held throughout chemistry. Despite the customary and convenient division of compounds into organic and inorganic, which originated because certain properties such as dissociation greatly predominated in only one of the divisions, they still maintained that electrically charged atoms really existed in both kinds of compounds.

On the other hand, many chemists regarded the application of the polar theory to both organic and inorganic compounds as an abuse of an idea

[1] John M. Nelson, Hal T. Beans, and K. George Falk, "The Electron Conception of Valence: IV. The Classification of Chemical Reactions," *J.A.C.S.* 35 (1913), 1811; Julius Stieglitz, "Molecular Rearrangements of Triphenylmethane Derivatives," *Proc. Nat. Acad. Sci.* 1 (1915), 196–210; Stieglitz, "The Electron Theory of Valence as Applied to Organic Compounds," *J.A.C.S.* 44 (1922), 1293–1313.

valid chiefly for inorganic compounds, more specifically for those compounds that actually yielded ions upon solution. William C. Bray and G. N. Lewis were among the first to make this criticism of the polar theory's all-embracing nature. By 1914 J. J. Thomson had also abandoned the unitary polar theory, and alternative theories appeared in the publications of William C. Arsem and Johannes Stark.

The Dualistic Valence Theory of Bray and Branch

William C. Bray belonged to the remarkable group of young physical chemists, including G. N. Lewis, Richard C. Tolman (1881–1948), and Charles A. Kraus, that Arthur A. Noyes gathered at the Massachusetts Institute of Technology in 1905. He remained there until 1912, leaving to join G. N. Lewis, who had become dean of the College of Chemistry and chairman of the Chemistry Department at the University of California, Berkeley.

Bray first called attention to the considerable confusion that had arisen over the meaning of valence, for the term then meant an atom's total number of bonds or its electrical polarity or charge in a molecule. Organic chemists, he said, favored the former meaning while the latter was becoming increasingly popular among inorganic chemists. In his 1913 paper "Valence and Tautomerism," published jointly with Gerald E. K. Branch, Bray sought to eliminate the confusion. They introduced the expressions *polar number* (now called the oxidation number) to represent an atom's positive or negative charge and *total valence number* to indicate the total number of bonds that the atom formed.[2]

Bray and Branch believed that either the use of directive valences or Fry's positive-negative symbolism would show the difference. In ammonium chloride, the polar number of nitrogen, N, was −3 (chlorine, Cl = −1; hydrogen, H = +1), but its total valence number was 5:

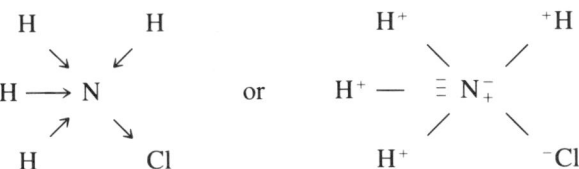

[2] William C. Bray and Gerald E. K. Branch, "Valence and Tautomerism," *J.A.C.S.* 35 (1913), 1440–41.

Similarly, the nitrogen atom in ammonia

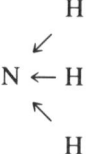

had a polar number of -3, and differed from the nitrogen atom in ammonium chloride only in having a total valence number of 3 instead of 5.

Bray and Branch did not assign polar numbers to all chemical compounds, however. They could not accept Falk and Nelson's claim that their research on the addition reactions of unsaturated hydrocarbons demonstrated the opposing directions of the two valences of the carbon-carbon double bond and thus established each carbon atom's electrical polarity. Nor could they accept Fry's argument that his investigation of benzene's substitution reactions revealed that adjacent carbon atoms in the six-membered ring were of opposite polarity. Indeed, not only Bray and Branch, but other chemists, among them G. N. Lewis, Arnold F. Holleman (1859–1953), Stuart J. Bates (1887–1961), and Roger F. Brunel (1881–1924),[3] questioned Fry's and Falk and Nelson's results.

Bray and Branch also pointed out that the polar theory had not solved the problem of correctly determining the polar numbers of carbon atoms in saturated hydrocarbons and their derivatives. This fact alone, they said, seemed to nullify its significance. One encountered striking difficulties upon assigning directive valences to the formulas of methane, $C \overset{\scriptstyle =}{\underset{\scriptstyle =}{}} H_4$, and carbon tetrachloride, $C \overset{\scriptstyle \rightarrow}{\underset{\scriptstyle \rightarrow}{}} Cl_4$. For no such profound difference in properties existed between them as existed between the corresponding nitrogen compounds ammonia, $N \overset{\scriptstyle =}{\underset{\scriptstyle =}{}} H_3$, and nitrogen trichloride, $N \overset{\scriptstyle \rightarrow}{\underset{\scriptstyle \rightarrow}{}} Cl_3$, and certainly not the difference expected between compounds containing carbon atoms with polar numbers -4 and $+4$. Bray and Branch argued that the bond between carbon and chlorine in carbon tetrachloride was not the same as that between sodium and chlorine in sodium chloride; otherwise, the compounds would show a greater similarity and not obvious differences,

[3]G. N. Lewis, "Valence and Tautomerism," *J.A.C.S.* 35 (1913), 1448–55; Arnold F. Holleman, "Substitution in the Benzene Nucleus," *J.A.C.S.* 36 (1914), 2495–98; Stuart J. Bates, "The Electron Conception of Valence," *J.A.C.S.* 36 (1914), 789–93; Roger F. Brunel, "A Criticism of the Electron Conception of Valence," *J.A.C.S.* 37 (1915), 709–22.

such as their ability to liberate their ions. Indeed, they claimed that the carbon atom's polar number was actually zero in both methane and carbon tetrachloride, as were those of hydrogen and chlorine. The polarity of these compounds, therefore, warranted no further consideration.[4]

Thus it seemed more reasonable to Bray and Branch to admit the likelihood of the organic chemists' nonpolar bond, and they consequently favored adopting the organic chemists' structural formulas, altering only those bonds that showed definite evidence of polarity. This design would permit the presence of polar and nonpolar bonds in the same compound, such as organic acids and bases and the organic derivatives of nitrogen and sulfur.[5]

In effect, Bray and Branch proposed two distinct types of unions between atoms: "polar, in which an electron has passed from one atom to the other, and non-polar, in which there is no motion of an electron [from one atom to the other]." Though these two types were the limiting cases, they believed that most compounds were actually intermediate in character, "in the sense that for each pair of atoms in a compound, the electron has actually passed from one atom to the other in a definit fraction of the total number of molecules [and] implies, of course, that the electrons are continually in motion within the molecules."[6]

Bray and Branch's dualistic interpretation of valence resulted in two general classes of chemical compounds, depending on whether the valence bonds were chiefly polar or nonpolar. The following shows the properties of the two classes:

	Polar	Nonpolar
Dielectric constant	High	Low
Degree of ionization	High	Low
Chemical reactivity	High	Low

Since their classification corresponded roughly to the division into organic and inorganic compounds, Bray and Branch also considered the ability of an atom, namely the carbon atom, to form chain compounds and stable isomers to be a further property of the nonpolar type. In fact, the far-

---

[4]Bray and Branch, "Valence and Tautomerism," p. 1443.
[5]Ibid.
[6]Ibid. Phonetic spelling was somewhat popular during the Progressive Years, 1890–1920, thus the use of *definit* instead of *definite*.

reaching character of this classification was to them a strong argument in favor of a nonpolar and a polar union between atoms.

G. N. Lewis and Dualism

There appeared simultaneously with Bray and Branch's article a second publication by G. N. Lewis favoring a dualistic interpretation of valence. Lewis believed that Bray and Branch had performed an important service to theoretical chemistry in differentiating between valence number and polar number, which he defined, respectively, as "the number of positions, or regions, or points (bond-termini) on the atom at which attachment to corresponding points on other atoms occurs" and "the number of negative electrons which an atom has lost." Yet he was certain at this time that the independence of "valence number" and "polar number" was even more complete than Bray and Branch had indicated. "These two conceptions are radically distinct," Lewis wrote, "and therefore we must recognize the existence of two types of chemical combination which differ, not merely in degree, but in kind." [7] A salt such as potassium chloride illustrated the first or polar type, and a paraffin hydrocarbon such as methane, the second or nonpolar type.

Lewis suggested that chemists could clearly express the difference between these two types if they called them mobile and immobile, instead of polar and nonpolar. The immobile compounds, he said, had a frame structure, "a fixed arrangement of the atoms within the molecule, which permits us to describe accurately the physical and chemical properties of a substance by a single structural formula." On the other hand, "the change from the nonpolar to the polar type may be regarded, in a sense, as the collapse of this framework. The nonpolar molecule, subjected to changing conditions maintains essentially a constant arrangement of the atoms; but in the polar molecule the atoms must be regarded as moving freely from one position to another, falling now into one place, now into another, like the bits of glass in a kaleidoscope." [8]

In the well-known cases of organic tautomerism the compound behaved like a mixture of two different substances in mobile equilibrium. This same phenomenon, Lewis added, appeared in more exaggerated form

[7] Lewis, "Valence and Tautomerism," p. 1448.
[8] Ibid., p. 1449.

in almost all inorganic compounds in which the molecule varied, perhaps among dozens of possible forms, thus accounting for the remarkable failure of structural formulas in inorganic chemistry. The cause of the mobility in polar compounds was the freedom of "the atom of electricity," or electron, to move from one position to another, and its motion was responsible for the polar molecules' electrical properties, such as the tendency to form ions and to have dielectric constants with high values.[9]

Lewis summarized in the lists below the properties of polar and nonpolar compounds.[10] Like Bray and Branch, he believed that a given molecule might be polar in one part and nonpolar in another and thus a given molecule did not correspond completely and at all times to either type.

Polar	Nonpolar
Mobile	Immobile
Reactive	Inert
Condensed structure	Frame structure
Tautomerism	Isomerism
Electrophiles	Non-electrophiles
Ionized	Not ionized
Ionizing solvents	Not ionizing solvents
High dielectric constant	Low dielectric constant
Molecular complexes	No molecular complexes
Association	No association
Abnormal liquids	Normal liquids

According to Lewis, it was, therefore, impossible to represent by a single structural formula a polar compound whose molecules were supposedly alternating among numerous tautomeric forms. Indeed, Lewis doubted whether any of the structural formulas currently in use represented even a single one of a molecule's momentary polar states. For this reason, he took issue with Bray and Branch for adopting Falk and Nelson's directive valences in their discussion on polar compounds. Their formulas, Lewis said, might be satisfactory enough in cases no more complicated than potassium chloride, $K \rightarrow Cl$, but they failed entirely when applied to compounds involving several atoms and electrons. The arrow, according to the theory of directive valences, indicated the two atoms between which an electron had passed. But since all electrons were alike, and presumably left

[9] Ibid.

[10] Ibid. In the original publication Lewis accidently interchanged the words *reactive* and *inert*. I have corrected them here.

no trail behind them, one could not argue for the hypothetical compound $A^+B^-C^+D^-$ that atom A lost an electron to atom B or that atom C lost one to atom D. The only conclusion possible was that atoms A and C each lost an electron, and atoms B and D each gained one.[11]

Thus the formula that Bray and Branch suggested as a possible structure for hydrogen peroxide,

$$H \rightarrow O \rightleftarrows O \leftarrow H,$$

seemed to have no meaning to Lewis.[12] He responded: "It indicates that one atom has given an electron to another, but has received one in return. Even supposing that the electron which returns goes to another part of the oxygen atom than that from which it departed, the formula would still mean nothing more than the formula,

$$H \rightarrow O \Updownarrow O \leftarrow H,$$

where the vertical arrows would indicate that an electron in each oxygen atom has moved from one position to another, so that each acquires thereby a positive charge at one point and a negative charge at another; but when written in this way, the molecule is not bonded together!"[13]

Actually, Lewis must have known, even if he did not agree, that the directive valence, from the time of its introduction by J. J. Thomson in 1904 to the later publications of Falk and Nelson, had always represented an electrostatic tube of force and was the physical link connecting a pair of atoms. Looking at the oxygen-oxygen bond of Bray and Branch's hydrogen peroxide formula, $H \rightarrow O \rightleftarrows O \leftarrow H$, a tube of force originated on each oxygen atom, terminated on the other, and physically connected the two

[11] Ibid., p. 1452.

[12] Actually, Bray and Branch ("Valence and Tautomerism," p. 1446) referred to this formula as one that did not deserve serious attention. They preferred

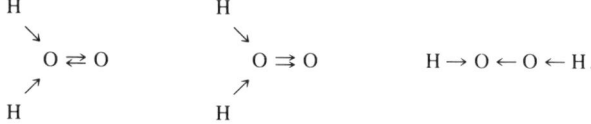

The last two were identical with those Nelson and Falk gave in "The Electron Conception of Valence: III. Oxygen Compounds," *Original Communications to the Eighth International Congress of Applied Chemistry* 6 (September 1912), 218.

[13] Lewis, "Valence and Tautomerism," p. 1452.

atoms. In Lewis's formula, $H \rightarrow O \Updownarrow O \leftarrow H$, the tube of force originated and terminated on the same oxygen atom, providing no connection at all between the two oxygen atoms. Clearly, the two formulas were not equivalent as Lewis implied in his attack on directive valences.

But, even to concede that the arrows indicated those atoms bound together by electrical forces, Lewis still maintained that this interpretation was equally misleading. A "positive charge," he said, "does not attract one negative charge only, but all the negative charges in its neighborhood. Consider the compound, H_2F_2; it seems extremely likely that it must assume, occasionally at least, the symmetrical form

$$
\begin{array}{ccc}
 & F^- & \\
H^+ & & H^+ \\
 & F^- &
\end{array}
$$

and yet such a form can in no way be expressed in terms of 'polar bonds.' " [14]

Here we see the influence of Alfred Werner's ideas on Lewis. In *New Ideas on Inorganic Chemistry*, Werner had given what Lewis called "a masterly presentation of many of the arguments against the valence formulas of polar compounds." [15] Werner did not regard chemical affinity as an attractive force that fixed the valence bonds in specific directions. Instead, he believed that the attractive force originated at the center of a spherical atom and acted uniformly toward all parts of the atom's surface. [16] Lewis adopted Werner's nondirectional valence force in his criticism of directive valences in hydrogen fluoride.

Werner's influence appeared again in Lewis's objection to the use of directive valences in compounds such as sulfuric acid:

$$
\begin{array}{ccc}
H \rightarrow O & & O \\
 & \nwarrow \quad \nearrow\nearrow & \\
 & S & \\
 & \swarrow \quad \searrow\searrow & \\
H \rightarrow O & & O\,.
\end{array}
$$

[14] Ibid., pp. 1452–53.

[15] Ibid., p. 1452. Werner's arguments are in *New Ideas on Inorganic Chemistry*, pp. 57–60 (first German edition of Werner's book was 1905; second edition, 1908).

[16] Werner, *Inorganic Chemistry*, p. 73; idem, "Valency," *Chemical News* 96 (13 Septem-

This formula, Lewis argued, was fundamentally erroneous because it led to artificial distinctions in groups of entirely analogous compounds: "Thus, just as the oxide of hydrogen combines with other oxides to form acids, so the hydrogen halides combine with other halides to form similar acids. For example, completely analogous to boric and silicic acids [meta-boric, HBO_2; orthoboric, H_3BO_3; disilicic, H_2SiO_5; metasilicic, H_2SiO_3] are hydrofluoboric and hydrofluosilicic acids, $HFBF_3$ and $(HF)_2SiF_4$ [HBF_4 and H_2SiF_6], but the latter compounds can in no way be represented by valence formulas. It is best to use a less pretentious formula such as H_2SO_4, or the old Berzelius form, H_2OSO_3, rather than one that leads to such unjustified distinctions."[17]

Practically the same argument had appeared earlier in Werner's *New Ideas*. Werner had pointed out that the organic chemists' valence bond theory failed to show the relation between certain oxides and double chlorides, which his coordination theory and the use of principal and auxiliary valences made evident:

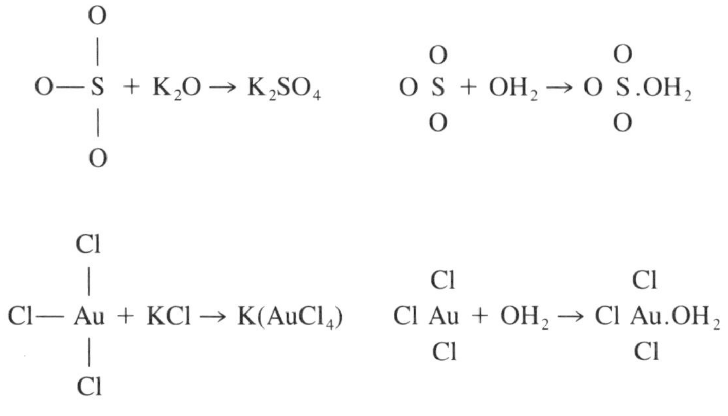

$$O-S + K_2O \rightarrow K_2SO_4 \qquad O\,S + OH_2 \rightarrow O\,S.OH_2$$

$$Cl-Au + KCl \rightarrow K(AuCl_4) \qquad Cl\,Au + OH_2 \rightarrow Cl\,Au.OH_2$$

He recognized, as Mendeleev did in *The Principles of Chemistry* (1868), the inherent connection among the various classes of so-called molecular compounds—the metal-ammonia salts, hydrates, and double salts. With his coordination theory Werner attempted to demonstrate these constitu-

---

ber 1907), 130. Werner in fact opposed the use of valence formulas in both polar (inorganic) and nonpolar (organic) compounds, though he based his arguments on examples chosen from polar compounds only.

[17] Lewis, "Valence and Tautomerism," p. 1453.

tional similarities. Indeed, Lewis believed that Werner's coordination theory and not the theory of directive valences was "the most important principle at present available for the classification of polar compounds."[18]

J. J. Thomson Adopts a Dualistic Theory of Valence: The Physical Evidence Supporting Nonpolar Chemical Bonds

At the same time that Lewis, Bray, and Branch were rejecting the theory of directive or polar valences on chemical grounds, at Cambridge J. J. Thomson, one of the theory's chief architects, also abandoned it. Thomson now maintained that in addition to electrostatic bonds, which he said were present only in "ionic molecules," there existed a second type of bond found in many molecules, whose atoms were not electrically charged. Thomson's arguments supporting a dualistic theory of valence appeared in several publications in 1911–14 and rested essentially on his investigations of the physical properties of gases.[19]

We consider first the evidence Thomson obtained from studying the "rays" of positive electricity that resulted from the electrical decomposition of a gas contained in a discharge tube. When the gas in the tube was carbon monoxide, CO, Thomson showed that, after passing through an opening in the cathode, it dissociated into charged atoms of carbon and oxygen, giving deflections $28/12 = 2.3$ and $28/16 = 1.7$ times that of the undissociated molecule. If the carbon atom in the undissociated molecule had a positive charge and the oxygen atom was negative, Thomson believed that when the impact of the rapidly moving electrons split carbon monoxide into atoms there would be a tendency for each atom to retain its original charge. Thus an analysis of the positive rays should have revealed the presence of a far greater number of carbon than oxygen atoms. Instead, Thomson found the ratio of carbon to oxygen atoms in the positive rays passing through the cathode to be nearly equal, 11 : 9. The approximate equality of this ratio, he argued, made it highly unlikely that the carbon

[18] Werner, *Inorganic Chemistry*, p. 24; Dmitri Mendeleev, *The Principles of Chemistry*; Lewis, "Valence and Tautomerism," p. 1454.

[19] J. J. Thomson, "The Structure of the Atom," *Engineering* 95 (1913), 328–30, 346–47, 397–98; idem, *Rays of Positive Electricity*; idem, "Rays of Positive Electricity," *Phil. Mag.* 21 (1911), 225–49; idem, "Further Experiments on Positive Rays," *Phil. Mag.* 24·(1912), 209–53; idem, "The Forces between Atoms and Chemical Affinity," *Phil. Mag.* 27 (1914), 757–89.

and oxygen atoms in the carbon monoxide molecule originally had opposite charges.[20]

Thomson's conclusion was in agreement with conclusions that Paul J. Kirkby (1869–1932) at Oxford reached after studying the effects of an electrical discharge on a gas. In discussing the forces in the oxygen molecule and those holding the oxygen and hydrogen atoms together in the water molecule, Kirkby had written: "The considerations prove definitely that the atoms of oxygen, however they come to be dissociated previous to their combining with hydrogen, are during their dissociate state uncharged. Hence their union with hydrogen cannot be attributed to electrostatic forces."[21]

Kirkby's observation that the most frequent effect of bombarding the oxygen molecule with electrons was its decomposition into two uncharged atoms rather than ions seemed to suggest a symmetrical distribution of the oxygen atoms' valence electrons. This was obviously irreconcilable with the electrostatic theory of molecular formation resulting from the attraction of oppositely charged ions.

Other studies on the electrical conductivity of gases, according to Thomson, led to the conclusion that the atoms in many molecules were not electrically charged. A good example, he said, was an equilibrium mixture of a gas and its dissociation products, a system in which new molecules of the gas were continually forming while others were breaking up. If the atoms in these molecules carried a positive or a negative charge and retained their charges upon dissociation, then the dissociated gas, containing a larger number of charged atoms, would be a good conductor of electricity. Taking nickel carbonyl, $Ni(CO)_4$, a gas that dissociated at temperatures well below the boiling point of water and was, consequently, very easy to experiment with, Thomson demonstrated the contrary: its conductivity when dissociated was no greater than that of any other gas. Like all gases, whether dissociated or not, nickel carbonyl showed some traces of conductivity, but certainly not enough, Thomson believed, to support the presence of charged atoms or molecules in the gaseous mixture.[22] Thom-

[20]Thomson, "The Forces between Atoms and Chemical Affinity," p. 760; idem, "The Structure of the Atom," p. 329; idem, *Rays of Positive Electricity*, pp. 63–64.

[21]Paul J. Kirkby, "A Theory of Chemical Action of Electrical Discharge in Electrolytic Gas," *Proc. Roy. Soc.* A 85 (1910), 163.

[22]Thomson, "Further Experiments on Positive Rays," p. 250; idem, "The Structure of the Atom," pp. 328–30, 346–47, 397–98; idem, *Rays of Positive Electricity*, pp. 64–65.

son also established that the formation of hydrogen chloride from hydrogen and chlorine under the influence of sunlight had no effect on the conductivity of these gases. As a result, he concluded that the atoms in the diatomic hydrogen and chlorine molecules had no electric charge.[23]

Léon Bloch's studies on conductivity supported Thomson's conclusions. Bloch (1876–1947) in Paris demonstrated in 1911 that an increase in electrical conductivity did not accompany the dissociation of arsine, AsH_3. He showed that many other chemical reactions involving gases— among them the oxidation of nitric oxide and ether vapor and the reaction between chlorine and arsenic at low temperatures—had little or no effect on conductivity.[24] These reactions, Thomson argued, provided clear evidence that no transfer of electric charge from one atom to another occurred.

Thomson's positive ray investigation called into question another fundamental postulate of the polar theory of valence—the belief that an atom's maximum positive valence was equal to the maximum number of electrons it lost. According to the polar theory, rare gases did not lose electrons and had zero valence, but Thomson discovered that helium, argon, and krypton were present in discharge tubes as He^{+2}, Ar^{+3}, and Kr^{+4} and Kr^{+5}.[25]

Thomson also found evidence of positively and negatively charged hydrogen, oxygen, sulfur, and chlorine atoms but only positive charges on the nitrogen and mercury atoms. Mercury apparently lost from one to eight electrons. Gases such as methane and carbon dioxide that Thomson expected to yield carbon atoms carrying four negative and four positive charges, respectively, instead gave carbon atoms with a maximum of two charges.[26] Positive ray analysis thus seemed to indicate the lack of any relation between an atom's maximum positive valence and the number of electrons lost, leading Thomson to write: "The maximum number of charges carried by a multiply charged atom does not seem to be related to any

[23]Thomson, "The Structure of the Atom," p. 329; idem, *Rays of Positive Electricity*, p. 65.

[24]Léon Bloch, "Recherches sur les actions chimiques et l'ionisation par barbotage." *Annales de Chimie et de Physique* 22 (1911), 370–417, 441–95, and 23 (1911), 28–144.

[25]Thomson, *Rays of Positive Electricity*, p. 53.

[26]J. J. Thomson, "Rays of Positive Electricity," *Proc. Roy. Soc.* A 89 (1913), 1–20; idem, "Rays of Positive Electricity," *Phil. Mag.* 21 (1911), 239.

chemical property of the atom such as its valency, but to depend mainly on the atomic weight." [27]

Thomson's statement clearly stood in total opposition to perhaps the most fundamental postulate of the polar theory of valence. If correct, it could have had a revolutionary effect on the development of later electron theories of valence. Yet, his statement evoked little response either favorable or unfavorable from chemists investigating the problems of atomic structure and valence. Probably, most of them were unaware of Thomson's research on positive rays or assumed that his publications were of interest only to physicists and not to chemists. In fact, as H. S. Fry correctly noted, though at a much later date, the evidence from positive rays and other physical investigations was largely irrelevant anyway because Thomson carried out his experiments under isolated conditions, such as high vacuum and high voltages. These conditions, Fry said, were not comparable with those prevailing in the great majority of chemical reactions, and consequently both the results and the conclusions were bound to differ. [28]

Actually, much of the physical evidence Thomson presented as contrary to the theory of polar valences soon appeared in an article that Stuart J. Bates at the University of Illinois published in 1914 in the *Journal of the American Chemical Society*. His publication was necessary, Bates believed, because most chemists did not know the theory's weaknesses, only the evidence in favor of it. Physicists, but not chemists, he pointed out, had already begun to reconsider whether the chemical bond between two atoms resulted from the transfer of an electron from one atom to the other: "Even Sir J. J. Thomson, who proposed and first developed the electron theory of valence, has, chiefly as a result of his investigation of positive rays, changed his opinion regarding the mechanism of the attraction between atoms." [29]

Quoting Thomson, Bates wrote: " 'We are led by these results to regard the electrical forces which keep the atoms in a molecule together as due not to one atom being charged positively and the other negatively but to the displacement of the positive and negative electricity in each atom. Thus each atom acts like an electrical doublet, and attracts another atom in much the same way that two magnets attract each other.' " [30]

[27] Thomson, *Rays of Positive Electricity*, p. 53.

[28] Harry S. Fry, *The Electronic Conception of Valence and the Constitution of Benzene*, p. 283.

[29] Bates, "The Electron Conception of Valence," p. 789.

[30] Ibid. (Thomson's statement is from *Rays of Positive Electricity*, p. 66).

Bates concluded, however, that chemistry contributed the most satisfactory evidence in favor of the polar theory. That the theory could explain simply and clearly the oxidation and reduction of electrolytes in solution seemed sufficient justification for its application to these reactions. But at the same time, he cautioned that an electrolyte's ability to separate into oppositely charged ions was not conclusive evidence that in the original molecule the atoms or atomic groups were already electrically charged. Ionization, he said "may consist essentially in the passage of an electron from one atom or group to another, these then separate from one another forming the ions." [31]

The ease with which the transfer of the electron took place, Bates argued, appeared to depend on the solvent as well as on the dissolved substance. Thus, the distinction between electrolytes and nonelectrolytes depended largely on the extent to which the environment acted on the substance. "In the pure state as solid, liquid or gas, the molecule of an electrolyte differed but little in its behavior from that of other substances," Bates said. [32]

Bates criticized Nelson, Beans, and Falk's claim that the electron theory of polar valences enabled them to classify all chemical reactions into three general types: (1) oxidation-reduction; (2) onium compound formation, $NR_4^+X^-$, such as the formation of ammonium chloride from ammonia and hydrogen chloride; and (3) metatheses or simple replacement or rearrangement. From the definitions of these reaction types and their breadth of classification, it seemed to him that Nelson, Beans, and Falk had removed all discussion of the electronic changes taking place during reaction from dependence upon experimental data. Their classification neither supported nor refuted their valence theory for it was impossible "to discover or even imagine any reaction which could not be classified and explained on such a basis." [33]

Nevertheless, Bates believed that Nelson, Beans, and Falk's classification of compounds and reactions certainly had value. What he doubted was the necessity of their fundamental premise, a complete electron transfer in bond formation. Some other theory, such as Thomson's, Johannes Stark's, or Niels Bohr's, which assumed an electrical attraction between two atoms

---

[31] Bates, "The Electron Conception of Valence," p. 792.

[32] Ibid.

[33] Bates, "The Electron Conception of Valence," p. 792; Nelson, Beans, and Falk, "Valence: Classification of Reactions," p. 1815.

without any electron transfer, would lead to a similar classification of chemical compounds and reactions.[34]

Thomson's Electronic Interpretation of the Dualistic Theory of Valence

In his 1914 publication, "The Forces between Atoms and Chemical Affinity," Thomson introduced new evidence in addition to positive ray analyses that, he said, made it essential to replace the polar theory with a dualistic theory of valence.[35] He obtained this evidence from specific inductive capacity measurements of compounds, and it was, according to Thomson, particularly strong and easily interpreted, especially when the compounds were gases.

The gases hydrogen chloride and ammonia and vaporized methyl and ethyl alcohol had abnormally high specific inductive capacities, but the diatomic molecules of the elements hydrogen, oxygen, nitrogen, and chlorine gave quite normal values. The high values found in the former group of molecules, Thomson believed, were due to the presence of large dipole moments, which resulted from the intramolecular ionization of these molecules.[36] When oriented under the influence of the electric field the dipoles contributed to the specific inductive capacity. Thus, intramolecularly ion-

---

[34] Bates, "The Electron Conception of Valence," p. 793; Thomson, "The Forces between Atoms and Chemical Affinity," pp. 757–89; Johannes Stark, "Zur Energetik und Chemie der Bandenspektra," *Phys. Zeit.* 9 (1908), 85–94; Stark, "Die Valenzlehre auf atomistisch elektrischer Basis," *Jahrbuch der Radioaktivität und Elektronik* 5 (1908), 124–53; Stark, *Prinzipien der Atomdynamik: Part II. Die elementare Strahlung*; Stark, "Folgerung aus einer Valenzhypothese: I. Bandenspektrum und Valenzenergie," *Jahrbuch der Radioaktivität und Elektronik* 9 (1912), 15–27; Stark, "Über das dreiatomige Wasserstoffmolekül nach Hypothese und Erfahrung," *Zeit. Elektrochemie* 19 (1913), 862–63; Niels Bohr, "On the Constitution of Atoms and Molecules," *Phil. Mag.* 26 (1913), 1–25.

[35] Thomson, "The Forces between Atoms and Chemical Affinity," pp. 760–64.

[36] *Specific inductive capacity* is defined as the ratio of the capacitance (the ratio of the charge Q to the voltage V) of a condenser with a given substance as dielectric to the capacitance of the same condenser with air or a vacuum as dielectric. A *dielectric* is an insulator, a body that does not conduct electricity. A dielectric permits the passage of the lines of force of an electrostatic field but does not conduct the electric current. *Dipole moment* is a measure of the extent to which an electric field separates the positive and negative charges in a molecule thus giving rise to a "pole," one end of which is positive, the other negative. *Intramolecular ionization* is a term that Thomson introduced in his 1914 paper to indicate the presence of charged atoms in highly polar molecules. These molecules were significantly polar due to the large values of their dipole moments.

ized molecules, since they consisted of charged atoms, tended to have abnormally large dipole moments and large specific inductive capacities.

Because the orientation of electrical dipoles required rotation of the entire molecule, Thomson argued that this movement was relatively slow and probably unaffected by rapid vibrations such as those of light waves in the visible region of the spectrum. The rotation of polar molecules, therefore, would not influence the value of the molecule's refractive index,[37] though as discussed above it would have an observable effect on a molecule's specific inductive capacity.

Clerk Maxwell had already shown that for many molecules the simple equation $K = n^2$ related specific inductive capacity K and refractive index n. The relation held, Thomson concluded, only because no intramolecular ionization took place within these molecules. Conversely, its failure when applied to other molecules indicated the presence of dipole moments and hence of intramolecular ionization. Taking the breakdown of Maxwell's $K = n^2$ rule as his criterion for the presence of intramolecular ionization, Thomson divided molecules into two bonding types, intramolecularly ionized (Group II) and nonintramolecularly ionized (Group I).[38] This division is outlined below:

I		II	
H_2	CO_2	H_2O	CH_3Cl
O_2	CS_2	NH_3	$CHCl_3$(slight)
N_2	CCl_4	SO_2	
He	C_6H_6	HCl	
Cl_2	CH_4	CH_3OH	
CO	N_2O	C_2H_5OH	

The dualism of Thomson and of Lewis and Bray and Branch received criticism from Roger Brunel at Bryn Mawr College. In an analysis of contemporary electrical theories of valence, Brunel claimed that Lewis had not really made clear his conception of an electron bond that represented all gradations between the extreme polar and nonpolar states. How could he explain, Brunel asked, a double decomposition reaction between a polar and a nonpolar compound. Of course, Brunel accepted the atom's electron constitution and believed that attractions between atoms were electrical,

---

[37] The *refractive index* is the ratio of light's velocity in a vacuum to its velocity in a substance.

[38] Thomson, "The Forces between Atoms and Chemical Affinity," p. 764.

but he favored a valence theory that did not require the existence of charged atoms in an undissociated molecule.[39]

Brunel thus opposed a dualistic theory. At the same time, he rejected completely the arguments put forward supporting the unitary theory of polar valences. The polar theory, he said, required the constant use of assumptions rendering it so elastic that its assumptions neither proved nor disproved the theory. The accepted existence of electromers was a good example. These structures, according to Brunel, permitted the distribution of electrons "largely in accordance with the exigencies of the particular equation in hand" and had developed "into merely a new system of nomenclature, desirable, if correct, but as far without direct support. This complex array of hypothetical electromeric substances [was not] an acquisition to the science which is desirable in itself," especially when there was not a single well-established case of their existence.[40]

Brunel believed that those chemists who supported the polar theory appeared imbued with the idea that all reactions were ionic just because some were. Such a claim, he added, was no more desirable than that of electromers. Indeed, it seemed to Brunel that the theory had already developed "into a complex mass of hypotheses, both fundamental ones dealing with the existence of electromeric substances, etc., and supplementary ones regarding the electronic formulas of particular substances."[41] Many statements made categorically in support of this theory, he maintained, were more correctly only hypothetical.

Among other things, Brunel criticized Fry's application of the polar theory to the benzene substitution problem. Fry's rule governing the position of substituents, he said, was "so simple as to arouse distrust at once." Brunel cited reactions involving phenylsulfonic acid and Arnold Holleman's studies on the nitration of benzene-halide compounds as instances in which Fry's rule failed.[42] In short, Brunel's 1915 publication was a point-by-point refutation of much of the experimental evidence supposedly favorable to the polar valence theory.

[39] Brunel, "A Criticism of the Electron Conception of Valence," p. 709, 719.
[40] Ibid., p. 718.
[41] Ibid., p. 721.
[42] Brunel, "A Criticism of the Electron Conception of Valence," p. 718; Harry S. Fry, "Interpretations of Some Stereochemical Problems in Terms of the Electronic Conception of Positive and Negative Valences: Part I. Anomalous Behavior of Certain Derivatives of Benzene," *J.A.C.S.* 36 (1914), 248, 252; Hollemann, "Substitution in the Benzene Nucleus," pp. 2495–98.

In concluding his criticism, Brunel raised two important questions that he believed any acceptable electron theory of valence had to answer, namely: (1) "What fundamental principle controls this tendency of atoms to gain and lose electrons? That is, what is the fundamental principle on which we are to determine the electronic formulas of substances?" and (2) "Do atoms which have not gained or lost electrons attract each other? If so, by what means?"[43]

In answer to the first question, many chemists and physicists—Abegg, Thomson, Ramsay—had used the rule of eight, the belief that an atom tried to acquire a maximum of eight valence electrons, as a guiding principle in assigning chemical formulas to compounds. This rule took on a greater significance only after Lewis used it in conjunction with his cubic atom to propose electron structures that accounted for the inertness and hence the stability of the noble gases. Other atoms, therefore, reacted to achieve a stable structure similar to one of the noble gases by gaining, losing, or sharing electrons.

An answer to Brunel's second question had already appeared in the electronic explanation of nonpolar bonds that Thomson offered in his 1914 paper. Thomson now assumed that the electrons in an atom were of two kinds. Those involved in bond formation were "mobile" electrons and arranged themselves in a ring near the atom's surface. The other electrons, the "fixed" electrons, remained in the one or more concentric layers or rings located within the outermost ring of mobile electrons. At the center of all the electron rings was the atom's dense positive core.[44]

For each kind of atom, the number of mobile electrons varied from zero to eight and was equal to the atom's group number (0, I, II, . . . VII) in the periodic table. Thus, the Group 0 elements, helium, neon, and argon, had no mobile electrons, or, looked at from another point of view, they had eight electrons in the outermost ring. According to Thomson, this made the ring so stable that the electrons were no longer mobile. Hydrogen and the Group I elements had one mobile electron and so on up to Group VII, in which each element had seven mobile electrons.

The process of forming a nonpolar bond, Thomson believed, required fixing a mobile electron from each of the combining atoms. The tube of force represented the physical link connecting atoms, and fixing occurred

---

[43] Brunel, "A Criticism of the Electron Conception of Valence," p. 721.

[44] Thomson, "The Forces between Atoms and Chemical Affinity," p. 781. Thomson left unchanged his interpretation of the bond in polar compounds.

when a tube of force starting from a mobile electron in one atom anchored itself to the positive core of a second atom. Alternatively, no fixing of a mobile electron occurred when its tube of force returned to the positive core in the same atom. These tubes possessed considerable mobility because the electron at one end of each could move freely about in the atom. If an atom had n mobile electrons, Thomson wrote,

> to fix these and thus saturate the atom, the n tubes of force which start from the n corpuscles [electrons] must all end on other atoms and not return to the original atom. Thus to ensure saturation from every free corpuscle in an atom, a tube of force must pass out of that atom and end on some other, and this must hold for every atom in the molecule. When the atoms are electrically neutral, *i.e.*, have no excess of positive over negative charge or *vice versa*, for each tube of force which passes out of an atom, another must come in; and thus each atom containing n corpuscles must be the origin of n tubes going to other atoms and also the termination of n tubes coming from other atoms. . . . With this arrangement every mobile corpuscle in the system is anchored by a tube of force to some other atom, and thus deprived of mobility: hence the system will be saturated.[45]

Thomson represented the tube of force emanating from each mobile or valence electron in an atom by an arrow pointing from the atom's symbol. The conversion of the organic chemists' structural formulas into his electron formulas amounted simply to doubling each of the lines representing a bond—one arrow indicated a tube of force going from the atom and the other, a tube returning to the atom. For a single, a double, and a triple bond, as in ethane, ethylene, and acetylene,[46] this gave

$$C \rightleftharpoons C \qquad\qquad C \lessgtr\gtrless C \qquad\qquad C \gtreqqless C.$$

Ethane   Ethylene   Acetylene

In each of these structures Thomson's bond was, therefore, a two-electron bond, though this was not at all a necessary postulate of his theory. Thomson's structures seemed to imply that if an atom A sent a tube of force to a second atom B it received one in return from B, but the only requirement was that the number of tubes that left A had to equal the number that returned. "It is not necessary," he wrote, "that the atoms from which A receives the tubes should be the same as those to which it sends them."[47]

[45] Ibid., p. 782.
[46] Ibid., p. 784.
[47] Ibid., p. 783.

Thomson's theory thus permitted a structure of benzene that had the carbon atoms joined to one another by three tubes of force, while a pair of tubes held each hydrogen atom to the carbon hexagon.[48] Indeed, his structure contained both "three-electron" and "two-electron" bonds. Because the same number of tubes of force originated and terminated on each atom (four for each carbon and one for each hydrogen), every atom in the molecule was electrically neutral, the benzene structure itself being nonpolar:

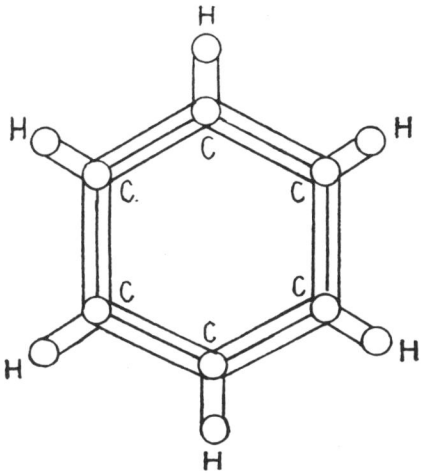

A similar application of the theory led Thomson to suggest the possible existence of H_3, a molecule for which he believed he had obtained experimental evidence in his positive ray study. He illustrated H_3 as follows:

As the diagram shows, each hydrogen atom had only to be the origin and termination of a tube of force (the one condition it had to fulfill) in order for this molecule to exist.[49]

The compounds $AgCl_2$, $AgCl_3$, $AgCl_4$, even $AgCl_n$ were also possible

[48] Ibid., p. 785.

[49] Thomson, *Rays of Positive Electricity*, p. 121; idem, "The Forces between Atoms and Chemical Affinity," p. 783.

additions to the one known silver chloride structure AgCl, though like H_3 their existence was not compatible with established ideas on valence. Such compounds, Thomson believed, would be mainly ring or long chain compounds, reminiscent of the structures once proposed for many metallic salts and oxyacids in an effort to maintain a constant rather than a variable valence for an atom.[50] Unless they contained carbon atoms, they were unlikely to have any great stability.[51]

In describing the bonding in both inter- and intramolecularly ionized compounds, Thomson's ideas remained unchanged from those found ten years earlier in his 1904 and 1907 publications.[52] The chemical bond was still an electrostatic attraction that resulted when the force between combining atoms was sufficiently great to remove an electron from the more electropositive atom and attach it to the electronegative one. Each electron upon leaving the one atom transferred its entire tube of force to the other. After the transfer, no tubes left the electropositive atom and none entered the electronegative atom. There were, therefore, no mobile electrons left on the electropositive atom. Since the number of electrons in the outer ring of the electronegative element would always have increased to eight, forming a rigid ring, it also had no mobile electrons.

Thus, according to Thomson's dualistic theory of valence, the division into two types of compounds, polar and nonpolar, depended solely on whether a molecule contained charged atoms. The valence conditions for the nonpolar were far more elastic than those for the polar, because the nonpolar type permitted twice as many bonds in a molecule and allowed for numerous compounds not possible in a unitary theory of polar bonds. Indeed, as Joseph W. Mellor (1869–1938) pointed out in 1916, the chief problem with Thomson's theory of valence was that it explained too much.[53]

Criticisms of the Dualistic Theory of Valence

Falk and his associates, John M. Nelson and Hal T. Beans, of course, opposed a dualistic valence theory. They argued that such a view intro-

[50] See Alexander Smith, *General Chemistry for Colleges*, p. 199.

[51] Thomson, "The Forces between Atoms and Chemical Affinity," p. 788.

[52] J. J. Thomson, *Electricity and Matter*; idem, *The Corpuscular Theory of Matter*.

[53] Thomson, "The Forces between Atoms and Chemical Affinity," pp. 787–88; Joseph W. Mellor, *Modern Inorganic Chemistry*, p. 874.

duced "complications in the theoretical treatment of chemical reactions" that chemists could largely avoid if they adopted the unitary polar theory. Their electronic theory was superior, they said, because it enabled chemists to classify all chemical reactions into three main types: (1) oxidation-reduction; (2) "onium" compound formation, the formation of ammonium and its derivatives $NR_4^+X^-$; and (3) simple replacement or rearrangement reactions such as hydrolysis, the mixing of salt solutions, and precipitation reactions.[54]

With the polar theory, chemists could apply to inorganic and organic compounds the same definitions of oxidation and reduction, the loss and the gain of electrons, respectively. In a theory that permitted polar and nonpolar bonds in the same compound they could not use these definitions without leading to difficulties.[55]

Falk and Nelson cited organic reactions in which a halogen atom such as chlorine replaced a hydrogen atom in a hydrocarbon and a hydroxyl group replaced the halogen. The experimental evidence supplied by hydrolysis, they pointed out, indicated that the chlorine atom and the hydroxyl group's oxygen atom when combined with carbon had a negative charge.[56] To accept this identity of charge, they said, implied that the substitution of a negative halogen or hydroxyl for a positive hydrogen had to be an oxidation reaction.[57] On the other hand, according to the dualistic theory, the hydrogen and the chlorine atom combined with carbon using nonpolar valences. But because hydrolysis indicated no change in chlorine's charge when a hydroxyl group replaced chlorine, it followed that the obviously polar bond between carbon and the hydroxyl group's oxygen atom of hydroxyl was also nonpolar.

Falk and Nelson, therefore, argued that if chemists admitted the existence of nonpolar bonds, "the phenomenon of oxidation in organic com-

[54] K. George Falk and John M. Nelson, "The Electron Conception of Valence: V. Polar and Non-Polar Valence," *J.A.C.S.* 36 (1914), 214. Nelson, Beans, and Falk in "Valence: Classification of Reactions," pp. 1810–21, discuss the classification scheme.

[55] Falk and Nelson, "Valence: Polar and Non-Polar," p. 210.

[56] Feodor Selivanov, "Beitrag zur Kenntnis der gemischten Anhydride der unterchlorigen Säure und analoger Säuren," *Berichte* 25 (1892), 3617–23; Julius Stieglitz, "On the Beckmann Rearrangement: I. Chlorimidoesters," *Amer. Chem. Journal* 18 (1896), 756; William A. Noyes, "An Attempt to Prepare Nitro-Nitrogen Trichloride, an Electromer of Ammono-Nitrogen Trichloride," *J.A.C.S.* 35 (1913), 769.

[57] Nelson, Beans, and Falk, "Valence: Classification of Reactions," p. 1816; Falk and Nelson, "Valence: Polar and Non-Polar," p. 210. See also Fry, "Interpretations of Some Stereochemical Problems," pp. 248, 262.

pounds would possess an entirely different significance from that which it does in inorganic reactions. On the non-polar view, chlorination consists of a simple replacement of hydrogen by chlorine, oxidation would then consist of a simple replacement of hydrogen by hydroxyl [or oxygen], and the series of transformations methane→methyl alcohol→formaldehyde→formic acid→carbon dioxide, for example, would represent replacements with no change in the charge of the carbons." They regarded this as untenable because it introduced an artificial differentiation between the reactions of organic and inorganic compounds. The dualistic theory resulted in "the absolutely separate characterization of reactions involving the oxidation for organic and inorganic substances as distinct and unrelated phenomena." [58]

The Grignard reaction, below, revealed another weakness of the dualistic theory:

$$RI + Mg = Mg \genfrac{}{}{0pt}{}{\nearrow I}{\searrow R}$$

$$Mg \genfrac{}{}{0pt}{}{\nearrow I}{\searrow R} + R_1I = MgI_2 + RR_1$$

According to the theory, RI and R_1I, two compounds with nonpolar bonds, reacted with the element magnesium to form nonpolar RR_1 and magnesium iodide, MgI_2, an obviously polar compound. Clearly, a change in bond type occurred during the reaction, but, as Falk and Nelson pointed out, when it took place was certainly not apparent from the reaction's equations. That magnesium iodide behaved as an electrolyte did not indicate when the change occurred, for its behavior was only an effect of the reaction and not the cause.[59] Using polar or directive valences to describe the reaction, they said, eliminated this confusion:

[58] Falk and Nelson, "Valence: Polar and Non-Polar," p. 211.
[59] Ibid.

$$R^+ \rightarrow {}^-I + Mg^0 = Mg^{+-}\overset{\nearrow\ I^-}{\underset{\nwarrow\ R^+}{}} \rightleftharpoons Mg^{++}\overset{\nearrow\ I^-}{\underset{\searrow\ R^-}{}}$$

$$Mg^{++}\overset{\nearrow\ I^-}{\underset{\searrow\ R^-}{}} + R_1^+ \rightarrow {}^-I = Mg^{++}\overset{\nearrow\ I^-}{\underset{\searrow\ I^-}{}} + R_1^+R^-$$

In the initial step, R^+I^- combined with a neutral magnesium atom, giving the intermediate product $Mg^{+-}\langle{}^{I^-}_{R^+}$. An intramolecular oxidation-reduction converted this hypothetical intermediate into the first definite product, $Mg^{++}\langle{}^{I^-}_{R^-}$. A simple metathesis then followed, resulting in the final products, magnesium iodide and $R_1^+R^-$. There was no need to assume a change in bond type anywhere in the reaction, and for this reason Falk and Nelson found it "much simpler and more satisfactory to follow the transformations from this point of view than with the combination of the two kinds of valence, polar and non-polar." [60]

Falk and Nelson maintained that the Wurtz synthesis provided one more example of the dualistic theory's weakness: $2RI + 2Na = RR + 2NaI$. As in the Grignard reaction, it could not show when the iodine atom changed its bond from nonpolar to polar. In the polar theory the synthesis was simply another oxidation-reduction reaction in which one of the R^+ groups took a single electron from each sodium atom:

$$2R^+I^- + 2Na^0 = R^+R^- + 2Na^+I^-.$$

Indeed, Falk and Nelson believed that the only difference between organic and inorganic oxidation-reduction reactions was that in the former the changes in electrical charge were not as apparent and as easily followed. Fundamentally, there was no difference. Chemists could interpret all organic reactions, "simply and satisfactorily" and bring "them into relationship with the reactions of inorganic chemistry by using polar valences." The unitary theory assumed that all chemical changes depended

[60] Ibid., p. 212.

on the atoms' electrical charges. It "show[s] a simpler and more satisfactory classification than does the dualistic view which leads to contradictions or to an arbitrary separation of phenomena fundamentally similar."[61]

William A. Noyes expressed the same sentiments regarding the unitary polar theory of valence. The magnitude of electrical forces, he said, "makes it seem quite improbable that the atoms of compounds such as hydrochloric acid, hydriodic acid and water are held together by electrical forces while those of methane and ethane are held by gravitation or some other non-electrical force." Edward C. Franklin's study of nonaqueous solvents convinced Noyes that ammonia contained positive hydrogen and negative amide, NH_2, just as water contained positive hydrogen and negative hydroxyl. There was also considerable evidence, Noyes wrote, that acetylene, $H—C\equiv C—H$, ionized to yield positively charged atoms. If this were true, he said, ethane, $H_3C—CH_3$, which reacted to give acetylene, most likely contained positively charged hydrogen atoms.[62]

Arsem's One-Electron Theory of Valence

The unitary polar theory, according to its proponents, accounted very well for the formation of all chemical compounds. Lewis and Bray and Branch argued on the contrary that the facts of chemical union demanded a dualistic theory of polar and nonpolar bonds for their complete explanation.

William C. Arsem (1880–1971), a research chemist and engineer with the General Electric Company in Schenectady, New York, offered a third possibility. In his 1914 paper on valence, Arsem sought to avoid classifying the elements as positive or negative and valence theories as unitary or dualistic. His theory had originated about ten years earlier, in 1904, he claimed, and had been in the process of development and extension since that time. But the recent publication of ideas that he thought resembled his own convinced Arsem that the time had now come to publish his theory. It was in harmony with the trend of scientific speculation.[63]

[61] Ibid., pp. 213, 214.

[62] William A. Noyes, "The Nature of the Forces Holding Atoms in Combination," *J.A.C.S.* 36 (1914), 214.

[63] William C. Arsem, "A Theory of Valency and Molecular Structure," *J.A.C.S.* 36 (1914), 1656. He was referring to Ramsay who believed the electron served as the bond of union between atoms in a compound and especially to Stark, who assumed in his publications (1908–13) that the chemical bond resulted from the simultaneous attraction of atoms for the same electron.

In Arsem's theory, chemical combination between a pair of atoms occurred whenever a single valence electron from one atom altered its direction of motion so that it oscillated periodically between the two atoms. This electron—the bonding electron—was therefore common to both atoms, and its motion though described as oscillating actually included any kind of motion such as "revolution in an orbit or motion in a complex curvilinear path." Arsem developed his theory on the assumption that each bond corresponded to a single valence electron, but he agreed that a group or system of electrons might perform the same function.[64]

Valence thus became "the property, or power, which an atom possesses of sharing a certain number of electrons with one or more other atoms in such a way that the atoms so united form a *complete* or *perfect* electron system which is electrically neutral." That a chemical bond resulted from the permanent transfer of an electron from one atom to another was unacceptable to Arsem. The two atoms in a diatomic molecule like hydrogen or chlorine would have opposite charges and no conclusive experimental evidence supported this hypothesis. Making use of an argument so often raised against the unitary polar theory, Arsem wrote: "the fact that many substances can be dissociated into oppositely charged atoms or groups does not prove that these atoms or groups are permanently charged in the same way when in the undissociated compound."[65]

According to Arsem, a single valence electron sufficed for the union of monovalent atoms, as in the hydrogen and halogen molecules:

The black dot in his diagrams indicated the valence electron and the dotted line its path of oscillation. Arsem represented the atoms in the molecule by spheres in contact at a single point, though he believed that the union was really much more intimate. In fact, as Arthur A. Noyes implied in 1908, Arsem thought that the separate atoms probably lost their identity when combined in the molecule.[66]

Arsem's structures for monovalent diatomic molecules also showed that they could not dissociate symmetrically, for upon dissociation, one of the

[64] Ibid., p. 1659.

[65] Ibid., pp. 1657, 1659.

[66] Ibid., p. 1659; Arthur A. Noyes, "The Conductivity and Ionization of Salts, Acids, and Bases in Aqueous Solutions at High Temperatures," *J.A.C.S.* 30 (1908), 351–52.

atoms retained the valence electron. Since the other atom no longer had a share of this electron, the two atoms were now oppositely charged ions: HeH $\rightleftarrows$ H + eH. Unit valence for each ion, unlike other electron theories of valence, corresponded to an excess or deficiency of only one half of the electron's charge, since in the neutral molecule each atom originally possessed only one half of the electron. Using Arsem's symbol E for the electron charge, each ion thus had a "unit ionic charge" of $\pm E/2$.[67]

Hydrogen chloride, its hydrogen and chlorine atoms held together by a single valence electron, resembled structurally the hydrogen and halogen molecules. Its synthesis from hydrogen and chlorine appears below in (I). The structures in (II) illustrate Arsem's reaction mechanism.[68] The same mechanism, he said, held for all monovalent atoms and explained the phenomena such as conduction in metals and in electrolytes, keto-enol tautomerism, and the association of simple molecules:

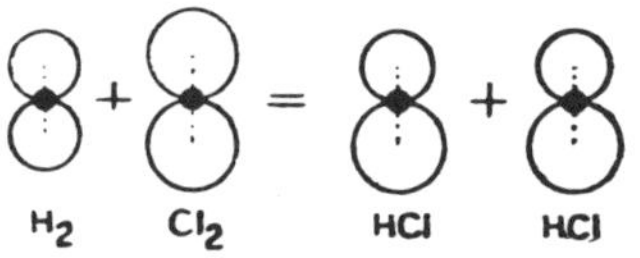

(I) 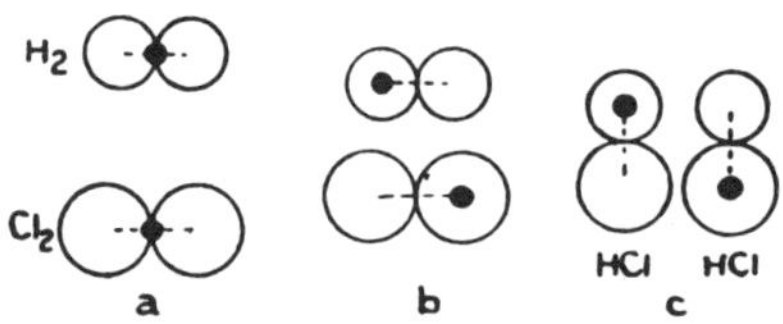   (II)

Part a of (II) showed the single valence electron of the hydrogen and chlorine molecules in the middle of its path of oscillation. In part b they were at opposite ends of their paths, and each hydrogen atom was now adjacent to an oppositely charged chlorine atom. When the valence electrons were in this relative position at the moment of collision, combination occurred. In c, the valence electrons oscillated in new paths, forming hydrogen chloride, HCl, a more dynamically stable electron system than either hydrogen or chlorine. According to Arsem, a chemical reaction meant "nothing more than the readjustment of the paths of oscillation of valence-electrons between the atoms of contiguous molecules."[69]

Arsem applied his valence theory to divalent atoms such as the oxygen atoms in the diatomic oxygen molecule. His model required two valence electrons, one from each atom, to form the molecule's two bonds:

[67] Arsem, "A Theory of Valency," pp. 1659, 1673.
[68] Ibid., pp. 1660, 1666–72.
[69] Ibid., p. 1667.

Dissociation of the double bond occurred in two different ways. First, each atom retained one of the valence electrons and remained electrically neutral. It simultaneously lost and gained one-half of an electron charge:

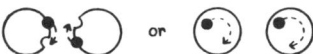

Second, one of the atoms retained both of the valence electrons:

Each atom then had a charge twice that of the unit ionic charge, $E/2$, or a charge equal to the electronic charge, $2(E/2) = E$. One atom's charge was $+E$, the other's $-E$.[70]

Like most valence theories of that time—Fry's, Ramsay's, and Falk and Nelson's—Arsem's theory was what Fry called a "formulative hypothesis." In other words, it was another theory in which a one-electron bond replaced each valence bond of the preelectronic valence theory, and which did not attempt to explain chemical affinity or the atom's structure. But Arsem added that each chemical bond between a pair of atoms resulted from the periodic oscillations of a single electron between them. His theory thus differed from the generally held polar interpretation of the chemical bond, which required the complete transfer of a valence electron or electrons between the combining atoms.

Arsem, therefore, did not use an electrostatic tube of force to represent an atom's valence electron. The tube of force not only provided the physical link between combining atoms, but by being directive it also indicated which of the atoms donated the valence electron. Arsem's structures never showed which of the combining atoms provided the oscillating valence electron; they also ignored the presence of all other electrons in the atom except those forming bonds. In this respect his theory was quite vague when compared with the polar theory, which clearly indicated the origin of

[70] Ibid., p. 1662.

each bonding electron. But one of Arsem's aims was to eliminate the classification of elements as either electropositive or electronegative, that is, as electron donors or electron acceptors. From this point of view, his theory of electron sharing was perhaps an improvement over the polar theory, especially when applied to compounds like the hydrocarbons, which had never given any evidence that they contained electrically charged atoms.

Within a few years after Arsem published his theory, it became clear that he had gone too far in assuming a shared one-electron bond even in compounds that clearly consisted of ions. His theory was no longer tenable once X-ray studies showed the presence of ions rather than molecules in many inorganic crystalline salts such as sodium chloride. Nor was a one-electron bond in agreement with magnetic studies on atoms and molecules in the early 1920s. These studies indicated an electron arrangement in most compounds that allowed the magnetic effect associated with each electron's motion to neutralize itself. Most chemical compounds were, therefore, not magnetic (diamagnetic), but according to Arsem's theory, they would exhibit a magnetic behavior (paramagnetic) because of the single valence electron constituting the chemical bond.

The Electroatomic Theory of Johannes Stark: An Attempt to Avoid Both a Dualistic and a Unitary Theory of Valence

The electroatomic theory of Johannes Stark, professor of physics at the Technische Hochschule in Aachen, Germany, was a second valence theory based on electron sharing. Stark, who regarded his theory as a revival of Berzelius's electrochemical theory, used electron sharing to avoid having to distinguish between dualistic and unitary valence theories. He attributed the different valences of atoms to a difference in the electric field of force on each atom's outer surface.[71]

At the time of Stark's publications on valence (ca. 1908–15), physicists and chemists believed that atoms contained only positive and negative units of electricity. Stark accepted this, though in putting forward his valence theory he disregarded dealing with the atom's inner structure. It was only necessary, he said, to assume that the atom's surface was a three-

---

[71] Dorothy A. Hahn and Mary E. Holmes, "The Valence Theory of J. Stark from a Chemical Standpoint," *J.A.C.S.* 37 (1915), 2613. This paper is an abstract of Paul Ruggli's treatise *Die Valenzhypothese von J. Stark vom chemischen Standpunkt.*

dimensional arrangement of positive and negative electrical charges. Allowing for the greater difference in "size" between the fundamental units of positive and negative electricity, Stark's atomic surfaces consisted of spheres or distorted spherical zones of positive electricity. Between or even above these were the comparatively small, point-like, negative valence electrons.

Physicists usually represented an electric field's structure by the arrangement and density of its lines of force, and Stark believed that lines of force radiated in all directions from the negative electrons in an atom. Many of the lines terminated on the atom's extended or distorted positive zones. Stark did not distribute the positive electricity uniformly over the atom's surface but concentrated it at certain points on the atom. The attraction between positive spheres and neutralizing valence electrons was strongest at these points, and the electrons therefore occupied definite positions on the atom's surface.[72]

Stark illustrated the arrangement of the positive surface and the valence electrons with numerous atomic models.[73] The models below are for a monovalent atom:

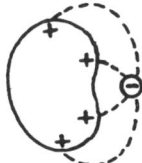   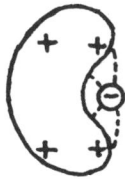

Monovalent metal (sodium)   Monovalent nonmetal (chlorine)

They make clear that the difference between metallic and nonmetallic atoms was the shape of the positive surface and the arrangement of the electron's field of force on the surface. Indeed, with his variable fields of force Stark hoped to avoid the difficulties of both the unitary and the dualistic valence theories. Thus, the protruding valence electron on the metallic atom's surface and the nonmetallic atom's receding valence electron,

[72] Stark, "Zur Energetik und Chemie der Bandenspektra," pp. 85–94; Hahn and Holmes, "The Valence Theory of J. Stark," p. 2613.

[73] Johannes Stark, "Anwendung einer Valenzhypothese auf Erscheinungen der Fluoreszenz," *Zeit. Elektrochemie* 17 (1911), 515; idem, *Atomdynamik: II*, p. 104; Hahn and Holmes, "The Valence Theory of J. Stark," p. 2625.

giving prominence to the positive sphere, produced the apparent polarity of these atoms. But no gain or loss of electrons actually occurred.[74]

The following models are for multivalent atoms.[75] They are, from left to right, a divalent electropositive atom (magnesium), a divalent electronegative atom (oxygen), and a trivalent atom (nitrogen). The last model shows the extreme case of a noble gas. Its valence electrons were so deeply embedded in the positive sphere that their lines of force could no longer unite with the positive spheres of other atoms to form a chemical compound. This gave the appearance of having zero valence:

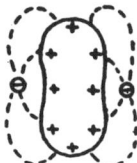  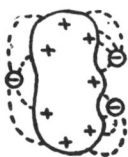 

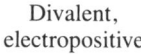

| Divalent, electropositive | Divalent, electronegative | Trivalent | Zero valence |

A chemical bond between two atoms resulted, he said, whenever one of the atoms detached some of the lines of force emanating from a valence electron in the other atom and attached them to its positive sphere. The united atoms did not actually interpenetrate; their lines of force held them together. As the negative electron of one atom approached the positive sphere of another, there was also, according to Stark, a shortening of the lines of force or a contraction of the field of force between the atoms. Stark's theory thus led to the conclusion that a change in volume accompanied chemical reaction. The work of Theodore W. Richards (1868–1928) at Harvard corroborated Stark's hypothesis. Richards had shown experimentally in 1902 that a negative heat of reaction, which indicated the likelihood of a given reaction's taking place, corresponded to a net decrease in the final volume compared with that of the original reactants.[76]

[74] Stark, *Atomdynamik: II*, pp. 102–6; Hahn and Holmes, "The Valence Theory of J. Stark," p. 2625.

[75] Stark, *Atomdynamik: II*, pp. 102–6.

[76] Stark, "Zur Energetik und Chemie der Bandenspektra," pp. 86–87; idem, *Atomdynamik: II*, pp. 102–6; Hahn and Holmes, "The Valence Theory of J. Stark," p. 2613; Theodore W. Richards, "Die mögliche Bedeutung der Änderung des Atomvolums," *Zeit. phys. Chemie* 40 (1902), 169–84; idem, "Die Bedeutung der Änderung der Atomvolume," *Zeit. phys. Chemie* 40 (1902), 597–610.

Stark illustrated the bonding of a monovalent electropositive and a monovalent electronegative atom, a halogen and an alkali metal, with the diagram:

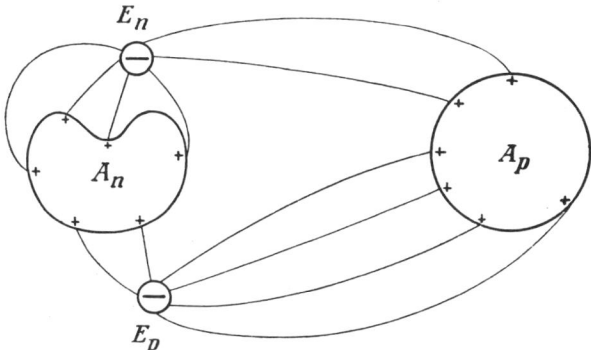

where A_n = electronegative atom; A_p = electropositive atom; E_n = electron from A_n; and E_p = electron from A_p. The rearrangement of the lines of force showed the chemical bond existing between the two atoms and that the electron E_p, belonging to the electropositive atom A_p, was now closely associated with A_n, giving the appearance of polarity, $A_n^- A_p^+$, to the compound.[77]

Other diagrams represented the bonding between two identical (I) and two different (II) monovalent electronegative atoms (below).[78] They demonstrate clearly the absence of any polarity in these molecules:

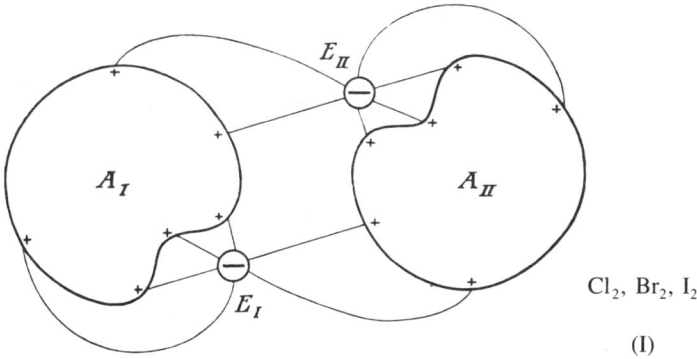

Cl_2, Br_2, I_2

(I)

[77] Johannes Stark, *Prinzipien der Atomdynamik: Part III. Die Elektrizität in chemischen Atom*, p. 75.
[78] Ibid.

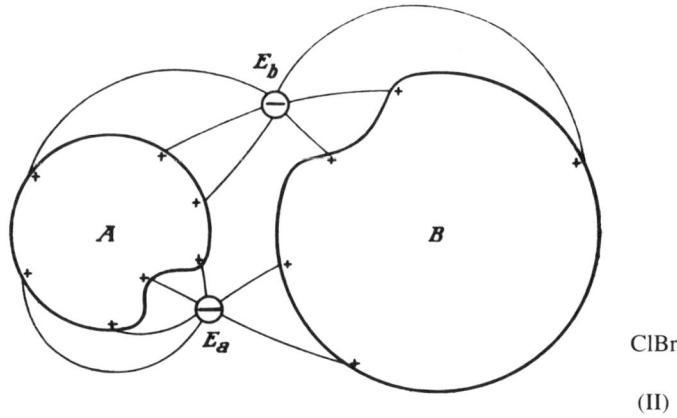

ClBr

(II)

Using lines of force, Stark also offered an explanation for the formation of molecular compounds that in some respects resembled the Oliver
Lodge and Percy Frankland hypotheses of 1904. Suppose, Stark suggested, that at one or more points on a stable molecule some of the lines of
force emanating from saturated valence electrons or from positive spheres
protruded considerably beyond the molecule's contour and that similar
lines of force protruded from a second molecule. An electrical union might
take place because the positive spheres of one molecule attracted the negative electrons of the other. But Stark believed that a saturated valence electron used only a very small fraction of its lines of force in this way and that
the resulting attraction was, consequently, quite weak.[79]

The same explanation, according to Stark, accounted for the formation
of crystalline structures. Under the influence of the electrical fields of force
that existed on the surfaces of molecules, adjacent molecules oriented
themselves so that the positive spheres and negative electrons of one molecule were opposite those of the other. If the energy due to the molecular
orientations were sufficient to withstand the disruptive tendency that accompanied the molecules' thermal motion, a regular molecular arrangement or crystalline structure resulted. Hence, the difference between molecular compounds and crystalline structures was only a difference of
degree.

[79] Oliver Lodge, "Residual Affinity," *Nature* 70 (23 June 1904), 176; Percy Frankland,
"Residual Affinity," *Nature* 70 (7 July 1904), 222–23; Hahn and Holmes, "The Valence Theory of J. Stark," p. 2617.

Stark's Electron Pair Bond

In the period 1913–14, Bray and Branch, G. N. Lewis, and J. J. Thomson turned to a dualistic valence theory, requiring polar and nonpolar bonds, as a feasible alternative to the unitary polar theory. With the exception of Thomson, none of them developed his theory to any extent. William Arsem also rejected the unitary view. But he found the dualistic theory equally unsatisfying, suggesting instead that the chemical bond consisted of a single oscillating shared electron. Like the unitary theory, his theory failed to explain why chemical compounds easily fit into two general classes obviously differing in properties, one of which was, in fact, polarity.

On the other hand, Stark, represented the chemical bond in each of his molecular structures given above with a shared pair of electrons. He had shown with his structures how the position of the electron pair could account for the presence or absence of polarity in a molecule. With the electron pair he had clearly overcome the chief weakness of the unitary polar theory, namely, of postulating the presence of charged atoms in every chemical compound.

Yet, Stark's shared electron pair bond was not essential to his valence theory. It followed directly from his assumption that a valence electron was saturated only after attaching its lines of force to another atom's positive sphere in addition to its own sphere. Indeed, his bond resembled closely the electron bond Thomson described in his 1914 paper on valence. Hence, in Stark's binary halogen compounds, the lines of force from each valence electron attached themselves to both atoms, giving rise to a two-electron, or electron pair, bond.

Stark, therefore, represented the hydrogen molecule with the electron structure below:

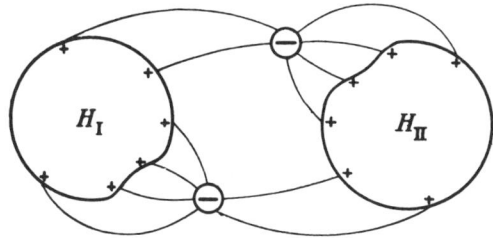

But because the electron pair bond had no special significance in his theory, Stark, like Thomson, did not rule out the possible existence of H_3 or "triatomic" hydrogen. A single electron sufficed in this case to bond adjacent hydrogen atoms in a ring-like structure.[80] He illustrated it as follows:

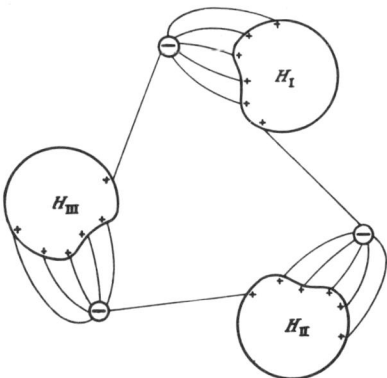

According to Stark's diagram, the "valence field" between the hydrogen atoms was less concentrated in H_3 than in H_2, suggesting an unstable triatomic hydrogen molecule. This was obviously the reason, Stark believed, why physicists and chemists had not established its existence. Quite understandably, he did not predict the molecules H_4, H_5, . . . H_n, though structures for them would have followed simply by expanding the ring structure of the H_3 molecule.

Stark's structures for the carbon-carbon single bond and the carbon-hydrogen bond in saturated hydrocarbons also required a pair of electrons.[81] In ethane, H_3CCH_3, each carbon atom contributed a single electron to the carbon-carbon bond, while to form each of the six carbon-hydrogen bonds each carbon and hydrogen atom provided one electron.

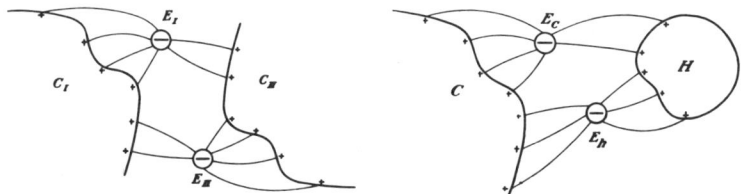

[80] Stark, "Folgerung aus einer Valenzhypothese: I," p. 19; idem, *Atomdynamik: III*, pp. 120, 121.

[81] Stark, *Atomdynamik: III*, p. 81.

Four electrons, two from each carbon atom, and six electrons, three from each carbon atom, represented, respectively, a double and a triple bond between carbon atoms in an unsaturated hydrocarbon. Stark's electron structures thus were in agreement with the carbon atom's well-established tetravalence as well as hydrogen's monovalence. But clearly Stark never could have predicted the valence of carbon, hydrogen, or any other atom with his electron structures. He applied them only to known atomic valences.

In his structures of the ammonia and water molecules we might have expected Stark to use once again an electron pair bond. Instead he assumed, and without explanation, that the nitrogen-hydrogen and the oxygen-hydrogen bonds were essentially single electron bonds.[82] Stark's structure of the ammonia molecule clearly illustrates the electron bonds:

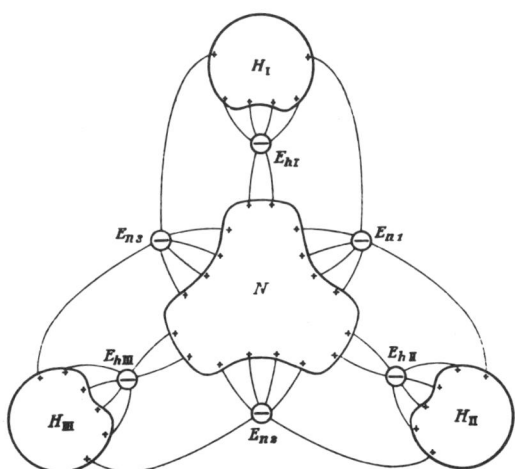

Conclusion

Stark's electroatomic theory was another attempt to overcome the inadequacies of the unitary theory of polar valences. Indeed, Lewis and Thomson introduced their dualistic theories for precisely this reason. They believed that in order to account for the obviously different properties of polar and nonpolar compounds a theory with two kinds of valence, polar

[82] Ibid., pp. 98–99.

and nonpolar, was necessary. Stark, on the other hand, maintained that the apparent polarity of many compounds resulted simply because valence electrons moved away from the positive surface of one atom toward the positive surface of another. The second atom's greater attraction for the valence electrons produced the movement. A nonpolar compound followed if the valence electrons remained more or less intermediately between the bonded atoms. His theory was a unitary theory but not one requiring complete electron transfer between the combining atoms.

According to Stark, an atom's position in the periodic table determined the direction of its valence electrons' movement. Those atoms located on the left side of the table, the metals, held their valence electrons loosely. Thus, other elements easily pulled the valence electrons away from the positive surface of these atoms. Conversely, the nonmetals lying on the right side of the table, tightly held their own valence electrons and pulled those of other atoms toward their positive surfaces. The carbon atom presented a unique case. Because it occupied a position about halfway across the periodic table, its valence electrons apparently had an equal tendency to move toward or away from its positive surface. Hence, in forming a bond with another atom, especially a hydrogen or a second carbon atom, Stark seemed to assume that the tendency of the carbon atom's valence electron to move away from its surface was as great as that for an electron from the other atom to move toward its surface. This resulted in the combining atoms' holding an electron pair rather than a single electron between them.

In Stark's and in Thomson's valence theories a pair of electrons sometimes constituted a chemical bond. Both of them almost always restricted the electron pair to certain kinds of compounds or made it an incidental feature of their theories. The electron pair bond was never the basis or the starting point from which they developed their valence theories. Indeed, this is precisely what differentiated all earlier valence theories from the Lewis valence theory. In his 1916 paper "The Atom and the Molecule," G. N. Lewis assumed that a chemical bond always consisted of two electrons shared between the combining atoms. With this idea he laid the foundation of the modern electron theory of valence.

7. G. N. Lewis and the Shared Electron Pair Bond

Introduction

In 1916 G. N. Lewis published for the first time his theory of the cubic atom. The theory had some rather obvious shortcomings, but out of it evolved a fundamental idea—the shared electron pair bond. In its simplified representation as a pair of dots, the shared electron pair gave a satisfying picture of polar and nonpolar molecules and of the polarity present to some degree in many essentially nonpolar organic molecules. According to Lewis, chemists and physicists no longer needed to invent such hypothetical species as intramolecular ions and electromers. Lewis could never really explain why the two electrons forming the bond remained paired. The great serviceability of the shared electron pair bond provided for its success.

Nearly simultaneously with the appearance of Lewis's paper, the German physicist Walther Kossel published a theory of atomic structure and valence that offered an interpretation of the polar bond similar to the Lewis theory. But Kossel's work had no bearing on the main idea developed in this chapter, the theory of the shared electron pair bond. To complete the historical picture, I have therefore discussed Kossel's paper in an historiographical section near the end of the chapter. A discussion on the establishment of atomic numbers that occurred in this same period also appears in that section.

G. N. Lewis: Early Life

Gilbert Newton Lewis was born in Weymouth, Massachusetts, near Boston, in 1875. At age nine he moved with his family to Lincoln, Nebraska. He received little formal education in either state until 1889, when he enrolled in the University of Nebraska's preparatory school. After attending the University of Nebraska for two years, Lewis transferred to Harvard College in 1893, graduating in 1896. He taught for a year at Phillips Academy at Andover before returning to Harvard for graduate work, receiving an M.A. degree in 1898 and a Ph.D. under Theodore W. Richards in 1899.

After remaining at Harvard for one year as an instructor in chemistry, Lewis went abroad on a traveling fellowship. He spent a semester at Leipzig with Wilhelm Ostwald and another at Göttingen with Walther Nernst, at that time the recognized leader in physical chemistry. For the next three years, 1901–4, Lewis was again an instructor at Harvard, and then served for a year as superintendent of weights and measures in the Philippine Islands and chemist in the Bureau of Science at Manila.

In 1905, Lewis returned to the United States, joining the group of physical chemists that Arthur A. Noyes brought together in the Research Laboratory of Physical Chemistry at the Massachusetts Institute of Technology. Chemists were just beginning to recognize the importance of physical chemistry at this time. Indeed, Noyes's laboratory became the first center for physical chemists established in the United States and the center from which modern physical chemistry spread throughout the entire country.

Lewis remained at M.I.T. for seven years, publishing over thirty papers, among them his epoch-making "Outlines of a New System of Thermodynamic Chemistry" and "The Free Energy of Chemical Substances." They were among the most important in a long series of papers on the experimental determination of free energies and ultimately led in 1923 to the publication of Lewis's great work *Thermodynamics and the Free Energy of Chemical Substances.*[1]

By 1912, Lewis held the rank of full professor. He left M.I.T. that same year to become dean of the College of Chemistry and chairman of the Chemistry Department at the University of California, Berkeley. Berkeley was an institution then rapidly rising under the able leadership of its president, Benjamin Ide Wheeler. Lewis remained at Berkeley until his death in 1946.

During his Harvard period of 1901–4 Lewis originated the cubic atom. This model illustrated the electronic union of polar compounds but could not deal adequately with nonpolar molecules such as the hydrocarbons. His

---

[1]G. N. Lewis and Merle Randall, *Thermodynamics and the Free Energy of Chemical Substances.*

theory of atomic structure and valence was therefore incomplete, and, though Lewis discussed it with his colleagues and in his classes, he withheld it from publication chiefly for this reason.[2]

There was a second reason why Lewis refrained from publishing his theory of structure and valence at that time. Chemistry was still highly descriptive at the turn of the century, and only a few physical chemists tried to give their science a more theoretical base. Many chemists were strongly critical of this movement. Among the American chemists, Edgar Fahs Smith (1856–1928) at the University of Pennsylvania said chemists had only to recognize, not believe in physical chemistry. He barely tolerated Arrhenius's theory of ionization. Louis Kahlenberg at the University of Wisconsin was a vigorous skeptic of the ionic and the electron theories.[3] At Harvard, Richards, though certainly not opposed to the study of physical chemistry, instead concerned himself primarily with establishing accurate atomic weights.

Thus, at the turn of the century chemists were unwilling to give a theoretical idea such as the cubic atom a favorable hearing. Lewis, himself, pointed out this fact in a letter that he wrote several years later to the physicist Robert Millikan at the University of Chicago:

I went from the Middle-west to study at Harvard, believing that at that time it represented the highest scientific ideals. But now I very much doubt whether either the physics or the chemistry department at that time furnished real incentive to research. . . . A few years later [1902] I had very much the same ideas of atomic and molecular structure as I now hold, and I had a much greater desire to expound them, but I could not find a soul sufficiently interested to hear the theory. There was a great deal of research work being done at the university, but as I see it now the spirit of research was dead.[4]

The Theory of the Cubic Atom

Lewis originally conceived of the cubic atom in 1902 while attempting to explain the periodicity of atomic properties, the repetition occurring in

---

[2] There are two other brief notes by Lewis that show the cubical atom: one from Manila, the other from M.I.T. Neither one is dated, but both must be prior to 1913—the year Lewis was at Berkeley (Lewis Papers, Office of the Dean, College of Chemistry, University of California, Berkeley).

[3] Joel H. Hildebrand, "Chemistry, Education, and the University of California," an interview conducted by Edna Tartaul Daniel (Berkeley: Regional Cultural History Project, General Library, University of California, 1962), p. 34; Louis Kahlenberg, *Outlines of Chemistry*, p. 432.

[4] Lewis to Millikan, October 28, 1919, Lewis Papers.

the atoms' chemical behavior that the periodic table conveniently summarized. He wanted also to account for the striking fact that the difference between an element's maximum negative and maximum positive valence, or polar number, was frequently eight and in no case more than eight.[5] The latter observation led Richard Abegg at about the same time (1902–4) to propose his well-known rule of eight or law of normal and contravalence. Indeed, the Lewis cubic atom appeared to be a geometric representation of Abegg's arithmetical rule:

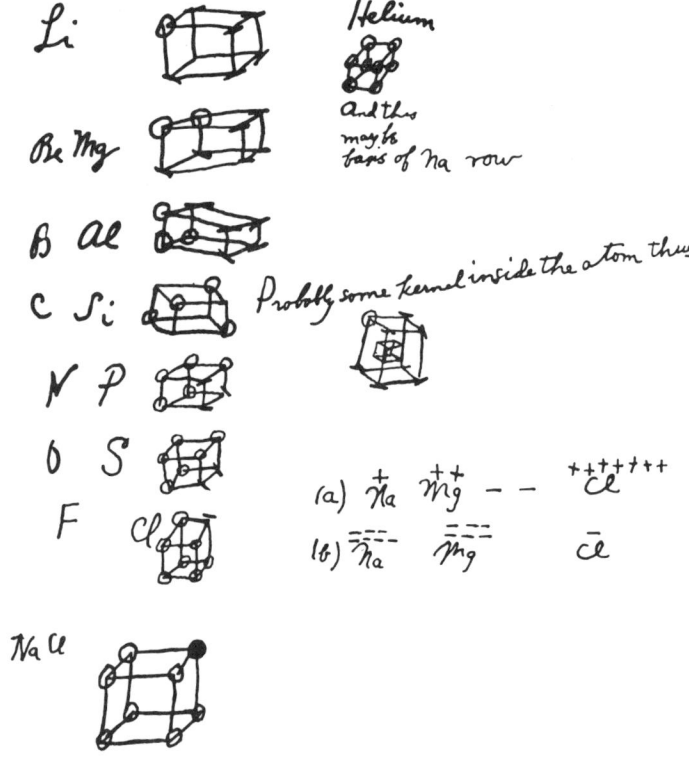

On the origin of the cubic atom and its main features, Lewis wrote:

In the year 1902 (while I was attempting to explain to an elementary class in chemistry some of the ideas involved in the periodic law) becoming interested in the new

[5]G. N. Lewis, *Valence and the Structure of Atoms and Molecules*, p. 29.

theory of the electron, and combining this idea with those which are implied in the periodic classification, I formed an idea of the inner structure of the atom, which, although it contained certain crudities I have ever since regarded as representing essentially the arrangement of electrons in the atom. . . . The main features of this theory of atomic structure are as follows:

(1) The electrons in an atom are arranged in concentric cubes.

(2) A neutral atom of each element contains one more electron than a neutral atom of the element next preceding.

(3) The cube of 8 electrons is reached in the atoms of the rare gases, and this cube becomes in some sense the kernel about which the larger cube of electrons of the next period is built.

(4) The electrons of an outer incomplete cube may be given to another atom, as in Mg^{++}, or enough electrons may be taken from other atoms to complete the cube, as in Cl^-, thus accounting for "positive and negative valence."[6]

For metallic elements, Lewis thus assumed the number of electrons in an atom's outermost cube to equal the atom's positive electrical charge in electrochemical reactions. Since sodium, magnesium, and aluminum had the charges Na^+, Mg^{++} and Al^{+++}, the outermost cube of each contained, respectively, one, two, and three electrons.

Nonmetals, on the other hand, usually had negative charges and presented a more difficult problem. Electrochemical reactions could, in some cases, show how many electrons a nonmetal gained, but this method did not, of course, indicate the number of electrons originally present in the neutral atom's outermost cube. The halogen atoms (Group VII), since they always gained a single electron forming an ion with a charge of -1, probably had seven electrons in each atom's outermost cube and therefore required only one additional electron to complete the cube.

Lewis relied mainly on chemical formulas established from combining weight and volume relations to arrive at the number of electrons in a nonmetal's outermost cube. Carbon and silicon (Group IV), which formed a maximum of four bonds, had four outermost electrons; nitrogen and the Group V elements had five; the oxygen group (Group VI), six; the halogens (Group VII), seven; and the rare gases, including helium, because they formed no compounds, had eight electrons in the outermost cube, or

[6]Ibid., pp. 29–30. Regarding the location of the positive charge that balanced that of the electrons, Lewis described his ideas as "very vague," though he "inclined at the time toward the idea that the positive charge was also made up of discrete particles, the localization of which determined the localization of the electrons." The positive charge and the electron were spatially related to one another, perhaps coupled, at each of the corners in the cubes.

in other words a completed cube.[7] In general, for a nonmetal the number of electrons in its outermost cube was 8 minus the value of its negative valence.

Compound formation occurred, Lewis believed, because every atom, with the exception of the rare gases, attempted to complete its outermost or valence cube by gaining electrons if that cube was already more than half complete, or losing electrons if it was less than half complete. In either case, electron transfer took place until the valence cube was either totally filled or totally empty. In this way Lewis accounted for negative and positive valence.[8] The formation of polar binary compounds such as sodium chloride thus followed simply by assuming that the sodium atom gave up

[7] The cube of eight electrons present in the unreactive gases neon and argon thus represented a stable structure having zero valence. For this reason Lewis originally accepted Mendeleev's idea that hydrogen was the first member of a full period and erroneously assumed that helium contained a single cube of eight electrons. For Mendeleev's observations, see his "Versuche eines systems der Elemente nach ihren Atomgewichten und chemischen Funktionen," *J. prakt. Chemie* (1869); "Essai d'un système des éléments d'après leurs poids atomiques et propriétés chimiques," *Journal of the Russian Physical and Chemical Society* 1 (1869), 60–77; and "The Periodic Law of the Chemical Elements," *J. Chem. Soc.* 55 (1889), 634–56 (Faraday Lecture for 1889). Wilhelm Ostwald published German versions of some of Mendeleev's early works as "Das natürliche System der chemischen Elemente," in *Ostwald's Klassiker der exakten Wissenschaften* 61–68, pp. 18–118. Included in this selection were a translation of Mendeleev's first memoir of 1869, "Die Beziehungen zwischen den Eigenschaften der Elemente und ihren Atomgewichten"; a short German abstract of that first memoir, "Über die Beziehungen der Eigenschaften zu den Atomgewichten der Elemente," which had first appeared in *Zeitschrift für Chemie* 5 (1869), 405–6; and a German translation of his second memoir of 1871, "Die periodische Gesetzmässigkeit der chemischen Elemente," first published in *Annalen der Chemie* Supplementband 8 (1871), 133–229. An English translation of the second (1871) memoir appeared as "The Periodic Law of the Chemical Elements," *Chemical News* 40 (November–December, 1879), and 41 (January–March, 1890). Lothar Meyer's article "Die Natur der chemischen Elemente als Funktion ihrer Atomgewichte" appeared shortly after Mendeleev's first memoir, in *Annalen der Chemie* Supplementband 7 (1870), 354–64, and also in *Ostwald's Klassiker* 61–68, pp. 9–17.

[8] Actually, the first publication that pointed out the positive-negative valence relation and the stability of the group of eight electrons was Richard Abegg's "Versuch einer Theorie der Valenz und der Molekülarverbindungen," *Christiania Videnskabs Selskabets Skrifter* 12 (1902) (abstract in *J. Chem. Soc.* 84 (1903), Part II, 536), in which he introduced his theory of normal and contravalences. Abegg greatly developed the same idea in a 1904 publication: "The sum of 8 of our normal and contravalences possesses therefore simple significance as the number which for all atoms represents the points of attack of electrons, and the group number or positive valence indicates how many of the 8 points of attack must hold electrons in order to make the element electrically neutral" ("Die Valenz und das periodische System: Versuch einer Theorie der Molekülarverbindungen," *Zeit. anorg. Chemie* 39 [1904], 379–80).

the single electron present in its outermost cube to the chlorine atom. This atom, which already had seven electrons in its outermost cube, could acquire one and only one more electron:

Despite the cubic atom's success in accounting for the formulas of simple polar compounds, essentially those of inorganic chemistry, Lewis recognized a serious limitation of his model. It left unexplained the vast number of nonpolar or organic compounds that clearly did not consist of ions and hence required a different interpretation of chemical union.

Numerous papers dealing with the atom's electron structure and the electron theory of valence appeared within the next ten years. Lewis published several papers on electrochemistry and thermodynamics during this period but nothing on the atom and valence. Indeed, in his first publication on valence, his 1913 paper "Valence and Tautomerism," Lewis did not mention the cubic atom, choosing instead to discuss the difference in chemical behavior between polar and nonpolar compounds. Falk and Nelson, he said, blurred the difference with their theory of polar valences, and even his Berkeley colleagues, Bray and Branch, in their 1913 paper on valence, failed to distinguish clearly between the two. Polar compounds, according to Lewis, were reactive, ionized, and had high dielectric constants; nonpolar structures were unreactive, not ionized, and had low dielectric constants.

Lewis discussed the electrostatic force that resulted from electron transfer and united the atoms in simple polar compounds. But his 1913 paper gave the electron no role in the bonding of nonpolar compounds. "We need only assume," he wrote, "that upon each atom there are definit regions, or points, at which direct connection to similar points on other atoms may be made, and that the number of occupied regions on a given atom is the valence number of that atom. We thus visualize the structure in which each atom is tied directly to one or more atoms at one or more points. One such point or region can be attached to one and only one point of another atom or atoms." Lewis maintained in 1913 that there were two distinct types of chemical bonds, polar and nonpolar. They differed not merely in degree but in kind.[9]

By 1916, however, Lewis had abandoned his earlier position. In a paper entitled "The Atom and the Molecule," he introduced the revolutionary idea that polar and nonpolar bonds were essentially the same, a pair of electrons shared by the two bonded atoms.[10] If the two bonded atoms shared the pair of electrons more or less equally, the resulting bond was nonpolar; but if one atom had a stronger attraction for the electron pair than the other and displaced the pair, the bond was polar. In extreme cases, one atom might actually transfer an electron to the other, forming a purely polar compound.

We turn now to a detailed analysis of Lewis's 1916 paper, which not only contains some of the rich consequences of his new valence theory but describes further the development of his own work prior to that date.

The Lewis Theory of Atomic and Molecular Structure in 1916

In the historical part of his 1916 paper, Lewis published for the first time his theory of the cubic atom. By including it Lewis showed how he arrived at a theory of chemical combination that held for both polar and nonpolar compounds. Indeed, in no other way could he make clear that the shared electron pair bond originated in the sharing of electrons by the incomplete shells of cubic atoms. Lewis also introduced at this time the electron dot notation, with which he conveniently symbolized the shared electron pair bond without having to illustrate the actual sharing of rather cumbersome cubes.

Thus when Lewis abandoned the cubic atom in this same 1916 paper (for reasons discussed below) it was only after he had justified sufficiently its introduction with lengthy exposition: the cubic atom was the theoretical model out of which evolved the shared electron pair bond and ultimately the Lewis dot notation. Though the cubic atom disappeared almost as quickly as it had appeared, the shared electron pair bond and the dot notation became a permanent part of modern valence theory. In the remainder of his paper, Lewis illustrated how they enabled him to unravel such long-

[9]G. N. Lewis, "Valence and Tautomerism," *J.A.C.S.* 35 (1913), 1448, 1451. Like his colleagues Bray and Branch, Lewis used *definit* instead of *definite*.

[10]G. N. Lewis, "The Atom and the Molecule," *J.A.C.S.* 38 (1916), 762–85.

standing problems as the electron structures of ammonium chloride and the oxyacids, and the different strengths of substituted organic acids.

The Lewis cubic atom in 1916 appears below for the atoms lithium through fluorine; they represent the elements in Groups I–VII.[11] As in his cubic atom of 1902, each circle indicated a valence electron located at a corner of the cube:

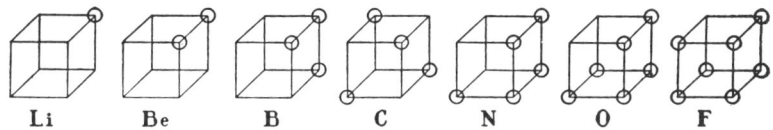

Li   Be   B   C   N   O   F

The main postulates of his theory in 1916 were:

(1) In every atom is an essential *kernel* which remains unaltered in all ordinary chemical changes and which possesses an excess of positive charges corresponding in number to the ordinal number of the group in the periodic table to which the element belongs.

(2) The atom is composed of the kernel and an *outer atom or shell*, which in the case of the neutral atom, contains negative electrons equal in number to the excess of positive charges of the kernel, but the number of electrons in the shell may vary during chemical change between 0 and 8.

(3) The atom tends to hold an even number of electrons in the shell and especially to hold eight electrons which are normally arranged symmetrically at the eight corners of a cube.

(4) Two atomic shells are mutually interpenetrable.

(5) Electrons may ordinarily pass with readiness from one position in the outer shell to another. Nevertheless, they are held in position by more or less rigid constraints, and these positions and the magnitude of the constraints are determined by the nature of the atoms and of such other atoms as are combined with it.

(6) Electric forces between particles which are very close together do not obey the simple law of inverse squares which holds at greater distances.[12]

The first two postulates show that Lewis considered the atom to consist of two parts: a positively charged kernel and an outer shell containing the valence electrons. Lithium and the other Group I elements, including hydrogen, each had a kernel with a single positive charge and one valence electron in the outer shell. At the other extreme of the periodic table, each

[11] Ibid., p. 767.
[12] Ibid., p. 768.

Group VI element had a kernel with a charge of $+6$ and six valence electrons; a Group VII element's kernel charge was $+7$ with seven valence electrons. The same relation between kernel and outer shell electrons held for the other groups in the table but not for the transition elements. In these elements the atomic kernel probably varied during chemical change.[13]

Lewis maintained that the rare gases (Group 0), except for helium, had a kernel with a $+8$ charge and a completed outer cube of eight electrons.[14] These gases were extremely inert, forming no compounds, and for this reason Lewis considered the outermost cube to be part of the unreactive kernel.

To illustrate the correctness of his third postulate, Lewis pointed out that when atoms entered into combination, "among the tens of thousands of known compounds" only a few of them did not have an even number of electrons (E) in their valence shells. In every compound in which each element used either its highest or its lowest valence or polar number, E was a multiple of eight.[15] Using Lewis's notation,

$$NH_3 = NH_3E_8 \qquad MgCl_2 = MgCl_2E_{16} \qquad NaNO_3 = NaNO_3E_{24}$$
$$H_2O = H_2OE_8 \qquad\qquad\qquad\qquad K_2CO_3 = K_2CO_3E_{24}$$
$$KOH = KOHE_8 \qquad\qquad\qquad\qquad Al(OH)_3 = Al(OH)_3E_{24}\,.$$

For compounds whose elements had polar numbers intermediate between the highest and the lowest:

$$SO_2 = SO_2E_{18} \qquad\qquad C_2H_2 = C_2H_2E_{10}$$
$$NaClO = NaClOE_{14} \qquad C_6H_5OH = C_6H_5OHE_{36}\,.$$

The following, among them the Group I and VII elements when in the free state, were exceptions to the above rules. But each was highly reactive and tended to form a structure with an even number of valence electrons:

$$Na = NaE \qquad NO = NOE_{11} \qquad ClO_2 = ClO_2E_{19}$$
$$I = IE_7 \qquad NO_2 = NO_2E_{17} \qquad (C_6H_5)_3C = (C_6H_5)_3CE_{91}\,.$$

According to Lewis, NO_2E_{17}, when pure and in the liquid state, conducted an electrical current that suggested the presence of the ions $NO_2^+ =$

[13] Ibid., p. 769.

[14] The reason Lewis gave for changing his earlier opinion regarding helium appears later in this chapter in the section "The Introduction of Electron Dot or Lewis Formulas."

[15] Ramsay had adopted somewhat similar formulas in his 1908 paper "The Electron as an Element," *J. Chem. Soc.* 93 (1908), 774–88.

NO_2E_{16} and $NO_2^- = NO_2E_{18}$. Similarly, ClO_2E_{19} apparently dissolved in solution, giving the ions $ClO_2^-E_{20}$ and $ClO_3^-E_{26}$. Lewis called such molecules "odd molecules," and to account for their reactivity he assumed that only weak electrical constraints (forces) held an odd electron in the molecule. Placing it in a medium that weakened further the constraints, namely a polar medium, completely removed the odd electron.[16]

In postulate four—that two atomic shells were mutually interpenetrable—Lewis gave the mechanism for electron pair sharing. Because of shell interpenetrability, an electron from one atom could form part of another atom's shell but not belong exclusively to either atom. Hence, neither atom had lost or gained the electron. Only in purely polar molecules was the transfer complete and the distinction applicable. Further discussion of electron pair sharing follows throughout this chapter.

Lewis's fifth postulate resulted from his conviction that isomers of the intra-atomic type did not exist. He argued that "if the electrons of the atomic shell could at one time occupy one set of positions and at another time another set, and if there were no opportunity for ready transition from one of these sets of positions to another, we should have a large number of isomers differing from one another only in the situation of electrons in the atomic shell."[17] Because chemists had not demonstrated the existence of these isomers, Lewis believed that the electrons had considerable freedom to change from one arrangement in the shell to another.

Finally, by abandoning Coulomb's law for particles at small distances, as he did in his sixth postulate, Lewis allowed two negatively charged electrons to occupy adjacent positions in an atom. Each electron, he said, behaved as a magnet, and the attraction of these magnetic electrons or magnetons when properly oriented accounted for the stability of paired electrons.

Some Illustrations of Electon Pair Sharing

In his 1916 paper, Lewis showed how the cubic atom led to a theory of chemical combination that accounted for both polar and nonpolar molecules. Despite the great difference between typical members of the two classes, he said, chemical molecules passed "from the extreme polar to the

---

[16] Lewis, "The Atom and the Molecule," p. 771.
[17] Ibid., p. 772.

extreme nonpolar form, not *per saltum*, but by imperceptible gradations."
Chemists only had to recognize that an electron could become the common
property of two atomic shells.[18]

Lewis had illustrated in 1902 that a polar molecule resulted whenever
an atom with a few electrons in its outermost or valence shell lost these
electrons to an atom that already had several valence electrons, and tended
to increase this number to eight. The electrostatic attraction of the oppositely charged ions produced the polar bond. Sodium and calcium,
whose neutral atoms contained one and two valence electrons, respectively, lost electrons and formed positive ions, Na^+ and Ca^{+2}, with an
empty valence shell. The ions of chlorine, sulfur, and nitrogen, Cl^-, S^{-2},
and N^{-3}, were negative; each had eight electrons at the corners of its outermost cube or completed valence shell. The formation of sodium chloride,
an obviously polar molecule, appears below:

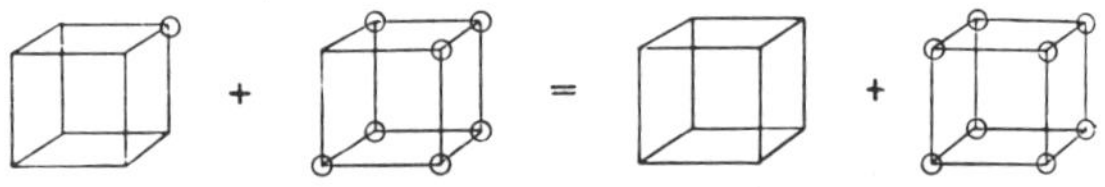

To account for nonpolar molecules Lewis applied his fourth postulate,
the interpenetrability of atomic shells.[19] With this one fundamental idea, he
showed how an atom in a nonpolar molecule acquired an outermost shell or
group of eight electrons by sharing one or more pairs of electrons with another atom or atoms. Each shared electron pair, according to Lewis, was a
chemical bond and replaced the electrostatic or polar bond in nonpolar
molecules. Thus the single bond in the nonpolar diatomic halogen molecules resulted from the atoms' sharing a pair of electrons along an edge of
each cube:

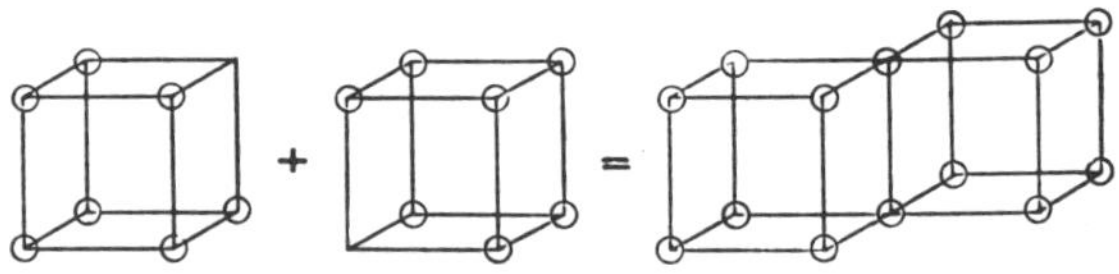

The double bond followed whenever two atoms shared two electron
pairs, or a face of each cube as in the oxygen molecule:

[18] Ibid., p. 775.
[19] Ibid.

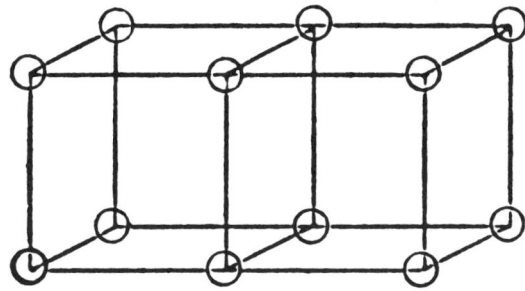

The nonpolar structures given above for the halogen and oxygen molecules were those idealized for single- and double-bonded molecules. Indeed, Lewis believed that each represented only one of numerous possible electronic structures, which he called tautomers and which resembled the electromers of the earlier polar theory and to some extent the resonance forms of later electron valence theories. Thus, in a binary compound, such as *AB*, one of the atoms might hold the electron pair more closely than the other and the asymmetry would result in an electrically polar molecule. The molecule's chemical environment might produce an even greater separation of the charged atoms, increasing its polarity until it completely ionized. Between the perfectly symmetrical and nonpolar tautomer and the completely polar or ionized structure there was, Lewis said, "an infinity of positions representing a greater or lesser degree of polarity." [20]

Because of the double bond's pronounced tendency to break, Lewis maintained that a molecule with a double bond also existed as a single-bonded tautomer with an unpaired electron in the shell of each bonded atom.[21] In assuming this structure, it would thus exhibit all the properties of an odd molecule (a molecule with one or more unpaired electrons), such as paramagnetism and color. Lewis cited the behavior of the double-bonded diatomic oxygen molecule. At low temperatures most reactions of oxygen first produced a peroxide, which showed, he said, that the oxygen molecule behaved to some extent as an odd molecule having an unpaired electron on each atom. It could add directly to other atoms and was similar to ethylene, which formed addition compounds. Molecules with double bonds, like single-bonded structures, existed in numerous tautomeric forms

[20] Ibid., p. 776.

[21] The double bond could break in a highly polar environment and both electrons go to one of the atoms, giving oppositely charged atoms within the molecule.

ranging from the double-bonded structure to the paramagnetic single-bonded structure.[22]

Fundamental Shortcomings of the Cubic Atom

The cubic atom, while able to illustrate the union of atoms by single and double bonds, offered no structural model for the triple bond present in numerous organic compounds such as acetylene, H—C≡C—H, and in the nitrogen molecule, N≡N. Lewis had represented carbon and nitrogen, atoms with four and five electrons in their valence shells, as cubes lacking four and three electrons, respectively, but there was no way for either cube to share three electron pairs with another carbon or nitrogen atom.

The cubic arrangement of valence electrons also led to a rigid single bond, because two cubic atoms could not possibly rotate freely about their common edge. A rigid single bond produced structural isomers of substituted ethane,

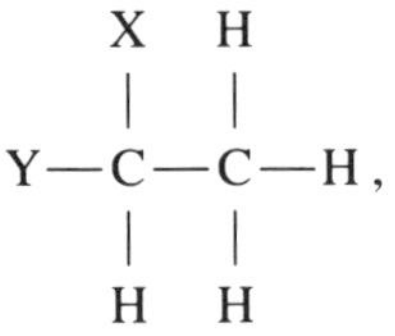

$$
\begin{array}{cc}
X & H \\
| & | \\
Y-C-C-H, \\
| & | \\
H & H
\end{array}
$$

just as a rigid double bond accounted for the isomers of a substituted ethylene:

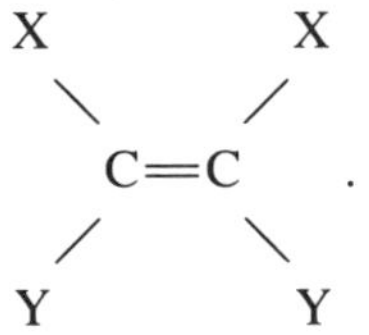

$$
\begin{array}{ccc}
X & & X \\
\diagdown & & \diagup \\
& C=C & \\
\diagup & & \diagdown \\
Y & & Y
\end{array}
$$

Yet, chemists had established the free rotation of single-bonded carbon atoms in the late nineteenth century.

Thus there was, Lewis wrote, "a need to assume a somewhat different arrangement of the group of eight electrons, at least in the case of the nonpolar substances whose molecules are as a rule composed of atoms of

[22] Lewis, "The Atom and the Molecule," pp. 778–79.

small atomic volume." [23] This new arrangement, the tetrahedron, emphasized the shared electron pair. A magnetic force or other force acting strongly at small distances drew together the electrons that occupied the corners of the cube. This force caused the electrons to arrange themselves in pairs at the four corners of a tetrahedron:

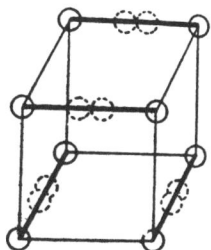

Lewis's new tetrahedral atom brought his hydrocarbon structures in agreement with those of organic chemists. Two tetrahedra, attached by one or by two corners of each, represented, respectively, the single and the double bond. In the former, two atoms held one pair of electrons in common, in the latter, two atoms held two pairs. Two tetrahedra joined at three corners or on one face represented a triple bond. They shared three electron pairs—a sharing that the cubic atom could not indicate.

The Introduction of Electron Dot or Lewis Formulas

In order to express conveniently and clearly his theory of the shared electron pair bond, Lewis in his 1916 paper introduced the now well-known electron dot or Lewis formulas. The use of dots to represent bonds in chemical formulas was not an entirely new idea, however. Organic chemists, long before the discovery of the electron and its application to valence theory, had represented single, double, and triple bonds with one, two, and three dots, respectively. In the Lewis theory the important innovation was that each dot no longer represented a valence bond of unknown nature but an electron, and each pair of dots or colon symbolized the electron pair constituting the chemical bond. [24]

[23] Ibid., p. 780.
[24] Ibid., pp. 776–79.

For the single bond in the nonpolar chlorine molecule Lewis thus wrote Cl : Cl. To indicate polarity in a molecule he placed the pair of dots closer to the more negative element, for example, the chlorine atom in iodine monochloride, I : Cl. The dot notation made it possible, though certainly not practical, to illustrate some of the numerous electronic tautomers for a given formula by placing the electron pair at different positions between the bonded atoms.

The evidence from radioactive phenomena and from Moseley's X-ray spectral analysis led by 1916 to general agreement among physicists and chemists that helium contained two electrons and was the only element between hydrogen and lithium. Helium was also inert and resembled the members of the rare gas or Group 0 family. According to Lewis, for this element the electron pair played the same role as the group of eight in the heavier elements. Indeed, Lewis believed that in row one of the periodic table, consisting of hydrogen and helium, the "rule of two" replaced the "rule of eight."

Thus, hydrogen with a single electron in its valence shell not only gave up this electron and like lithium or sodium yielded a +1 ion, it gained in other reactions a second electron, forming a stable pair just as fluorine or chlorine acquired an electron to attain the stable group of eight.[25] Lewis structures for hydrogen's polar compounds, the nonpolar hydrogen molecule, and a few other single-bonded molecules appear below:

$$H \; :Cl \qquad\qquad H \; : \; H \qquad\qquad Na \; :H$$

$$H:\overset{..}{\underset{..}{O}}:H \qquad\qquad H: \overset{..}{\underset{..}{I}}: \qquad\qquad :\overset{..}{I}:\overset{..}{\underset{..}{I}}:$$

The second row of structures, unlike the first row, shows all the valence electrons belonging to each atom, including those not involved in bond formation. Lewis placed them in pairs, though he had not yet discussed in his paper such an arrangement of nonbonding valence electrons.

Each double bond in a Lewis structure required two electron pairs (two pairs of dots):

$$:\overset{..}{O}::\overset{..}{O}: \qquad\qquad H: \overset{H}{\overset{..}{C}} :: \overset{H}{\overset{..}{C}} :H$$

But Lewis believed that molecules with double bonds also existed in a second tautomeric form:

[25] Ibid., p. 774.

$$: \overset{..}{O} : \overset{..}{O} :$$

$$\underset{\overset{.}{H} : C : C : H}{\overset{..}{H} \quad \overset{..}{H}}$$

Changes in the position of each unpaired electron once again gave rise to numerous forms between the single- and double-bonded molecules.[26]

Lewis assumed that triple-bonded structures were tautomeric. Either one or two of the electron pairs making up the triple bond could break, leaving a molecule with an unpaired electron on each of the bonded atoms, or one in which each bonded atom had a pair of unshared electrons. Using the dot notation, Lewis's tautomers were:

$$H : C ::: C : H , \qquad H : \overset{.}{C} :: \overset{.}{C} : H , \qquad H : \overset{.}{C} : \overset{.}{C} : H$$

It is clear from the examples given that the Lewis bond, though requiring a pair of electrons held in common by two atoms, corresponded to the valence bond used in graphic or structural formulas. Replacing the dash representing the single bond by a pair of electrons sufficed to convert a structural formula into a Lewis formula. But Lewis intended his formulas to be more than an alternative means of expressing structures. Since each dot indicated one valence electron, Lewis's formulas completely described the valence shell of each atom in a molecule. They also solved a problem that had proved extremely embarrassing to a number of valence theories, namely, establishing the structure of ammonia, NH_3, and of ammonium chloride, NH_4Cl.

Due to the polar character of ammonia and the hydrogen ion, chemists had sometimes regarded the ammonium ion as a loose complex resulting from the electrical attraction of the two polar substances. Yet, substituting methyl, ethyl, or other organic radicals for the hydrogen atoms left no doubt that four groups were attached directly to the nitrogen atom and held firmly enough to permit the isolation of numerous stereochemical isomers. Victor Meyer's studies in 1876 on the mixed methylethyl ammonium derivatives showed that the same compound resulted upon reacting dimethylamine with ethyl iodide or diethylamine with methyl iodide.[27] His findings clearly demonstrated the equivalence of the four nitrogen valences that held the alkyl groups.

On the other hand, Abegg, Friend, and Arrhenius believed that the ni-

---

[26] Ibid., p. 779.

[27] Victor Meyer and Marco T. Lecco, "Untersuchungen über die Konstitution der Ammoniumverbindungen und des Salmiaks," *Annalen der Chemie* 180 (1876), 173–91.

trogen atom, which had a polar valence of -3 in ammonia, formed five bonds in ammonium chloride, the result of using its oppositely charged contravalences or residual valences:

$$
\begin{array}{l}
\text{H} + - \\
\text{H} + - \quad \mathbf{N} \quad - + \text{H} \\
\text{H} + - \quad \quad + - \text{Cl}
\end{array}
$$

Their contravalences or residual valences thus accounted for the widely held view that nitrogen was pentavalent in ammonium chloride, and their hypothesis received experimental support from the fact that perfectly dry ammonium chloride vaporized without dissociating into ammonia and hydrogen chloride gas.[28]

A third alternative was Werner's assumption of a tetravalent nitrogen atom in the ammonium ion:

$$\text{H}_3\text{N} \ldots \text{H.Cl}$$

Friend criticized Werner's structure, claiming that Werner used a divalent hydrogen atom and treated three of the hydrogen atoms in ammonium chloride differently from the fourth.[29] But Werner's belief in a tetravalent nitrogen atom ultimately became a part of the presently accepted ammonium chloride structure that Lewis first proposed in his 1916 paper.

In the Lewis structure, the nitrogen atom used three of its five valence electrons to form single bonds with three of the hydrogen atoms. Each hydrogen atom contributed the second electron to the electron pair bond.[30] In forming the fourth bond, the nitrogen atom supplied both electrons but now to a positively charged hydrogen ion rather than a neutral hydrogen atom. Nevil V. Sidgwick named the fourth nitrogen-hydrogen bond a coordinate covalent or dative bond that, despite the different mechanism of formation, was identical to the usual two-electron bond:

$$
\begin{array}{cc}
\quad\text{H} & \quad\text{H} \\
\text{H:N:} + \text{H}^+ = & \text{H:N:H}^+ \\
\quad\text{H} & \quad\text{H}
\end{array}
$$

The ammonium ion carried a net charge of $+1$ because four electron pairs or eight negative charges held together the nitrogen atom with a $+5$ kernel

---

[28] Herbert B. Baker, "Influence of Moisture on Chemical Change," *J. Chem. Soc.* 65 (1894), 612; Frederick M. G. Johnson, "Der Dampfdruck von trocknem Salmiak," *Zeit. phys. Chemie* 61 (1908), 457–63.

[29] John Newton Friend, *The Theory of Valency*, p. 120.

[30] Lewis, "The Atom and the Molecule," p. 778.

charge and the four hydrogen atoms, each with a $+1$ kernel charge, or a total kernel charge of $+9$. The positive charge belonged not to any specific hydrogen atom but to the ammonium group as a whole. Thus, in ammonium chloride there was no direct attachment of the chloride ion to nitrogen. An electrostatic force held the two ions together:

$$\overset{\displaystyle H}{\underset{\displaystyle H}{H:\overset{..}{N}:H^+}} + :\overset{..}{\underset{..}{Cl}}:^-$$

Lewis's electron dot structures corrected a second assumption generally accepted in 1916—that a divalent atom always used two bonds to combine with another atom or remained unsaturated. Lewis showed how the divalent oxygen atom in many of its compounds shared only one pair of electrons with another atom and yet the compound was perfectly saturated. In the oxyacid structure

$$\overset{\displaystyle :\overset{..}{O}:}{\underset{\displaystyle :\overset{..}{O}:}{:\overset{..}{O}:X:\overset{..}{O}:}}$$

every atom had a completed shell of eight electrons. If X was silicon, phosphorus, sulfur, or chlorine, the above structure represented all of the following ions: silicate, SiO_4^{-4}; orthophosphate, PO_4^{-3}; sulfate, SO_4^{-2}; and perchlorate, ClO_4^-. Indeed, the success of his electronic notation when applied to the structure of the oxyacids and ammonium chloride led Lewis to write: "While the two dots of our formulae correspond to the line which has been used to represent the single bond, we are led through their use to certain formulae of great significance which I presume would not occur to anyone using the ordinary symbols."[31]

The Use of a Magnetic Force to Account for the Attraction between Electrons

Because Lewis had arranged the electrons in an atom or a molecule in pairs, he clearly required some kind of additional force to overcome the intrinsic repulsion of their negative charges. As a way out of this dilemma, Lewis assumed in the final postulate of his cubic theory that Coulomb's law did not hold for electric particles at small distances. He suggested,

[31] Ibid.

though only very briefly in his 1916 paper, that perhaps a magnetic force or some other force acting at small distances caused the electron attraction.[32]

One year earlier, William Ramsay and Alfred L. Parson (b. 1889) had published theories of atomic and molecular structure in which they assumed that a magnetic force produced the electronic attractions in atoms and molecules. They believed that the magnetic force resulted from the rotation of an atom's electrons and that the attraction between the rotating valence electrons of different atoms led to a chemical bond between the atoms. Lewis acknowledged Parson's work in his paper on valence but did not mention Ramsay's publication.

Ramsay had discussed the electron's magnetic attraction in a series of lectures he delivered in 1915 at the opening of the Rice Institute in Houston. Physicists had shown, he said, that two parallel conductors each carrying a stream of electrons in the same direction attracted one another, but the conductors repelled one another when the electron streams moved in opposite directions.[33] What applied to a large number of electrons probably applied to single electrons and to rotary (circular) as well as linear motion. Indeed, according to Ramsay, the contributions of Zeeman and Lorentz clearly established the permanent rotation of electrons in atoms and molecules. The rotating electrons, he believed, were really the tubes of force that Thomson and others had introduced to account for the chemical bond.

Ramsay's valence electrons rotated within a spherically shaped atom, not around a circumference but in a small circle parallel to a fixed equatorial plane. In metals and the hydrogen atom they rotated in a clockwise direction, but in nonmetals the motion was counterclockwise. Thus, in the sodium-chlorine and hydrogen-chlorine reaction, magnetic attraction occurred between a valence electron on each atom, and combination took place when the electron planes of rotation in the two atoms were parallel.[34] Ramsay said nothing about the bond in the elementary diatomic molecules such as chlorine or oxygen. Clearly, their formation followed only if he assumed that the electron rotations were clockwise in half of the atoms but counterclockwise in the others:

[32] Ibid., p. 780.

[33] William Ramsay, "Compounds of Electrons," *Rice Institute Pamphlet* 1 (July 1915), 422–24. Ramsay's 1915 paper had the directions reversed, but this was corrected in his 1916 paper, "A Hypothesis of Molecular Configuration in Three Dimensions of Space," *Proc. Roy. Soc.* A 92 (1916), 451–62.

[34] Ramsay, "Hypothesis of Molecular Configuration," p. 453; Ramsay, "Compounds of Electrons," pp. 422–24.

Parson called his magnetic theory of atomic and molecular structure the magneton theory.[35] His magneton was actually an electron, not a tiny spherical electron, but a unit negative charge distributed in a very thin ring of radius about 1.5×10^{-9} cm. It did not orbit within the atom, as in Ramsay's theory, but rotated on its axis with nearly the velocity of light. Parson's magneton thus combined the electrostatic properties of the electron with the magnetic properties of a ring carrying a current.

The magnetic force between magnetons, he said, led to their arrangement in cubes lying side by side (not concentrically) within a sphere of positive electricity. This arrangement, which clearly resembled Thomson's positive sphere model of the atom, was lowest in magnetic energy. The magnetons' cubic arrangement, he pointed out, obviously agreed with the elements' distribution in the periodic table. Indeed, rotating rings of electrons, like those in physicist Niels Bohr's structural atom, could not "possibly harbor any essential peculiarity that could explain the definite system of 'octaves' which is the predominating feature of the Periodic Scheme." [36]

In Parson's magneton theory, polar bond formation differed from the electronic interpretation only in that a magnetic force brought about the transfer of a magneton and hence the attraction of the oppositely charged atoms.[37] But Parson also believed that very often the magnetic attraction was not great enough to draw a magneton from one atom to another. This low attraction level resulted in magneton sharing between the atoms and the formation of a nonpolar bond.

Actually, Parson thought that most molecules, for example HF or NH_3, were mixtures of polar and nonpolar structures with one or the other structure predominating. In both structures the more negative atom tried to complete a group of eight whenever possible, either by outright magneton transfer or by sharing one or more pairs of magnetons with the more positive atom.[38]

The diatomic halogen, oxygen, and nitrogen molecules also consisted

[35] Alfred L. Parson, "A Magneton Theory of the Atom," *Smithsonian Miscellaneous Publication* 65 (1915), 1–80.

[36] Ibid., pp. 12–13.

[37] Ibid., p. 29.

[38] Ibid., pp. 29–33.

of polar and nonpolar structures. But in each nonpolar structure the two atoms shared all of their valence electrons, giving what Parson called a "condensed" group of fourteen for the chlorine molecule (seven from each chlorine atom), a group of twelve for the oxygen molecule (six from each oxygen atom), and a group of ten for the nitrogen molecule (five from each nitrogen atom). Only for the diatomic hydrogen molecule did he represent the nonpolar bond by a single pair of electrons shared between the two hydrogen atoms and eliminate the accompanying polar form.

Parson's magneton theory received the attention of physicists David L. Webster (b. 1888) and Arthur Holly Compton (1892–1962) in the United States and chemist Herbert Stanley Allen (1873–1945) in Scotland.[39] Lewis was also very much aware of Parson's work. Though Parson developed many of his ideas at Oxford, he spent some time at Harvard (1913–14) with Webster and then at Berkeley in Lewis's laboratory (1914–15), where he completed his studies on the magneton theory.

Lewis had therefore known of Parson's theory for some time before he published his own paper on valence. He found Parson's ideas extremely interesting and acknowledged that in several important points the two theories appeared to coincide. Parson considered the completed cube to be the most stable arrangement of the atomic shell. He represented the chemical bond in the hydrogen molecule by a shared pair of magnetons, and he arranged the magnetons in the valence shell of the simpler atoms in pairs.[40]

On the other hand, Parson never really showed how to relate the theory of cubic shells to an atomic model with unshared or free pairs of magnetons in the valence shell. It remained for Lewis to illustrate the relation upon converting the cube into a tetrahedron having the electrons or magnetons situated in pairs at the tetrahedron's four corners. Lewis introduced the new tetrahedral arrangement in 1916 and continued to use it in later

[39] David L. Webster, "Parson's Magneton Theory of Atomic Structure," *Physical Review* 6 (1915), 54; idem, "The Theory of Electromagnetic Mass of the Parson Magneton and Other Non-Spherical Systems," *Physical Review* 9 (1917), 484–99; idem, "The Scattering of α-Particles as Evidence of the Parson Magneton Hypothesis," *J.A.C.S.* 40 (1918), 375–79; Arthur Holly Compton and Oswald Rognley, "Is the Atom the Ultimate Magnetic Particle?" *Physical Review* 16 (1920), 464–76; Herbert Stanley Allen, "The Case for a Ring Electron," *Chemical News* 118 (21 March 1919), 137–39, and (28 March 1919), 149–51. See also Lars O. Grondahl, "Experimental Evidence for the Parson Magneton," *Physical Review* 10 (1917), 586–89.

[40] Lewis, "The Atom and the Molecule," p. 774; Parson, "A Magneton Theory of the Atom," pp. 30, 31, 33.

publications. He discussed in greater detail the magnetic theory of electron attraction in his 1923 text *Valence and the Structure of Atoms and Molecules* and in the paper "The Magnetochemical Theory," published the following year.[41]

Contemporary Developments in Atomic Structure and Valence: The Introduction of the Nuclear Atom and Atomic Numbers and Walther Kossel's Electron Theory of Valence

In 1902 Lewis had deduced correctly from purely chemical evidence the number of valence electrons in many of the atoms; but not until 1914, with the establishment of atomic numbers from both physical and chemical evidence, had physicists resolved the total number of electrons in different atoms. Lewis, therefore, could construct electron shell structures in agreement with the known atomic numbers in his 1916 paper. In 1902 he thought the helium atom contained a single completed cube of eight electrons and eight neutralizing positive charges. But by 1916, demonstrations of the radioactive decay series and identification of the alpha particle as a positively charged helium atom, He^{+2}, with a relative mass of four units strongly suggested that the helium atom had two rather than eight electrons. Indeed, this evidence clearly agreed with the conclusions of Rutherford's alpha and beta scattering experiment of 1911.[42]

Rutherford's scattering experiment showed that the total positive charge in an atom occupied a very small and dense volume of radius $10^{-12} - 10^{-13}$ centimeters at the atom's center, called its nucleus, and not a surface of radius 10^{-8} centimeters as in Thomson's positive sphere model. This startling discovery resulted, of course, in Rutherford's theory of the nuclear atom. But his experiment also demonstrated that the number of electrons surrounding an atom's nucleus was nearly equal to one-half the

[41] G. N. Lewis, "Color and Chemical Constitution," *Chemical and Metallurgical Engineering* 24 (1921), 869–75 (paper delivered by Lewis upon receiving the Nichols Medal for the best original paper published in any of the journals of the American Chemical Society for the year 1921); idem, *Valence*, pp. 57–59, 81, 82, 147–48; idem, "The Magnetochemical Theory," *Chemical Reviews* 1 (1924), 231–48.

[42] William Ramsay and Frederick Soddy, "Experiments in Radioactivity and the Production of Helium from Radium," *Proc. Roy. Soc.* 72 (1903), 204–7; Ernest Rutherford and Thomas Royds, "The Nature of the α-Particle from Radioactive Substances," *Phil. Mag.* 17 (1909), 281–86; Rutherford, "The Scattering of α and β Particles by Matter and the Structure of the Atom," *Phil. Mag.* 21 (1911), 669–88.

atom's weight. Since Rutherford had already proved that the alpha particle was really an ionized helium atom with a +2 charge and a mass of four, helium obviously contained only two electrons.

As a direct consequence of Rutherford's work, the Dutch physicist Antonius van den Broek (1870–1926) in 1911 proposed that the integral number with which physicists represented the positive charge on an atom's nucleus was the same as the number (the ordinal number) chemists used to indicate the sequence of elements in the periodic table. He called this number the atomic number and defined it as either the number of positive charges in an atom's nucleus or the number of electrons surrounding the nucleus of a neutral atom.[43]

Van den Broek's speculations were in agreement with results obtained from the study of radioactive disintegrations. According to the laws of disintegration proposed in 1913 chiefly by Frederick Soddy in Glasgow and Kasimir Fajans in Karlsruhe, an atom undergoing alpha (α) decay lost two positive charges and four units of mass.[44] It agreed in chemical behavior with the atoms in the group two places to its left in the periodic table. Radium in Group II changed to radon in Group 0, which in turn decayed to polonium in Group VI.

This transformation was as follows (the lower digits indicate the atomic number; the upper digits, the mass number of each atom):

$$^{226}_{88}\text{Ra} \xrightarrow{-\alpha} {}^{222}_{86}\text{Rn} \xrightarrow{-\alpha} {}^{218}_{84}\text{Po} , \qquad \text{where } \alpha = {}^{4}_{2}\text{He}$$

Group II   Group 0   Group VI

When an atom underwent beta (β^-) decay, it lost a negative charge— the equivalent of gaining a positive one—and was then identical in behav-

[43] Antonius van den Broek, "The Number of Possible Elements and Mendeleev's 'Cubic' Periodic System," *Nature* 87 (20 July 1911), 78. See also idem, "Die Radioelemente das periodische System und die Konstitution der Atome," *Phys. Zeit.* 14 (1913), 33–41; idem, "Intra-Atomic Charge," *Nature* 92 (27 November 1913), 372–73; idem, "Intra-Atomic Charge and the Structure of the Atom," *Nature* 92 (4 December 1913), 476–78; Frederick Soddy, "Intra-Atomic Charge," *Nature* 92 (4 December 1913), 399–400; Soddy, "The Structure of the Atom," *Nature* 92 (18 December 1913), 452; Ernest Rutherford, "The Structure of the Atom," *Nature* 92 (11 December 1913), 423; Niels Bohr, "On the Constitution of Atoms and Molecules," *Phil. Mag.* 26 (1913), 1.

[44] Frederick Soddy, "The Radio-Elements and the Periodic Law," *Chemical News* 107 (28 February 1913), 97–99; idem, "Intra-Atomic Charge," pp. 399–400; Kasimir Fajans, "Die Stellung der Radioelemente im periodischen System," *Phys. Zeit.* 14 (1913), 136–42; Fajans, "On a Relation between the Nature of a Radioactive Transformation and the Elec-

ior with the atoms in the group one place to its right in the periodic table. Lead in Group IV became bismuth in Group V, and bismuth disintegrated to polonium in Group VI.

$$^{210}_{82}\text{Pb} \xrightarrow{-\beta^-} {}^{210}_{83}\text{Bi} \xrightarrow{-\beta^-} {}^{210}_{84}\text{Po}, \qquad \text{where } \beta^- = {}^{0}_{-1}e$$

What became clear from beta decay was that the atomic weights of successive elements remained the same because of the beta particle's negligible weight. Thus, instead of mass it was really the "electrical content" or the atomic number that determined an atom's chemical properties and hence its position in the periodic table. For the series of atoms from thallium, atomic number 81, to uranium, atomic number 92, unit difference in charge corresponded to unit difference in position in the periodic table. The loss of an alpha particle was equivalent electrically to the gain of two beta particles.[45]

Henry Gwyn-Jeffreys Moseley in 1913–14 tested experimentally van den Broek's hypothesis.[46] Upon photographing the X-ray spectra of atoms from aluminum to gold, Moseley found that the wavelength and, accordingly, the frequency of the emitted radiation varied in a regular manner from one atom to the next when he followed the order of increasing atomic weight given in the periodic table. Taking the frequency of the α-line, the stronger of the two lines constituting the K series of each atom, he showed that this frequency was directly proportional to the square of an integer Q, and that, of each atom, Q's numerical value was one less than the atom's ordinal or atomic number. Indeed, except for the pairs cobalt-nickel, argon-potassium, and tellurium-iodine, his sequence of the atoms was identical with the order based on increasing atomic weight.

For the relation between the frequency, v, and Q, Moseley arrived at the following equation:

$$Q = \sqrt{\frac{v}{\frac{3}{4}v_0}};$$

trochemical Behavior of the Radioelement Involved," in Alfred Romer, *Radiochemistry and the Discovery of Isotopes*, pp. 198–206.

[45]Frederick Soddy, "The Radioelements and the Periodic Law," *B.A.A.S. Report* 83 (1913), 445–47; idem, "Die Radioelemente und das periodische Gesetz," *Jahrbuch der Radioaktivität und Elektronik* 10 (1913), 193.

[46]Henry Gwyn-Jeffreys Moseley, "The High Frequency Spectra of the Elements," pt. 1, *Phil. Mag.* 26 (1913), 1024–34; pt. 2. *Phil. Mag.* 27 (1914), 703–13.

where Q = an integer, one less than the atomic number; v = frequency of the spectral line; and v_0 = fundamental frequency of ordinary line spectra R_c, where R is the Rydberg constant and c the velocity of light. Since ¾ v_0 was a constant, the equation reduced to

$$v = AQ^2,$$

where A = ¾ v_0.

Because Q for the K_α line was equal to $N - 1$, with N the atomic number of the element producing the spectrum, Moseley obtained $v = A(N - 1)^2$ as his final equation. Plotting the atomic number against the square root of the corresponding K_α frequency for a large number of atoms gave nearly a perfectly straight line.[47]

On the basis of his investigation, Moseley concluded that the atom contained a fundamental quantity that increased in a regular manner upon passing from one atom to the next and that this quantity was the charge on the central positive nucleus. "We are therefore led by experiment," he wrote,

to the view that N is the same as the number of the place occupied by the element in the periodic system. This atomic number is then H 1 for He 2 for Li 3 . . . for Ca 20 . . . for Zn 30, etc. This theory was originated by Broek and since used by Bohr. We can confidently predict that in the few cases in which the order of the atomic weight A clashes with the chemical order of the periodic system, the chemical properties are governed by N; while A is itself probably a complicated function of N. The very close similarity between the x-ray spectra of the different elements show that these radiations originate inside the atom, and have no direct connection with the complicated light spectra and chemical properties which are governed by the structure of its surface.[48]

Moseley also pointed out that "if either the elements were not characterized by these integers, or any mistake had been made in the order chosen or in the number of places left for unknown elements, these regularities would at once disappear."[49] From the evidence of X-ray spectra alone, he realized that these integers, the atomic numbers, were characteristic of the atoms. They proved how fundamental the atomic number was and provided a means of determining each atom's relative nuclear charge. Since it was highly improbable that two different stable atoms would have the same X-

[47] Moseley, "The High Frequency Spectra of the Elements," pt. 1, p. 1030, and pt. 2, p. 709.

[48] Moseley, "The High Frequency Spectra of the Elements," pt. 1, p. 1031.

[49] Moseley, "The High Frequency Spectra of the Elements," pt. 2, p. 711.

ray spectrum and hence the same value of Q or N, Moseley's work settled definitely the total number of elements lying between hydrogen and uranium, the two extremes in the periodic table. Consecutive integers, the atomic numbers, represented all the positions, and they expressed an actual relation between atoms, namely, the nuclear charge.

The following discussion on Kossel's electron theory of valence will make clear his relative contribution, for too often his theory is treated as if it were in all respects identical to the Lewis theory. Kossel, who was in Munich serving as an assistant to Arnold Sommerfeld (1868–1951), published his article on atomic and molecular structure and valence in 1916, nearly simultaneously with the appearance of Lewis's paper.[50] The papers presented parallel pictures of atomic structure and of polar bond formation, but Lewis—not Kossel—introduced the theory of the shared electron pair to account for the nonpolar bond.

The electrons in the Lewis atom were relatively motionless and distributed in three dimensions, but Kossel, following Bohr, placed the electrons in a plane, in concentric rings revolving in orbits around the nucleus. Moseley's experiments had established the number of electrons in most atoms, and Kossel in his 1916 paper clearly recognized that an atom's electronic arrangement accounted for its chemical behavior and therefore its position in the periodic table. Thus hydrogen's ring contained only one electron; helium, the next lightest atom and the first member of the rare gases, had a ring with two. The ten electrons in neon, the second rare gas and the last atom of the second period, were in two rings, an inner ring of two electrons and an outer ring of eight or an electron configuration of 2-8. Similarly, the rare gas argon at the end of period three had three rings and a configuration of 2-8-8. Kossel, like Lewis, attached special significance in his electron structures to the number eight's recurrence at the end of each period, but his rings seemed less likely than the cube or tetrahedron to indicate the extraordinary stability associated with this number.

In assigning structures to other atoms, Kossel assumed an inner pair of

[50] Walther Kossel, "Über Molekülbildung als Frage des Atombaus," *Annalen der Physik* 49 (1916), 229–362. Kossel's article was received on 27 December 1915 and published on 7 March 1916. Lewis's paper was received on 26 January 1916 and published in April 1916. Despite the greater length of Kossel's paper, most of it is a general discussion of the state of inorganic chemistry. His diagrams of chemical bonding, with the exception of those on p. 359, where he showed electron ring sharing in several molecules, all illustrate polar bonds holding the compound together and put an emphasis on achieving a stable octet. Kossel did not attach any importance to the shared electron pair, which was precisely the point of Lewis's article. For a good description of Kossel's theory, see Alfred Stock, *The Structure of Atoms*.

electrons and a sufficient number of eight-membered rings surrounding the nucleus to give a configuration identical to the rare gas immediately preceding the atom in the periodic table. Then he added an outermost valence ring capable of holding from one to eight electrons. The atomic number and the group in the periodic table to which the atom belonged determined the number of electrons in the valence ring. For example, the electron structure of fluorine in Group VII, atomic number 9, was 2-7; sodium, Group I, atomic number 11, was 2-8-1; calcium, Group II, atomic number 20, was 2-8-8-2.

According to Kossel, all atoms in forming molecules tended either to gain or to lose electrons in order to acquire the electron structures of the unreactive Group 0 gases. Electron gain or loss, therefore, occurred most easily for atoms in the groups (I and VII) adjacent to the rare gases, though Kossel believed that any atom in the first three periods could lose all of its valence electrons. Thus chlorine was sometimes heptapositive and sulfur, hexapositive in a molecule.

In this respect, Kossel differed completely with Lewis, who resorted to electron sharing rather than transfer large numbers of electrons. Indeed, Kossel argued that the atoms in nonpolar molecules were present as intramolecular ions. They were able, just as Abegg had hypothesized in his theory of normal and contravalences, to acquire a positive or a negative charge. Carbon was in fact negative in methane and positive in carbon tetrachloride.

Two illustrations of Kossel's molecular structures appear below.[51] They show only the valence ring of eight electrons, which, he believed, rotated at right angles to a line connecting the two nuclei of a binary molecule:

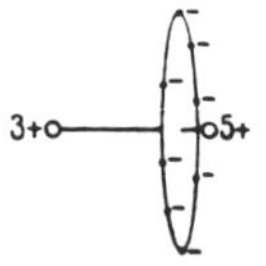

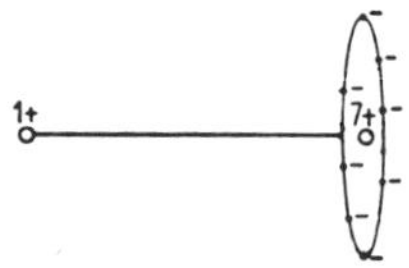

Kossel's structures for the elementary diatomic molecules did not require a ring of eight electrons, however. The nitrogen molecule N_2, in which each nitrogen atom had five valence electrons, had a ring of ten electrons rotating midway between the two atoms. The oxygen and fluo-

[51] Kossel, "Über Molekülbildung," p. 359.

rine molecules, O_2 and F_2, contained, respectively, rings of twelve and fourteen electrons.[52] Kossel's nitrogen structure appeared as:

Kossel, like Werner earlier, did not believe that an atom's bonds had fixed spatial orientations. Once again he differed with Lewis, who had abandoned his cubic atom for a tetrahedral model in order to account for the spatial orientation of the chemical bond. Indeed, Kossel believed that the tetrahedral orientation of the carbon atom's four bonds resulted simply because the atoms bonded to carbon assumed the form of a tetrahedron as the most stable structure. If the attractive force was sufficiently strong, the atoms could not change their positions. Kossel held this view despite conclusive evidence that showed that the carbon atom, even when surrounded by fewer than four atoms, had its bonds directed in space, as in the ethylene molecule H_2CCH_2.

Lewis and Kossel had presented similar pictures of the atom's structure. They assumed that the electrons surrounded the small positive nucleus in concentric groups and that the first group contained two electrons, the second, eight, and the third, eight. Then came groups of somewhat indeterminate character and finally an outer group holding from one to eight electrons. But while Kossel arranged the successive electron groups in concentric rings, Lewis placed them in concentric shells forming a three-dimensional structure about the nucleus.

In both theories, the electron groups attained the highest degree of symmetry and stability in the atoms of the rare gases with their groups of 2; 2,8; 2,8,8; and so on. Lewis and Kossel offered identical interpretations of polar compounds: atoms other than the rare gases had a strong tendency to gain or to lose electrons in order to acquire a rare gas structure. Each atom in the compound was, therefore, an ion. Here, the similarity between the two theories ends. For Lewis went beyond Kossel and provided an equally satisfying interpretation of the chemical bond in nonpolar molecules, his theory of the shared electron pair bond.

[52] Ibid.

Conclusion

With the publication of his 1916 paper "The Atom and the Molecule," Lewis offered an electron theory of valence capable of accounting for both polar and nonpolar bonds. His fundamental conception, the shared electron pair bond satisfied the long-established rule of eight, and its representation by the simple and pictorially satisfying electron dot formulas made his theory operable. Lewis readily wrote formulas and easily resolved many of the difficulties that had plagued theoretical chemistry for so many years: the formulas for ammonia and the ammonium compounds, the diatomic gases H_2, O_2, and N_2, and the oxyacids.

In spite of these achievements (the shared electron pair bond and the electron dot formulas became permanent parts of chemical tradition), Lewis's 1916 paper was a transitional one, containing residues of the past and unsolved problems for the future. The cubic atom had no place in the new theory; its limitations were too fundamental and too obvious. Yet Lewis used it in the paper as he must have used it in his own theory's development, as a lead-in to the shared electron pair bond. The cubic atom was a geometric representation that he easily adapted to the transfer of electrons and to the sharing of one or two pairs (but not three pairs) of electrons while preserving eight electrons as the complete atomic shell.

But electrons are not simply dots on paper. They have an experimental reality, consisting of real negative charge, and the question of how two such charges could form a stable nonpolar bond was a question Lewis really could not answer in 1916.

Lewis seemed to accept Parson's idea that each electron was not only an electric charge but also a small magnet. Thus, a magnetic force apparently was responsible for the pairing of electrons. However, a magnetic electron was of necessity a moving electron, whereas in the Lewis atom the electrons were essentially motionless. How Lewis eventually reconciled these opposing views to allow for the motion of electrons and yet retain his idea of electron pair sharing is the subject of the final chapter.

8. The New Theory of Valence

Introduction

In 1916 Lewis presented an atomic model that clearly explained the chemical bond, deriving it historically from his 1902 model of the cubic atom. In its most useful form, he condensed the eight electrons of a saturated cube to four pairs of electrons, with their positions oriented toward the vertices of a tetrahedron. Each electron pair constituted a bond when shared with another atom. Lewis offered no theoretical justification for this model; it was a chemist's model—pragmatic, convenient in use, widely applicable to a variety of chemical problems. It correlated to the latest fundamental physical knowledge of atomic structure only in utilizing the electron. In particular, Lewis's atom was static, consonant with its historical origin in a cubic geometry. It was in total contrast to Bohr's dynamic atom, whose dramatic success in accounting for the spectral lines of atomic hydrogen originated in an electron in orbit about a positive nucleus. Recognizing this apparent conflict, Lewis in 1917 wrote "The Static Atom," in which he attempted to justify the static model.

Ironically, just at this time when Lewis felt compelled to accommodate his static model to Bohr's highly successful dynamic atom, new evidence was appearing that made the cubic atom, which Lewis had abandoned, seem less arbitrary, more real.

The Arguments for a Static Atom

The Lewis static atom appeared first in the form of a cube, though in his 1916 paper Lewis had given two reasons for rejecting the cube and adopting a tetrahedral arrangement of electron pairs in an atom: (1) the cubic atom did not permit free rotation about a single bond, and (2) it could not show the formation of a triple bond as found in the nitrogen molecule.

Maurice Huggins (b. 1897), who was at Berkeley with Lewis and since 1919 had worked on extending the Lewis theory to the heavier atoms, also adopted a tetrahedral arrangement of the electron pairs. Huggins believed that whenever an atom formed a bond with a carbon or nitrogen atom, a closely spaced electron pair connected the two atoms. Thus, he found it very improbable that only two of the eight electrons in an atom's completed valence shell would form a pair while the others remained relatively far apart, as in the cubic arrangement. Again, when a molecule contained a carbon-oxygen or carbon-nitrogen double bond, Huggins argued that the four electrons not acting as bonds had paired off and were no longer at the corners of a cube. With Lewis, Huggins believed that in an atom the electron pair rather than the single electron was the basic unit of both bonding and nonbonding electrons.[1]

Almost at the same time that Lewis and Huggins were questioning the suitability of the cubic atom on chemical grounds, physical evidence appeared that tended to support a cubic electron arrangement. From the X-ray analysis of iron crystals, Albert W. Hull (1880–1966) at the General Electric Company in Schenectady, New York, concluded in 1917 that the electrons in the iron atom lay along the diagonals of a cube in four groups of 2, 8, 8, and 8 electrons. One electron was at each corner of three concentric cubes, with the remaining two electrons situated close to the nucleus. The electrons, according to Hull, had fixed positions in the atom though they moved about them in very small orbits.[2]

The studies of Max Born and Alfred Landé from 1918 to 1920 on the compressibilities of the alkali halides also supported a cubic atom. Born and Landé at Frankfurt am Main compared the observed with the calculated compressibility, which they defined as the ratio of the decrease in volume to the corresponding increase in pressure. The assumption that the ions in the cubic halide crystals consisted, as in Bohr's atom, of rings of electrons rotating in the same plane, gave poor agreement between the calculated and observed compressibilities. Assuming that the ions were tiny cubes led to much better agreement.[3]

[1]Maurice L. Huggins, "Electronic Structures of Atoms," *J. Phys. Chem.* 26 (1922), 602–3. Huggins wrote this paper in 1920, but it was not published immediately.

[2]Albert W. Hull, "The Crystal Structure of Iron," *Physical Review* 9 (1917), 86.

[3]Max Born and Alfred Landé, "Über die Berechnung der Kompressibilität regulärer Kristalle aus der Gittertheorie," *Verh. deut. phys. Ges.* 20 (1918), 210–16; Born, "Über kubische Atommodelle," *Verh. deut. phys. Ges.* 20 (1918), 230–39; Born, *The Constitution*

Of even greater significance, the investigations of Hull, Born, and Landé clearly suggested that the electrons in an atom were relatively motionless; each rotated only in a small orbit around its equilibrium position—the corner of the cube. Their atom was essentially a static atom and in agreement with Lewis's model, though it differed in the actual arrangement of the electrons. It was the very antithesis of the dynamic atom that Niels Bohr (1885–1962) in Copenhagen used quite successfully in 1913 to account for the spectral series of the hydrogen atom and the one-electron ions helium, He^+, and lithium, Li^{++}.[4]

Thus, as Lewis insisted in his 1916 publication, physicists had not actually offered any irrefutable arguments favoring a dynamic atom of the type proposed by Bohr. Indeed, Lewis again raised serious objections to Bohr's theory of the orbiting electron in a paper presented before a joint meeting of the American Association for the Advancement of Science, The American Physical Society, and the American Chemical Society held in New York on December 27, 1916.[5] Bohr's dynamic atom, he said, was unacceptable because it failed to account for an atom's physical and chemical properties. In view of the physicists' wide acceptance of Bohr's atom, Lewis intended to settle in his paper the question of whether the electrons in atoms and molecules were in rapid motion or essentially at rest.

His major criticism of the Bohr atom's physical properties was the implicit assumption that the electron's revolution continued even down to the absolute zero of temperature. If an electron were revolving about its nucleus and a charged particle approached it, Lewis argued, the electron, if it

---

of Matter, pp. 47–49; Born, "Dynamik der räumlichen Atomstruktur," *Zeit. Physik* 2 (1920), 83–86; Born, "Bemerkungen über die Grösse der Atome," *Zeit. Physik* 2 (1920), 87–89; Born, "Würfelatome, periodisches System und Molekülbildung," *Zeit. Physik* 2 (1920), 380–404. Landé had also developed a mathematical theory of the motions of eight electrons about points in an atom having cubic symmetry ("Dynamik der räumlichen Atomstruktur," *Verh. deut. phys. Ges.* 21 [1919], 2–12, 644–62). Wheeler P. Davey's extensive X-ray diffraction analyses of the lighter alkali halide crystals (lithium, sodium, and potassium chlorides) and of magnesium's and calcium's oxide and sulfide showed a few years later that the ions of these elements packed as if they were cubes but with well-rounded corners. Since each ion had the same number of electrons as the rare gas atom nearest to it in the periodic table, Davey believed that perhaps the atoms of the lighter rare gases also had a cubic shape. See Wheeler P. Davey, "The Cubic Shapes of Certain Ions as Confirmed by X-ray Crystal Analysis," *Physical Review* 17 (1921), 402–3; Davey, "Radiation," *Journal of the Franklin Institute* 197 (1924), 443–44.

[4]Niels Bohr, "On the Constitution of Atoms and Molecules," *Phil. Mag.* 26 (1913), 25.

[5]G. N. Lewis, "The Static Atom," *Science* 46 (28 September 1917), 298.

influenced the particle at all, would cause the particle to vibrate in sympathetic motion. The particle, therefore, received energy from an atom already *at absolute zero*, and this clearly was contrary to the most fundamental laws of thermodynamics.[6]

"Unless we are willing, under the onslaught of quantum theories, to throw overboard all of the basic principles of physical science," Lewis wrote, "we must conclude that the electron in the Bohr atom not only ceases to obey Coulomb's law, but exerts no influence whatsoever upon another charged particle at any distance. Yet it is on the basis of Coulomb's law that the equations of Bohr were derived." In Lewis's opinion, Bohr had invented a system not only inconsistent with the accepted laws of electromagnetics, but logically objectionable, "for that state of motion which produces no physical effect whatsoever may better be called a state of rest."[7]

Despite his objections to Bohr's dynamic atom, Lewis did not wish to minimize the broader significance of Bohr's work. Bohr had been the first to present an acceptable picture of the mechanism that produced the lines in atomic hydrogen's well-known spectral series and to trace a relation between the two natural constants R, Rydberg's constant of fundamental frequency, and h, Planck's constant. His theory provided interesting leads, but substituting a static atom for Bohr's dynamic atom led to similar relations. A static atom, Lewis said, was consistent with known chemical facts and did not require discarding the principal laws of mechanics and electromagnetics.[8]

Lewis, therefore, believed that he could explain the spectral series of hydrogen and other atoms by assuming that different "constraints" acting uniformly in all directions restricted each electron in an atom to a series of stable equilibrium positions with respect to the nucleus. Each electron vibrated in its position with a characteristic frequency determined solely by the magnitude of its constraints. Lewis intended to calculate the energy difference between successive equilibrium positions using the constants R, c, and h. Indeed, in his 1917 paper, Lewis sketched in rough outline an expression relating the frequencies of the spectral lines as a function of the electron's equilibrium position without using the newer quantum relations.[9]

[6] Ibid., p. 299.
[7] Lewis, "The Atom and the Molecule," *J.A.C.S.* 38 (1916), 772–73.
[8] Lewis, "The Static Atom," p. 299.
[9] Lewis, "The Atom and the Molecule," p. 773; idem, "The Static Atom," pp. 299–300.

Upon applying the Bohr atom to chemical phenomena, Lewis claimed that it failed to account for the chemical properties of even the simplest atoms. Chemists could hardly expect to solve the problems of chemical combination by assigning in advance definite laws of force between an atom's positive and negative constituents and then using these laws to construct mechanical models of the atom. Instead, Lewis argued:

We must first of all, from a study of chemical phenomena, learn the structure and the arrangement of the atoms, and if we find it necessary to alter the law of force acting between charged particles at small distances, even to the extent of changing the sign of that force, it will not be the first time in the history of science that an increase in the range of observational material has required a modification of generalizations based upon a smaller field of observations.

The existence of isomers presented a serious challenge to the acceptance of Bohr's dynamic atom because in these molecules the atoms always retained their relative positions. In one part of the structure, atoms replaced other atoms, and groups of atoms replaced other groups, without disturbing the arrangement of the atoms in another part. Hence, it seemed inconceivable to Lewis that electrons, since they obviously constituted the chemical bond uniting the atoms possessed any appreciable orbital or even chaotic motion. The electrons, he said, were in fixed equilibrium positions, about which they experienced minute oscillations under the influence of high temperature or electric discharge but from which they did not depart very far without altering the structure of the molecule.[10]

Other arguments against the dynamic atom appeared in several articles that J. J. Thomson published in 1919 and 1920 in the *Philosophical Magazine* and again in 1923 in his book *The Electron in Chemistry*.[11] While Thomson acknowledged that Bohr's arithmetical results happened to be in good agreement with experiment, he would not accept the physical basis of the Bohr atom. Indeed, Thomson claimed that Bohr's orbiting electron failed to explain the results of his experiments on the scattering of light by hydrogen molecules.[12]

[10]Lewis, "The Atom and the Molecule," p. 773; idem, "The Static Atom," p. 298. Isomers are compounds that are composed of the same number and kinds of atoms but differ in the arrangement of the atoms in relation to each other.

[11]J. J. Thomson, *The Electron in Chemistry*. This book contains the substance of five lectures delivered by Thomson at the Franklin Institute, Philadelphia, April 1923.

[12]J. J. Thomson, "On the Origin of Spectra and Planck's Law," *Phil. Mag.* 37 (1919), 419–66, especially 420–21; idem, "On the Scattering of Light by Unsymmetrical Atoms," *Phil. Mag.* 40 (1920), 393–413.

Thomson also pointed out that in Bohr's atom an electron in its orbit of lowest energy did not vibrate or rotate with a frequency corresponding to any of the lines found in the atom's spectrum. Bohr said the lines were due to the electron falling from a higher to a lower energy orbit after an atomic collision or an electric discharge had placed it in the higher orbit. But according to Thomson, considerable experimental evidence indicated that an electron in an unexcited atom vibrated with the frequencies of its spectral lines. Thomson cited as evidence studies on the absorption spectra of the alkali metal vapors, which showed that the comparatively cold vapors of these atoms absorbed light of the same frequency as that emitted in the spectrum of the luminous atom.[13]

Laws of Force Applied to the Static Atom

Bohr's atom achieved stability by balancing the centrifugal inertia of the rapidly moving electron and the coulombic attraction between it and the positive nucleus that kept the electron in orbit. In the static atom the electrons remained in relatively fixed positions. The atom, therefore, required some kind of repulsive force to prevent the electrons from falling into the nucleus, or alternatively a modification or even abandonment of Coulomb's law at intraatomic distances.

Abandoning Coulomb's law was certainly not an unrealistic alternative, especially when no direct experimental evidence verified the law at atomic dimensions. Indeed, most of the evidence supporting the law had come from experiments in which the distances were enormous compared with those of atoms. Clearly, it did not hold at intranuclear distances; otherwise, the repulsive force between the protons in the nucleus would cause the atom to disintegrate. Coulomb's law, Thomson said, might even reverse itself at distances that were small compared with the dimensions of an atom. Thus, "in considering the force which may exist in the atom, we must remember that we cannot assume that the forces due to the charges of electricity inside an atom are of exactly the same character as those given by the ordinary laws of electrostatics." [14]

Consequently, in his atomic model of 1913, Thomson replaced the coulombic force and centrifugal inertia of the Bohr atom with two new forces.

[13]Thomson, "Origin of Spectra," pp. 420–21.
[14]J. J. Thomson, "On the Structure of the Atom," *Phil. Mag.* 26 (1913), 793.

They were (1) a radial repulsive force diffused throughout the entire atom that varied inversely as the cube of the distance from the atom's center and (2) a radial attractive force varying inversely as the square of the distance from the center but restricted to a limited number of tubes of force in the atom.[15] Expressed mathematically, his law was

$$F = Ee\left(\frac{1}{r^2} - \frac{C}{r^3}\right),$$

where E = charge on nucleus; e = electronic charge; r = distance between charges e and E; and C = a constant varying from one kind of atom to another of the order 10^{-8} cm, measuring the distance at which the force changed from one of attraction to one of repulsion.

Thomson obtained some rather remarkable results with his law of force.[16] The electron configurations of atoms that were stable according to his calculations exhibited a chemical periodicity in agreement with the atom's position in the periodic table. He calculated the ionization potential and atomic radius of such simple atoms as lithium, beryllium, nitrogen, and oxygen; for the alkali metals he determined the frequency of the light at which each metal became photoelectrically active and measured their compressibilities.[17]

Wheeler P. Davey (1886–1959) at Pennsylvania State University found Thomson's arguments convincing. But the most obvious way to modify Bohr's equation, he said, was to add to Coulomb's inverse square law a sine or cosine law with a decrement. The decrement, if large enough, would reduce the equation to the inverse square law at points beyond the atom's radius and thereby satisfy all existing experimental data. Davey recognized that Thomson's equation was simpler and lent itself more readily to numerical calculation, however. A graph of Thomson's equation gave only one point of inflection and simplified to Coulomb's law at a distance determined by the constant C. "The correspondence between the various calculated values and the experimental results is such," Davey

[15] Ibid.

[16] Thomson gave only a statement of his force law in his 1913 paper "Structure of the Atom," p. 793. An equation appears in his 1919 paper "Origin of Spectra," p. 421, and in "On the Structure of the Molecule and Chemical Combination," *Phil. Mag.* 41 (1921), 514. See also idem, *The Electron in Chemistry*, p. 4.

[17] J. J. Thomson, "Application of the Electron Theory of Chemistry to Solids," *Phil. Mag.* 43 (1922), 721–57; idem, *The Electron in Chemistry*, chap. 1.

wrote, "as to give considerable of a feeling of confidence in the picture of atomic structure upon which the calculations are based." [18]

In his 1917 paper "The Static Atom," Lewis presented an alternative force law that, he said, enabled a static atom "to give at least as satisfactory an explanation of the phenomena of spectroscopy, and of the relationships between the natural constants which have been found in the study of radiation, as can be afforded by the orbital atom." [19]

Initially, Lewis's equation took the form

$$f = \frac{\varepsilon^2}{r^2}\, e^{\frac{r}{r_0}},$$

where f was the force acting on an equal positive charge at a distance r from the electron charge ε, e was the natural logarithmic base, and r_0 a characteristic distance that did not differ greatly from the radius of the spherical electron (10^{-8} cm).

This equation only approximated the true force relation, and Lewis suggested replacing the exponential term with a periodic function, such as a trigonometric function of $1/r$. The resulting plot would give a curve that intersected the r axis (the abscissa) an infinite number of times as r approached zero, showing that the force varied from one of attraction to one of repulsion.

At each intersection of the r axis, the slope of the curve df/dr increased toward a finite limit as r approached zero and, according to Lewis, its square root determined the electron's natural frequency of oscillation. For the hydrogen atom, the limiting value of df/dr represented the limiting frequency of its spectral series. The static atom, Lewis concluded, presented "a picture of a system which, consistently with recognized principles of mechanics and electromagnetics, would give a series of spectral lines analogous to the series which are known for various elements." [20]

Irving Langmuir (1881–1957), Albert Hull's colleague at the General Electric Company, introduced in 1921 yet another force law to justify a static arrangement of the electrons in an atom.[21] In Bohr's atom, the elec-

---

[18] Davey, "Radiation," p. 446.

[19] Lewis, "The Static Atom," pp. 301–2.

[20] Ibid., pp. 300–301.

[21] Irving Langmuir, "The Structure of the Static Atom," *Science* 53 (25 March 1921), 290–93. Langmuir's papers on atomic structure and valence are contained in *The Collected Works of Irving Langmuir*, ed. Guy Suits and Harold E. Way, vol. 6, *The Structure of Matter*.

tron remained in orbit, balancing the coulombic attraction with its own frictionless inertial motion. But since it made little difference to the chemist what balanced the coulombic force so long as some kind of balance existed, Langmuir introduced an ad hoc repulsive force, which he called a quantum force, and represented it by the following equation:

$$F_q = \frac{1}{mr^3}\left(\frac{nh}{2\pi}\right)^2,$$

where r = distance between the electron and the nucleus, m = mass of electron, h = Planck's constant, n = an integer denoting the quantum state (equilibrium position) of the electron.[22] Combining his quantum force and Coulomb's law, Langmuir then derived an equation for the energy of a static atom that had a single electron capable of vibrating in different equilibrium positions:

$$W = \frac{Z^2 W_0}{n^2},$$

where Z = charge on nucleus and W_0 = energy of lowest equilibrium position.

Langmuir's equation was identical with Bohr's relation for the energy of an electron in its different stationary states. It gave, therefore, identical values for the Rydberg constant, the Balmer, and the other spectral series without requiring any electron rotation around the nucleus.[23] When n, the quantum number of an electron vibrating in a stable position, decreased by one unit, the electron became unstable and fell to a new equilibrium position, where it then radiated its excess energy according to the laws of classical electrodynamics.

In comparing the two theories of atomic structure, Langmuir, of course, acknowledged the similarity between Bohr's stationary states and the static atom's equilibrium positions. But he believed that for atoms other than hydrogen, the static theory would "go much further," than the older dynamic theory. The determination of equilibrium positions using static forces was extremely simple compared with the corresponding dynamical problem, he argued, and "we are not troubled by mysterious quan-

[22] Langmuir, "The Structure of the Static Atom," p. 290.

[23] Ibid., p. 291. See also Langmuir, "Future Developments of Theoretical Chemistry," *Chemical and Metallurgical Engineering* 24 (1921), 553–57; idem, *Collected Works*, vol. 6, *The Structure of Matter*, pp. 124–27.

tum conditions which are theoretically applicable only to periodic orbits while the calculated orbits are not periodic." [24] Langmuir pointed out that, according to Bohr's calculations, a lithium nucleus surrounded by three electrons in a single ring was more stable than an arrangement with two electrons in an inner ring and an outer ring containing a single electron. Yet, the latter was a structure clearly fitting lithium's chemical behavior. Indeed, Bohr's calculations, in contradiction to experiment, showed that helium ionized more easily than lithium. [25] Bohr's theory gave no reason for the tremendous difference in properties between the helium atom, which had two electrons in a single ring, and lithium, with three. Based on coplanar, concentric circular electron orbits, his theory simply could not explain why a pair of electrons, or for that matter, a group of eight, exhibited unusual stability.

Bohr's dynamic atom, in spite of its impressive success in accounting quantitatively for atomic spectra, thus had serious limitations when applied to basic chemical problems. The work of Parson, Lewis, Langmuir, Born, and Landé all supported various forms of static atomic models, which were in some significant measure superior to the Bohr atom in dealing with these problems. [26]

Reconciliation of the Static and Dynamic Atoms

In developing their theories of atomic structure, physicists and chemists had in general proceeded along fairly independent paths. The physicists were searching for a theory that accounted for an atom's spectrum and adopted a dynamic atom, while chemists, hoping to understand the atom's chemical properties, especially its valence, chose a static model. But the two theories, despite some obvious differences, had the same fundamental objective: to unravel completely the atom's properties. Indeed, by the 1920s both physicists and chemists began to recognize that the dynamic and static atoms were not incompatible.

[24] Langmuir, "The Structure of the Static Atom," pp. 290–93; idem, *Collected Works*, 6:127.

[25] Bohr, "On the Constitution of Atoms and Molecules," p. 492; Irving Langmuir, "Theories of Atomic Structure," *Nature* 105 (29 April 1920), 261; Langmuir, *Collected Works*, 6:111.

[26] Langmuir, *Collected Works*, vol. 6; Born and Landé, "Über die Berechnung"; Born, "Über kubischje Atommodelle"; Landé, "Dynamik der räumlichen Atomstruktur"; Born, "Dynamik der räumlichen Atomstruktur"; Born, "Würfelatome, periodisches System und Molekülbildung."

The dynamic models Bohr developed for the hydrogen (H_2) and methane (CH_4) molecules in his 1913 papers clearly met the valence requirements of the carbon and hydrogen atoms. For the hydrogen molecule, a pair of electrons, one from each hydrogen atom, revolved in a single orbit about a line connecting the centers or nuclei of the two atoms. The electron pair constituted the chemical bond.[27] In the methane molecule, once again a pair of rotating electrons represented each carbon-hydrogen bond, but Bohr did not generally identify the electron pair as the chemical bond: "The configuration suggested by the theory for a molecule of CH_4 is of the ordinary tetrahedron type; the carbon nucleus surrounded by a very small ring of two electrons being situated in the centre, and a hydrogen nucleus in every corner. The chemical bonds are represented by 4 rings of 2 electrons each rotating around the lines connecting the centre and the corners."[28]

William Noyes was the first chemist to suggest a possible connection between the motion of the valence electrons and chemical combination between atoms.[29] In a 1917 publication, Noyes wrote:

Let us suppose that two atoms, which have an affinity for each other are brought close together. A valence electron which is rotating around a positive nucleus in the first atom may find a positive nucleus in the second atom sufficiently close so that it will include the latter in its orbit and it may then continue to describe an orbit about the positive nuclei of the two atoms. During that portion of the orbit within the second atom that atom would become, on the whole, negative while the first atom would be positive. During the other part of the orbit each atom would be electrically neutral, and the atoms might fall apart. When we remember, however, the tremendous velocity of the electrons and the relatively sluggish motions of the atoms it seems evident that the motion of an electron in such an orbit might hold two atoms together. In ionization the electron would, of course, revolve about the nucleus of the negative atom leaving the other atom positive. It seems impossible to explain ionization otherwise than on the supposition of the complete transfer of the electron. This complete transfer in ionization is one of the strongest arguments against the magneton theory as the only explanation of chemical combination.[30]

Noyes also believed his hypothesis could account for the "localization

[27] Bohr's trilogy of 1913, "On the Constitution of Atoms and Molecules," appears in Niels Bohr, *On the Constitution of Atoms and Molecules*, ed. Leon Rosenfeld (see p. 22).

[28] Ibid., p. 72; Bohr, "On the Constitution of Atoms and Molecules," p. 874.

[29] We could consider Parson's magneton theory to be a kinetic or dynamic theory of chemical combination, but in his theory the electron or magneton was the entire ring or orbit. A particlelike electron rotating in an orbit was not a feature of Parson's theory.

[30] William A. Noyes, "A Kinetic Hypothesis to Explain the Function of Electrons in the Chemical Combination of Atoms," *J.A.C.S.* 39 (1917), 881.

of affinities," that is, the directive valences of organic compounds such as methane or ethane. Arguing from the fact that radioactive elements ejected discrete, positively charged helium nuclei, he concluded that discrete nuclei were present in the remaining elements. The carbon atom, according to Noyes, had one at each corner of a tetrahedron, and a valence electron rotated around each nucleus. If a hydrogen atom was drawn inside the electron's orbit, a carbon-hydrogen bond resulted.[31] Clearly in his crude bonding scheme Noyes took into account neither the concentration of the atom's positive charge in a single nucleus, as shown in Rutherford's nuclear atom, nor Moseley's recent work on atomic numbers. Noyes never said what happened to the hydrogen atom's valence electron and was not prepared at this time to replace his one-electron bond with the Lewis shared electron pair.

Langmuir, like Noyes, maintained that the dynamic and static atoms were not totally incompatible. In his first papers on valence, published in 1919, he wrote that the electrons in an atom could be stationary "or rotate, revolve or oscillate about definite positions in the atom." The positions of the electrons shown in the structural diagrams could be the centers of their orbits.[32] Langmuir believed that the two theories were reconcilable if physicists and chemists assumed that the electrons did not revolve about the nucleus "but about definite positions symmetrically distributed in three dimensions with respect to the nucleus."[33]

Norman R. Campbell's notes in *Nature* expressed a similar point of view. According to Campbell, the essential truth of Bohr's theory, which now included Arnold Sommerfeld's modifications, was in 1920 beyond doubt. At the same time the structures of Lewis, Langmuir, Born, and Landé received experimental support from William L. Bragg's work on atomic radii in crystals and appeared extremely plausible.[34] If the two theories were really inconsistent, he said, the situation would be intolerable.

Consequently, in Campbell's interpretation of the Bohr theory, the elec-

[31] Ibid., p. 882.

[32] Irving Langmuir, "The Arrangement of Electrons in Atoms and Molecules," *J.A.C.S.* 41 (1919), 932. Brief excerpts of this paper also appeared in "The Arrangement of Electrons in Atoms and Molecules," *Journal of the Franklin Institute* 187 (1919), 359–62, and in "The Properties of the Electron as Derived from the Chemical Properties of the Elements," *Physical Review* 13 (1919), 300.

[33] Langmuir, "Theories of Atomic Structure," p. 261.

[34] Norman R. Campbell, "Atomic Structure," *Nature* 106 (25 November 1920), 408; idem, "A Static or Dynamic Atom," *Nature* 111 (28 April 1923), 569; William L. Bragg, "The Arrangement of Atoms in Crystals," *Phil. Mag.* 40 (1920), 169–89; Bragg, "The Dimensions of Atoms and Molecules," *Nature* 107 (24 March 1921), 107.

trons did not have to rotate when in their stable states but only to have energy equivalent to what they would have if they were actually rotating in definite orbits. He did not find it logically impossible to maintain that the electrons possessed this energy and yet were at rest. Bohr used his correspondence principle to predict the intensity and polarization of the hydrogen and helium spectral lines. His principle required the existence of orbits but at the same time assumed that the electrons were not moving in them. Campbell simply applied the same argument to the electron's energy.[35]

Campbell also pointed out that since Bohr now assigned each electron its own orbit instead of having several electrons rotating in the same orbit, the sharing of an electron in the Lewis theory was really equivalent to sharing an electron orbit. Thus, a chemical bond resulted, Campbell wrote, "when some of the electronic orbits, instead of surrounding one nucleus only, surround both, and therefore help to complete the quantum groups of both atoms."[36]

In 1923 Carl A. Knorr (1894–1960) in Munich and Nevil V. Sidgwick at Oxford suggested similar theories of inclusive orbits. For each chemical bond Knorr required a pair of electrons rotating around two atoms.[37] He illustrated the four distinct orbits constituting the four carbon-hydrogen bonds in the methane molecule as follows:

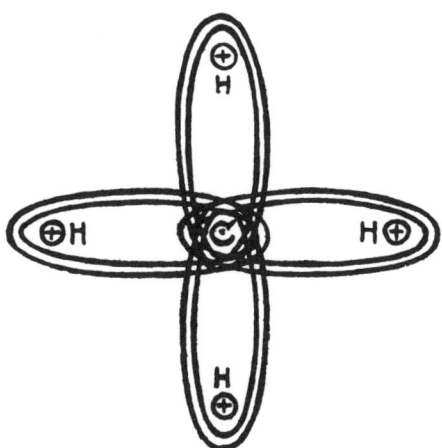

[35] Campbell, "Atomic Structure," p. 408.

[36] Campbell, "A Static or Dynamic Atom," p. 569; Niels Bohr, "Atomic Structure," *Nature* 107 (24 March 1921), 104–7.

[37] Carl A. Knorr, "Eigenschaften chemischer Verbindungen und die Anordnung der Elektronenbahnen in ihren Molekülen," *Zeit. anorg. Chemie* 129 (1923), 109–40.

Sidgwick outlined his ideas in a paper entitled "The Nature of the Nonpolar Link," which he presented at the Cambridge meeting of the Faraday Society held in July of 1923. The meeting was a general discussion on the electron theory of valence. In his paper Sidgwick discussed Bohr's theory as well as the "generally admitted idea that the nonpolar (as opposed to the polar) link consists in the 'sharing' of two electrons between the two linked atoms." [38] The only other necessary assumption, he said, was one requiring the orbit of each shared electron to include both of the connected nuclei. This assumption, as Norman Campbell had already pointed out, was an obvious extension of Bohr's theory to nonpolar bonds and accounted very well for the spectrum of molecular hydrogen.

Because the shared pair of electrons included both nuclei within their orbits, it was clear to Sidgwick why two electrons were absolutely essential for each chemical bond:

When one of the two [electrons] is near to, or on the far side of, one of the nuclei, its attraction on the other is negligible and it does nothing to prevent the two nuclei from separating: whereas if there are two, they may be so arranged in phase that one of them is always available to hold the nuclei together. It is an essential part of Bohr's theory that the valence electrons (with which we are concerned) pass very near the nucleus in the course of their motion, and that the powerful forces which come into play here are essential to the stability of the atom. Thus an orbit which surrounds and comes close to both of the two nuclei will form part of the constitution of both. [39]

Wheeler P. Davey's 1924 article in the *Journal of the Franklin Institute* was another of the publications aimed at minimizing the differences between the dynamic and static atoms. Davey pointed out that in the Bohr-Sommerfeld version of the dynamic theory, the orbits were ellipses rather than circles. If the nucleus were one focus of the ellipse, he argued, the electron's position in the static atom might well be the other focus. The two pictures of atomic structure, which at first seemed so widely different, were, therefore, practically the same though expressed in different words and from slightly different viewpoints. [40]

---

[38] Nevil V. Sidgwick, "The Nature of the Nonpolar Link," *Trans. Faraday Soc.* 19 (1923), 469. Sidgwick presented this paper as part of the general discussion on the electron theory of valency held by the Faraday Society on 13 and 14 July 1923 at Cambridge University.

[39] Ibid.

[40] Davey, "Radiation," p. 478.

G. N. Lewis and the Bohr Atom

At the 1923 meeting of the Faraday Society and shortly after in his monograph *Valence and the Structure of Atoms and Molecules*, G. N. Lewis discussed the complete reconciliation that had occurred between the dynamic and static theories of the atom. That Bohr and the physicists now assigned a separate orbit to each electron in an atom and oriented the orbits in space and not in a single plane removed, Lewis said, the last element of conflict between the two views. They had recognized at last that the orbit as a whole was important and not the electron's position in the orbit.

Lewis also pointed out that since each electron had its own orbit, this, in effect, fixed the orbits in space. The electron's location was its average position in its orbit and corresponded to the fixed position originally assumed in the static atom. Thus, Lewis could now reconcile the evidence of spectroscopy and the magnetic behavior of atoms, which were the two most important factors indicating the presence of rapidly moving electrons, with the facts of chemistry that required an atom with its electrons in a definite and permanent spatial orientation.[41]

In *Valence and the Structure of Atoms and Molecules* Lewis summarized the atom's structure with the following postulates:

1. . . . we shall adopt the whole of Bohr's theory insofar as it pertains to a single electron. There are no facts of chemistry which are opposed to this part of the theory. . . .
2. In the case of systems containing more than one nucleus or more than one electron, we shall also assume that the electron possesses orbital motion, for such motion seems to be required to account for the phenomenon of magnetism; and each electron in its orbital motion may be regarded as the equivalent of an elementary magnet or magneton. However, in the case of these complex atoms and molecules we shall not assume that an atomic nucleus is necessarily the center or focus of the orbits.
3. These orbits occupy fixed positions with respect to one another and to the nuclei. When we speak of the position of an electron, we shall refer to the position of the orbit as a whole rather than to the position of the electron within the orbit. With this interpretation we may state that the change of an electron from one position to another is always accompanied by a finite change of energy.
4. In a process which consists merely in the fall of an electron from one position to another more stable position, monochromatic radiant energy is emitted, and the

[41]Lewis, *Valence*, p. 56; idem, "Valence and the Electron," *Trans. Faraday Soc.*, 19 (1923), 452. Lewis's paper was the introductory address.

frequency of this radiation multiplied by h, the Planck constant, is equal to the difference in the energy of the system between two states.

5. The electrons of an atom are arranged about the nucleus in concentric shells. The electrons of the outermost shell are spoken of as valence electrons. The valence shell of a free (uncombined) atom never contains more than *eight electrons*. The remainder of the atom, which includes the nucleus and the inner shells, is called the kernel. In the case of the noble gases it is customary to consider that there is no valence shell and that the whole atom is the kernel.

6. In my paper on "The Atom and the Molecule," I laid much stress upon the phenomenon of pairing of electrons. I have since become convinced that this phenomenon is of even greater significance than I then supposed, and that it occurs not only in the valence shell but also within the kernel, and even in the interior of the nucleus itself.[42]

As his final postulate (7) Lewis adopted Bohr's idea

that the first shell is associated with a single energy level, and that this level can accommodate one pair of electrons, that the second shell contains two energy levels, each of which is capable of holding two pairs of electrons, making a maximum of eight electrons in the second shell. The third shell has three energy levels, each of which can hold three pairs of electrons, so that the maximum number of electrons in the third shell is eighteen. The fourth shell comprises four levels, each capable of holding four electron pairs, making a total of thirty-two electrons, and so on.[43]

Thus, by 1924 a truly static atom no longer had a part in the valence theories of Lewis and most other chemists, but the physicists had not ceased their criticism. Bohr attacked the static atom in his Nobel Lecture for 1922, maintaining that stationary electron positions in an atom were impossible if the inverse square law were to hold even approximately between electrical charges. Robert Millikan condemned it in his 1924 Faraday Lecture, "Atomism in Modern Physics," delivered before the London Chemical Society.[44]

Millikan christened the static theory the "loafer theory," one in which the chemist "imagined the electrons sitting around on dry goods boxes at every corner, ready to shake hands with, or hold onto similar loafer electrons in other atoms." He did not dispute the chemists' argument from classical electromagnetic theory that the electrons in the dynamic atom, if they continuously rotated at enormous speeds, would eventually dissipate

[42] Lewis, *Valence*, pp. 56–57.
[43] Ibid., p. 57.
[44] Niels Bohr, "The Structure of the Atom," supplement to *Nature* 112 (7 July 1923), 41; Robert Millikan, "Atomism in Modern Physics," *J. Chem. Soc.* 125 (1924), 1405–17.

all their energy. "God did not make electrons that way," he replied. Millikan denied that any physicist had "*ever advanced the theory that the electrons all rotate in coplanar orbits.* Localized valences," he claimed, "are probably just as compatible with the orbit theory when the electrons are properly distributed in space, as with the conception of stationary electrons." [45]

In 1923 and 1924 physicists and chemists had thus recognized the desirability of accommodating the atom's physical and chemical behavior within the same model, a dynamic one with directed valences. However, they accomplished this goal only after the introduction of quantum mechanics later in the decade. In the meantime physicists and chemists pursued their characteristic problems with objectives less lofty than the total reconciliation of their divergent theoretical needs. [46]

The Acceptance of the Electron Pair Theory of Valence

Following the introduction of the shared electron pair bond in 1916 and the publication of the static atom in 1917, Lewis did little to popularize his new valence theory. Service in World War I and interests in other areas of physical chemistry, chiefly thermodynamics, consumed most of his time. As a result, the chemists' acceptance of the shared electron pair was in part due to the efforts of Irving Langmuir, who between 1919 and 1921 wrote extensively and lectured widely in the United States and abroad on the Lewis theory. [47] Indeed, Langmuir received the Nichols Medal of the American Chemical Society in 1920 for his development of the Lewis theory of atoms and molecules. His interesting and convincing personality and his admirable methods of presentation clearly gave the theory considerable popularity.

In a 1919 paper Langmuir showed that atoms, molecules, or ions with the same number of electrons had the same electronic structure and usually similar physical properties, especially if the charge on each was identical. Langmuir called these structures "isosteres" and was probably the first chemist to support them with crystallographic data. [48]

[45] Ibid., p. 1411 (emphasis in original).

[46] As late as 1931 the static atom still appeared in the well-known text by Frederick Getman and Farrington Daniels, *Outlines of Theoretical Chemistry*, 5th ed., pp. 591–92. The added caption reminded readers not to take the theory too literally.

[47] Langmuir, *Collected Works*, vol. 6.

[48] Irving Langmuir, "Isomorphism, Isosterism and Covalence," *J.A.C.S.* 41 (1919),

Carbon dioxide, CO_2, and nitrous oxide, N_2O, each had 22 electrons arranged identically in the molecule. Their physical properties, such as density, critical temperature and pressure, refractive index, and dielectric constant, were nearly equal. They were, therefore, isosteres.[49] Carbon monoxide, CO, the nitrogen molecule, N_2, and the cyanide ion, CN^-, were also isosteres, each containing 14 electrons in identical electronic structures. But in this case only the neutral molecules, CO and N_2, had closely related physical properties: their freezing and boiling points, critical temperature and pressure, viscosity, and density.[50]

There are numerous cubic models of Langmuir's isosteres scattered throughout his 1919 paper.[51] But in his entire discussion, Langmuir never once employed the Lewis dot notation that clearly would have illustrated both the identity of the electronic structures and the sharing of electron pair bonds:

$$\overset{..}{:}\!O::C::\overset{..}{O}: \quad \text{and} \quad :N::N::\overset{..}{O}:$$

Langmuir also attempted to account for the behavior of the transition elements; Lewis had omitted them in his 1916 paper, and Langmuir hoped to include their electronic structures, indeed the electronic structures of all atoms and molecules, within his theoretical framework. He proposed an arithmetical rule that, he claimed, predicted the existence or nonexistence of chemical compounds, and in 1919 he coined the term *covalent bond* to replace the rather cumbersome *shared electron pair bond*.[52] The Lewis theory, particularly in England after Langmuir introduced it to the British

---

1543–59. *Isosteres* is Greek for "same structure." Today chemists call these structures isoelectronic.

[49] Ibid.; Langmuir, *Collected Works*, 6:43–44.

[50] Langmuir, "Isomorphism, Isosterism and Covalence," pp. 1543–59; idem, *Collected Works*, 6:44.

[51] Langmuir, *Collected Works*, 6:34, 38, 43.

[52] Langmuir, "Isomorphism, Isosterism and Covalence," p. 1543. See also Langmuir, "The Structure of Atoms and Its Bearing on Chemical Valence," *Journal of Industrial and Engineering Chemistry* 12 (1920), 386–89; idem, *Collected Works*, 6:32–35. Langmuir's rule was as follows: $p = \frac{1}{2}(8n - e)$, where e was the total number of available electrons in the valence shell of each combining atom; n, the number of octets formed by their combination; and p, the number of electron pairs held in common by the octets. For C_2H_6, $e = 14$, $n = 2$, and p therefore equals 1. Thus in C_2H_6 the two octets held one electron pair in common corresponding to a single bond between the two carbon atoms, $H_3C—CH_3$. Applying Langmuir's rule to C_2H_4 and C_2H_2 gave respectively double- and triple-bonded structures, $H_2C{=}CH_2$ and $HC{\equiv}CH$. All of Langmuir's structures were of course in agreement with each molecule's long-established structural formula.

chemists, often became the octet theory of valence or even Langmuir's theory of valence.

Lewis, of course, found this obvious neglect of his 1916 paper inexcusable. It led to a growing sense of embarrassment and to the beginning of a long-lasting grudge between the two men, for Lewis believed that Langmuir had never observed the established rules of priority regarding the origin of the cubic atom and the shared electron pair.[53] Langmuir's octet theory, Lewis pointed out, was identical to what he had concisely yet completely developed in his 1916 paper on valence. In a letter to Langmuir in 1919, Lewis wrote:

> To be perfectly candid I think there is a chance that the casual reader may make a mistake which I am sure you would be the last to encourage. He might think that you were proposing a theory which in some essential respects differed from my own, or one which was based upon some vague suggestions of mine which had not been carefully thought out. . . . It seems to me that the views which I presented were about as definite and concrete as was possible considering the condensed form of publication. I think if any confusion should arise it would be due perhaps to points of nomenclature. For example, while I speak of a group of eight, you speak of an octet.

To refer to the Lewis-Langmuir theory of valence, as chemists often did, seemed to imply some sort of collaboration between them, and to Lewis this was out of the question.[54]

Actually, even before Langmuir undertook his development of the Lewis theory, Lewis's associates were applying it to the problems of va-

---

[53]Lewis to A. B. Lamb, January 13, 1920, Lewis Papers, Office of Dean, College of Chemistry, University of California, Berkeley. See also Lewis to William A. Noyes, July 13, 1926, Lewis Papers. Langmuir seemed to give Lewis some priority when he wrote in 1919: "This theory, which assumes an atom of the Rutherford type, and is essentially an extension of Lewis's theory of the 'cubical atom' . . ." ("The Structure of Atoms and the Octet Theory of Valence," *Proc. Nat. Acad. Sci.* 5 [1919], 252). Again, in 1920 Langmuir wrote: "This work of Lewis has been the basis and the inspiration of my work on valence and atomic structure" ("Structure of Atoms and Its Bearing on Chemical Valence," p. 388). See also Langmuir, *Collected Works*, 6:98.

[54]Lewis to Langmuir, July 9, 1919, and Lewis to Lamb, January 13, 1920, Lewis Papers; Lewis, *Valence*, p. 87. In this regard see Joel Hildebrand's comparison of Lewis and Langmuir: "Lewis was an indoor man, an omniverous reader, much given to reflection, a scientific philosopher. Langmuir frequented the out-of-doors; he observed natural phenomena with a keen eye. In his Hitchcock lectures he told about lying on the ice of Lake George studying with a magnifying glass a curious pattern of bubbles he had noticed in the ice. I could not imagine Lewis doing anything of the kind. . . . Langmuir's mind was the more inductive; that of Lewis was more deductive" ("Irving Langmuir's Philosophy of Science, with an Introduction by J. H. Hildebrand," in Langmuir, *Collected Works*, 12:234).

lence and molecular structure. In 1918 F. Russell von Bichowsky (1889–1951), who had studied under Lewis at Berkeley, devised a theory of color for inorganic compounds based on the Lewis valence theory. He was also the first after Lewis to use the dot notation to represent the chemical bonds in a molecule. The following year, Joel Hildebrand (b. 1881), Lewis's longtime colleague at Berkeley, introduced the Lewis theory and electron dot notation in his textbook *Principles of Chemistry*.[55]

Within a short time other chemists in the United States, including Berkeley graduate Ermon D. Eastman (1891–1945), Ernest C. Crocker (1888–1964), Wallace H. Carothers (1896–1937), James B. Conant (1893–1978), and Howard J. Lucas (1885–1963), began applying the Lewis theory to the reactions of organic compounds. Some of the investigations they carried out were the study of the electronic structures of double- and triple-bonded molecules, the addition reactions of the double bond, the addition compounds of ketones, and the structures of aromatic molecules.[56]

In a publication on molecular volumes, Robert N. Pease (1895–1964) at Princeton analyzed the molecular volumes of water, ammonia, and methane according to the Lewis valence theory. At Berkeley, Wendell Latimer (1893–1955) and Worth Rodebush (1887–1959) used Lewis's theory when they introduced the idea of a hydrogen bond in their discussion on the association of liquids in 1920.[57] British chemists and physicists, among

[55] F. Russell von Bichowsky, "The Color of Inorganic Compounds," *J.A.C.S.* 40 (1918), 500–508; Joel H. Hildebrand, *Principles of Chemistry*, pp. 285–86.

[56] Ermon D. Eastman, "Double and Triple Bonds, and Electron Structures in Unsaturated Molecules," *J.A.C.S.* 44 (1922), 438–51; Ernest C. Crocker, "Application of the Octet Theory to Single-Ring Aromatic Compounds," *J.A.C.S.* 44 (1922), 1618–30; Wallace H. Carothers, "The Double Bond," *J.A.C.S.* 46 (1924), 226–36; James B. Conant, "Addition Reactions of the Carbonyl Group Involving the Increase in Valence of a Single Atom," *J.A.C.S.* 43 (1921), 1705–14; Howard J. Lucas and Archibald Y. Jameson, "Electron Displacement in Carbon Compounds: I. Electron Displacement versus Alternate Polarity in Aliphatic Compounds," *J.A.C.S.* 46 (1924), 2475–82; Howard J. Lucas and Hollis W. Moyse, "Electron Displacement in Carbon Compounds: II. Hydrogen Bromide and 2-Pentene," *J.A.C.S.* 47 (1925), 1459–62; Howard J. Lucas, Thomas P. Simpson, and James M. Carter, "Electron Displacement in Carbon Compounds: III. Polarity Differences in Carbon-Hydrogen Unions," *J.A.C.S.* 47 (1925), 1462–69; Howard J. Lucas, "Electron Displacement in Carbon Compounds: IV. Derivatives of Benzene," *J.A.C.S.* 48 (1926), 1827–38. Eastman was at Berkeley, Crocker at M.I.T., Carothers at the University of Illinois, Conant at Harvard, and Lucas at the California Institute of Technology.

[57] Robert N. Pease, "An Analysis of Molecular Volumes from the Point of View of the Lewis-Langmuir Theory of Molecular Structure," *J.A.C.S.* 43 (1921), 991–1004; Wendell M. Latimer and Worth H. Rodebush, "Polarity and Ionization from the Standpoint of the Lewis Theory of Valence," *J.A.C.S.* 42 (1920), 1430–32.

them Nevil V. Sidgwick, Ralph H. Fowler (1889–1944), Robert Robinson (1886–1975), William O. Kermack (1898–1970), and Thomas Lowry (1874–1936), gave a favorable reception to the shared electron pair.[58]

Lewis, of course, contributed to the theory's development and acceptance with his presentation at the Faraday meeting in 1923 and above all with the publication that same year of his monograph *Valence and the Structure of Atoms and Molecules*.[59] This rather brief volume of only 172 pages, which Lewis claimed in the preface would soon belong "to the ephemeral literature of science," instead became a classic in the history of valence theory. It contained not only all of the ideas Lewis had initially presented in his 1916 paper but also those which very likely would have appeared before 1923 had it not been for World War I. Its contents included the Lewis theory of acids and bases (the electron pair donor–electron pair acceptor theory), an examination of double and triple bonds in molecules, and a discussion on the source of chemical affinity, which Lewis still believed to be magnetic in origin.

In his treatment of acids and bases Lewis showed in considerable detail that the displacement of an electron pair or pairs accounted satisfactorily for a substituent's effect on the strength of an organic acid. The increase in acetic acid's ionization constant upon successively substituting chlorine atoms for hydrogen resulted, he said, because the more electronegative chlorine atom drew the bonding electron pair away from the carbon atom of the methyl group CH_3:

$$\begin{array}{cccc}
\text{H}\!:\!\ddot{\text{O}}\!: & \text{H}\!:\!\ddot{\text{O}}\!: & :\!\ddot{\text{C}}\text{l}\!:\!\ddot{\text{O}}\!: & :\!\ddot{\text{C}}\text{l}\!:\!\ddot{\text{O}}\!: \\
\text{H}\!:\!\text{C}\!:\!\text{C}\!:\!\ddot{\text{O}}\!:\!\text{H} < :\!\ddot{\text{C}}\text{l}\!:\!\text{C}\!:\!\text{C}\!:\!\ddot{\text{O}}\!:\!\text{H} < :\!\ddot{\text{C}}\text{l}\!:\!\text{C}\!:\!\text{C}\!:\!\ddot{\text{O}}\!:\!\text{H} < :\!\ddot{\text{C}}\text{l}\!:\!\text{C}\!:\!\text{C}\!:\!\text{O}\!:\!\text{H} \\
\text{H} & \text{H} & \text{H} & :\!\ddot{\text{C}}\text{l}\!:
\end{array}$$

---

[58] See "The Electronic Theory of Valency: A General Discussion," *Trans. Faraday Soc.* 19 (1923), 451–543, for the comments of these and other participants, including William A. Noyes, Robert Robinson, Charles R. Bury, J. J. Thomson, and William H. Bragg. Three other related articles are: Arthur Lapworth, "A Theoretical Derivation of the Principle of Induced Alternate Polarities," *J. Chem. Soc.* 121 (1922), 416–27; William O. Kermack and Robert Robinson, "An Explanation of the Property of Induced Polarity of Atoms and an Interpretation of the Theory of Partial Valences on an Electronic Basis," *J. Chem. Soc.* 121 (1922), 427–40; Martin D. Saltzman, "The Robinson-Ingold Controversy," *Journal of Chemical Education* 57 (1980), 484–88.

[59] Lewis was at first somewhat reluctant to write his monograph on valence because he was already committed to writing his textbook *Thermodynamics* (Lewis to William A. Noyes, September 30, 1919, Lewis Papers). Perhaps because Langmuir had almost completely overshadowed his original contribution to valence theory, he eventually wrote *Valence*.

This in turn displaced the electron pair between the hydrogen and oxygen atom toward the (now) more positive carbon of the methyl group and away from the acidic hydrogen of the carboxyl group COOH. A weakening of the oxygen-hydrogen bond followed, and hence an easier removal of the acidic hydrogen. The effect would, therefore, become more pronounced as the number of substituted chlorine atoms increased from one to three.[60]

Lewis's explanation of substituted organic acid strength thus differed from the interpretation K. George Falk and Harry Fry had given a few years earlier. These authors, in accordance with their theories of polar valences, had argued that the substituent's influence on acid strength resulted from its ability to donate or to receive a single electron when it formed a bond with one of the acid's carbon atoms, in particular with the alpha carbon atom, the carbon atom adjacent to the carboxyl group. An electropositive substituent, since it gave up an electron in bond formation, increased the concentration of negative charge around the carboxyl group and made it more difficult to remove the carboxyl group's positively charged hydrogen atom. The ionization constant, therefore, decreased. Conversely, an electronegative substituent, one that received an electron in bond formation, decreased the negative charge in the region of the carboxyl group and consequently increased the acid's ionization constant.[61]

Falk and Fry also tried to show that the carbon atoms in an open chain compound carried alternate positive and negative charges. Julius Stieglitz, Milton T. Hanke (1893–1961), and Karl K. Koessler (1880–1928) at Chicago supported their hypothesis. But most of the early evidence for alternate polarities appeared in a paper that Eustace Cuy (1898–1925) at Clark University published in 1920.[62]

Cuy based his arguments on the following: (1) the addition of hydrogen halides to ethylene homologs and allene homologs; (2) the rearrangement of alkyl bromides; and (3) what Cuy considered his strongest argument,

[60]Lewis, *Valence*, p. 139. Lewis treated this same topic briefly in his 1916 paper "The Atom and the Molecule," p. 782.

[61]K. George Falk, "The Electron Conception of Valence: II. The Organic Acids," *J.A.C.S.* 33 (1911), 1140–52; Harry S. Fry, "A Critical Survey of Some Recent Applications of the Electron Conception of Valence," *J.A.C.S.* 34 (1912), 664–73. See chapter 4.

[62]Julius Stieglitz, "The Electron Theory of Valence as Applied to Organic Compounds," *J.A.C.S.* 44 (1922), 1293–313, 1833–34; Milton T. Hanke and Karl K. Koessler, "The Electronic Constitution of Acetoacetic and Citric Acids and Some of Their Derivatives," *J.A.C.S.* 40 (1918), 1726–32; Eustace Cuy, "The Electronic Constitution of Normal Carbon Chain Compounds, Saturated and Unsaturated," *J.A.C.S.* 40 (1920), 503–14.

the alternating character of the melting point curves of hydrocarbon homologous series.

Howard J. Lucas and Archibald Y. Jameson (1902–35) at the California Institute of Technology showed in 1924 that the alternating character of the melting point curves actually resulted from differences in the crystal lattice structure and not from polarity within the molecule.[63] They demonstrated convincingly that the theory of electron pair displacement was far superior to alternate polarities in accounting for the addition and rearrangement reactions of organic molecules. Indeed, in the period 1924–26 Lucas and his collaborators provided the evidence that led to the extensive development and firm establishment of the electron pair displacement theory.[64]

Lewis used electron pair displacement to challenge another of the polar theory's fundamental postulates, the *real* existence of electromers. These were molecules that had an identical arrangement of atoms but a different distribution of the electronic charges. William A. Noyes in 1913 proposed the existence of two nitrogen trichloride electromers,

$$N^{+++}Cl_3^- \qquad \text{and} \qquad N^{---}Cl_3^+ ,$$

to account for the fact that acidic and basic hydrolysis produced, respectively, ammonia, NH_3, and nitrous acid, HONO, and he searched vainly for a number of years to isolate them. According to Lewis, nitrogen trichloride's behavior followed simply by assuming that the electron pair between nitrogen and chlorine, indeed between any pair of atoms, occupied a number of intermediate positions:

$$\overset{\cdot\cdot}{\underset{\cdot\cdot}{:Cl}}:\overset{\cdot\cdot}{N}:\overset{\cdot\cdot}{\underset{\cdot\cdot}{Cl}}:$$
$$\overset{\cdot\cdot}{\underset{\cdot\cdot}{:Cl}}:$$

In acidic hydrolysis, the nitrogen atom appeared more negative, just as it did in the ammonia molecule, because the electron pair remained with it.

[63] Lucas and Jameson, "Electron Displacement in Carbon Compounds: I," pp. 2475–82; Morris S. Kharasch and Frederick R. Darkis, "The Theory of Partial Polarity of the Ethylene Bond and the Existence of Electro-Isomerism," *Chemical Reviews* 5 (1928), 572–73.

[64] Lucas and Jameson, "Electron Displacement in Carbon Compounds: I," pp. 2475–82; Lucas and Moyse, "Electron Displacement in Carbon Compounds: II," pp. 1459–62; Lucas, Simpson, and Carter, "Electron Displacement in Carbon Compounds: III," pp. 1462–69; Lucas, "Electron Displacement in Carbon Compounds: IV," pp. 1827–38.

On the other hand, in basic medium the electron pair stayed with chlorine, and the nitrogen atom seemed to be more positive, as in nitrous acid.[65] A single Lewis formula with its shifting electron pairs rather than a pair of electromers sufficed to describe nitrogen trichloride's hydrolysis.

Instead of the electromers of benzenesulfonic acid

$$C_6H_5{}^+ \rightarrow SO_3{}^- \qquad \text{and} \qquad C_6H_5{}^- \leftarrow SO_3{}^+$$
$$\text{(I)} \qquad\qquad\qquad\qquad \text{(II)}$$

that Fry used to account for the acid's hydrolysis to phenol, C_6H_5OH, and sulfurous acid, H_2SO_3, in basic medium (I) and to benzene, C_6H_6, and sulfuric acid, H_2SO_4, in acid medium (II),[66] Lewis suggested the formula:

$$\ddot{\underset{\displaystyle :\ddot{O}:}{\overset{\displaystyle :\ddot{O}:}{C_6H_5 : \underset{}{S} : \ddot{O}:}}}$$

The bonding pair between phenyl, C_6H_5, and sulfur shifted toward one or the other group. In acid medium when the molecule broke at this point, the phenyl group, if it retained possession of the bonding pair, combined with a positive hydrogen ion, giving benzene, but upon losing possession of the pair it combined with a negative hydroxyl ion from the basic medium, forming phenol.[67] Thus, in the Lewis theory displacement of the two electrons constituting the single bond occurred within the molecule but always as a pair. Only in the tautomers of the double- and triple-bonded hydrocarbons did one of the bonding pairs split, leaving a single unpaired electron on each carbon atom.

Lewis had originally suggested in his 1916 paper that a bonding electron pair could occupy all possible positions between two atoms. However, by 1923 the idea of quantized electron orbits in an atom had gained considerable acceptance among physicists and chemists. In *Valence and the Structure of Atoms and Molecules* Lewis also assumed that the bonding

---

[65] William A. Noyes, "An Attempt to Prepare Nitro-Nitrogen Trichloride, an Electromer of Ammono-Nitrogen Trichloride," *J.A.C.S.* 35 (1913), 767–75; Lewis, *Valence*, pp. 85, 132; Lewis, "Valence and the Electron," p. 454.

[66] Harry S. Fry, "Positive and Negative Hydrogen, the Electronic Formula of Benzene, and the Nascent State," *J.A.C.S.* 36 (1914), 265; Lauder W. Jones, "The Beckmann Rearrangement of Hydroxamic Acids," *Amer. Chem. Journal* 48 (1912), 26.

[67] Lewis, *Valence*, p. 85; idem, "Valence and the Electron," p. 454.

pair occupied one of a finite though large number of specific positions between the two atoms. These different electronic structures or tautomers conceivably might resemble the electromers of Noyes, Falk, and Fry, he said, but they differed from them because the unequal sharing of an electron pair and not the transfer of a single electron produced the various electronic structures. Unlike electromers, which were physically distinct and hence separable, Lewis believed that his "tautomeric" structures were transient and had no real physical existence. In this regard they were very much like the molecular resonance forms that Linus Pauling introduced a few years later.[68]

Alfred Werner's coordination theory had long influenced Lewis's thinking on inorganic structures. It contributed greatly to his arriving at the electron dot structures of the ammonium ion and the oxyacids. While Lewis's 1916 paper only implicitly referred to Werner's work, in *Valence and the Structure of Atoms and Molecules* Lewis gave a much more thorough interpretation of Werner's coordination compounds using the shared electron pair. According to Werner, in the cobalt complex ion, $Co[NH_3]_6^{+++}$, the central cobalt atom held each of the six octahedrally arranged ammonia molecules with a secondary (un-ionized) valence. In the Lewis theory, Werner's secondary valence was simply a nonpolar bond, the result of the donation by each ammonia molecule, $:NH_3$, of its unshared electron pair to cobalt. In other coordination compounds containing groups such as H_2O, NO_2^-, or Cl^-, bond formation occurred in exactly the same way: each of the coordinated groups donated an unshared pair of electrons to the central atom.[69]

Nevil V. Sidgwick at Oxford University in the 1920s extended considerably the application of the shared electron pair to coordination compounds. He pointed out, as Lewis had, that the coordination number Werner assigned to the central atom was equal to the number of nonpolar bonds it formed. Werner's formulas followed, Sidgwick argued, from assuming (1) that only two kinds of bonds, polar and nonpolar, corresponding to Werner's primary and secondary valences, existed, and (2) that the charge on the coordinated group determined the compound's net charge. A negative monovalent group like nitrite, NO_2^-, or chloride, Cl^-, contributed a negative charge to the compound, but coordinated neutral molecules

[68] Lewis, *Valence*, p. 132; idem, "Valence and the Electron," pp. 456–67.
[69] Lewis, *Valence*, pp. 114–15.

such as NH_3 and H_2O contributed none. Since the coordinated group usually provided both electrons on forming the nonpolar bond, Sidgwick appropriately called this bond a coordinate covalent bond. The publication of Sidgwick's classic *The Electronic Theory of Valency* in 1927 provided both an excellent summary of his work and a masterly exposition of many of Lewis's ideas.[70]

There is little doubt that by 1923 Lewis's lucid and concisely written publications on electronic structures of atoms and molecules and their chemical reactions did much to convince the chemical community to acknowledge the superiority of his valence theory. William A. Noyes, a long-time advocate of a polar or electrostatic valence theory, accepted the Lewis theory in 1921.[71] As recently as 1920 and 1921, K. George Falk had published two monographs, *Chemical Reactions: Their Theory and Mechanisms* and *The Chemistry of Enzyme Actions*, in which he still adhered to a polar theory of valence. But by 1924, in the introduction to the second edition of *The Chemistry of Enzyme Actions*, Falk had also accepted the Lewis theory. The Lewis electron pair theory, he wrote, was proving to be extremely useful as a classifying principle for chemical reactions and in developing new lines of research.[72]

Indeed, by the mid-1920s, Lewis had placed the electron pair theory on a firm qualitative basis. The next step taken was its quantitative mathematical development. The application of wave mechanics to the chemical bond completed this final phase of the Lewis valence theory.

Conclusion

With the use of a single conception, the shared electron pair, Lewis offered a plausible explanation of the chemical bond in polar and nonpolar molecules. Originally without theoretical foundation when introduced in

[70] Nevil V. Sidgwick, "Coordination Compounds and the Bohr Atom," *J. Chem. Soc.* 123 (1923), 725–30; idem, "The Nature of the Nonpolar Link," pp. 472–75; idem, *The Electronic Theory of Valency*; Lewis, *Valence*, pp. 88–96, 137–42, 147–52; Lewis, "The Magnetochemical Theory," *Chemical Reviews* 1 (1924), 231–48.

[71] William A. Noyes, "An Attempt to Prepare Nitro-Nitrogen Trichloride: II. The Conduct of Mixtures of Nitrogen and Chlorine in a Flaming Arc," *J.A.C.S.* 43 (1921), 1774–82. As late as 1928, Fry had not abandoned the electrostatic theory (*Chemical Reviews* 5 [1928], 557–69).

[72] K. George Falk, *Chemical Reactions: Their Theory and Mechanism*; idem, *The Chemistry of Enzyme Actions*; idem, *The Chemistry of Enzyme Actions*, 2d ed., p. 17.

1916, the shared electron pair provided an accurate qualitative explanation and a predictive utility when applied to the polarity of substituted organic molecules—for example, the directive ability of groups attached to the benzene ring.

The shared electron pair enabled Lewis to develop a theory of acids and bases more universal than the Brønsted-Lowry proton donor–proton acceptor theory. With it he correctly accounted for the variation in organic acid strength upon substituting different atoms or groups in the acid.

The Lewis theory also removed the confusion that had long surrounded the meaning of Werner's primary and secondary valences in coordination compounds. Werner's primary valence was simply an ionic or polar bond; his secondary valence was a shared electron pair or nonpolar bond.

Because the Lewis shared electron pair offered such a satisfying electronic interpretation of chemical reactions and the structures of atoms and molecules, it quickly gained wide acceptance among chemists. Quantum mechanics has since provided the theoretical justification for the shared electron pair. But the shared pair remains today the core of practical chemistry, or the level of theory commonly used by chemists.

Epilogue

By 1923, after observing the behavior of atoms and molecules with an odd number of electrons, Lewis had concluded that the electrons in an atom were paired. The ionization potentials of odd-numbered atoms, he pointed out, were usually lower than those of adjacent even-numbered atoms, and all odd-numbered atoms were paramagnetic. Molecules containing an odd number of electrons were very few in number and were less stable than even-numbered molecules. They almost always reacted to form a structure with an even number of electrons. This last fact—the scarcity of molecules containing an odd number of electrons—was most important because it convinced Lewis that the nonpolar valence bond consisted of an electron pair shared between two atoms.

To account for the novel idea of two negative electrons attracting one another and hence pairing, Lewis assumed that a magnetic attraction existed between the electrons. But Lewis's hypothesis of electron pairing received no theoretical support until 1925, when Wolfgang Pauli (1900–1958) in Hamburg enunciated his famous exclusion principle.[1] According to this principle, no two electrons in an atom could have the same numerical value for each of the four quantum numbers used to describe an electron's energy state. In a pair of electrons having identical values for the n, l, and m quantum numbers, each had to have its spin axis, designated by the s quantum number, oriented in the opposite direction to the other. In other words, the two electrons were paired.

Pauli's principle limited the total number of electrons possible in an atom and permitted their correct distribution within each atom. Applied to the hypothesis of electron pairing in the chemical bond, it appeared equiv-

[1] Wolfgang Pauli, "Über den Zusammenhang des Abschlusses der Elektronengruppen im Atom mit der Komplexstruktur der Spektren," *Zeit. Physik* 31 (1925), 765–83; idem, "Zur Quantenmechanik des magnetischen Elektrons," *Zeit. Physik* 43 (1927), 601–23.

alent to what Lewis in 1924 had suggested in his magnetochemical theory. If the electrons in a nonbonded atom were paired magnetically, Lewis wrote, then a magnetic pairing also occurred between two electrons in different atoms upon forming a nonpolar bond. Indeed, Lewis's shared electron pair consisted of two electrons in identical energy states except for their opposing or paired spins.[2] Thus its saturation followed directly from Pauli's principle.

But while Lewis believed that the energy of the nonpolar bond was magnetic, Werner Heisenberg (1901–76) in Copenhagen and Paul Dirac (b. 1902) at Cambridge used the new quantum mechanics to show independently in 1926 that the electron interactions in a molecule resulted from a resonance or exchange effect. The following year, Walther Heitler (b. 1904) and Fritz London (1900–1954) in Zurich applied the Heisenberg-Dirac resonance or exchange effect to account for the energy of the electron pair bond in the hydrogen molecule.[3] Indeed, John H. Van Vleck (1899–1980) at the University of Wisconsin pointed out in 1928 that any magnetic attraction was almost entirely negligible in comparison with the electrical attraction, amounting to less than 0.1 percent of the electrical attraction in the hydrogen molecule.[4]

The resonance or exchange energy, according to Heitler and London, was the sum of the attractive and repulsive forces existing between the two positive hydrogen nuclei and the two negative electrons. It was due mainly to an exchange in position by the two electrons forming the bond and resulted in each electron's partially associating itself with one nucleus and partially with the other. Their interpretation of the bond led to approximate agreement between the theoretical and experimental values of the hydro-

[2]Linus Pauling, "The Shared-Electron Chemical Bond," *Proc. Nat. Acad. Sci.* 14 (1928), 359–62; G. N. Lewis, "The Magnetochemical Theory," *Chemical Reviews* 1 (1924), 231–48.

[3]Werner Heisenberg, "Mehrkörperproblem und Resonanz in der Quantenmechanik," *Zeit. Physik* 38 (June 1926), 411–26; Paul Dirac, "On the Theory of Quantum Mechanics," *Proc. Roy. Soc.* A 112 (August 1926), 661–77; Walther Heitler and Fritz London, "Wechselwirkung neutraler Atome und homöopolare Bindung nach der Quantenmechanik," *Zeit. Physik* 44 (1927), 455–72. Yoshikatsu Sugiura improved the mathematics of Heitler and London's work with "Über die Eigenschaften des Wasserstoffmoleküls im Grundzustande," *Zeit. Physik* 45 (1927), 484–92.

[4]John H. Van Vleck presented "The New Quantum Mechanics" at the Symposium on Atomic Structure and Valence, St. Louis Meeting of the American Chemical Society, April 1928. All of the papers given at the symposium are in *Chemical Reviews* 5 (1928), Van Vleck's on pages 467–507.

gen molecule's heat of dissociation, moment of inertia, and oscillational frequency.[5]

By 1928 the quantum mechanical treatment of the electron pair bond in the hydrogen molecule seemed beyond doubt. In that same year, London suggested that the resonance or exchange energy of two electrons, one from each of the atoms undergoing combination, was in every case the energy of the nonpolar bond. The two electrons forming the bond had paired their spins and were, therefore, incapable of further combination. Linus Pauling (b. 1901), at the California Institute of Technology, agreed that London's theory was in simple cases entirely equivalent to the Lewis theory. Quantum mechanics, he added, provided a theoretical justification for the rule of eight or octet theory at least for the first row elements:

The shared electron structures assigned by Lewis to molecules such as H_2, F_2, Cl_2, CH_4, etc., are also found for them by London. The quantum mechanics explanation of valence is, moreover, more detailed and correspondingly more powerful than the old picture. For example, it leads to the result that the number of shared bonds possible for an atom of the first row is not greater than four, and for hydrogen not greater than one, for, neglecting spin, there are only four quantum states in the L-shell and one in the K-shell.[6]

In 1931 Pauling gave a formal quantum mechanical proof to the Lewis idea that a pair of electrons constituted the nonpolar bond. Like Heitler, London, and John C. Slater (1900–1976), who was at the Massachusetts Institute of Technology, he believed the atoms in a molecule retained to some degree their electron configurations even when chemically bonded. Their interpretation of bond formation is called the valence bond theory.[7]

Shortly after the introduction of the valence bond theory, Friedrich Hund (b. 1896) in Leipzig, John E. Lennard-Jones (1894–1954) in Cambridge, and Robert Mulliken (b. 1896) at New York University continued the independent treatment of the hydrogen molecule that Edward Condon

[5]Sugiura, "Über die Eigenschaften," pp. 484–92. Hubert M. James and Albert S. Coolidge gave a thoroughly accurate treatment of the hydrogen molecule, showing that the quantum mechanical energy was in full agreement with the experimental value ("The Ground State of the Hydrogen Molecule," *J. Chem. Physics* 1 [1933], 825–35).

[6]Fritz London, "Zur Quantentheorie der homöopolaren Valenzzahlen," *Zeit. Physik* 46 (1928), 455–77; Pauling, "The Shared-Electron Chemical Bond," p. 360.

[7]Linus Pauling, "The Nature of the Chemical Bond: I. Application of Results Obtained from the Quantum Mechanics and from a Theory of Paramagnetic Susceptibility to the Structure of Molecules," *J.A.C.S.* 53 (1931), 1366–1400; John D. Slater, "Note on Hartree's Method," *Physical Review* 35 (1930), 210–11; Slater, "Directed Valence in Polyatomic Molecules," *Physical Review* 37 (1931), 481–89.

(1902–74) had published in 1927. Condon was at the time a National Research Fellow at Göttingen and Munich with Max Born and Arnold Sommerfeld. Like Condon, Hund, Lennard-Jones and Mulliken approached the problem of bond formation by assuming that the electrons of the combining atoms belonged to the molecule as a whole.[8] This is the molecular-orbital theory. It has been particularly important in understanding the bonding in aromatic compounds,[9] conjugated systems, and the oxygen molecule.

The two theories agree in most of their conclusions, though the molecular orbital theory is conceptually simpler. In either case, as Pauling wrote in the preface of *The Nature of the Chemical Bond*, the chemical bond still remains a shared pair of electrons, just as G. N. Lewis first proposed in his classic 1916 paper "The Atom and the Molecule."[10]

[8] Edward U. Condon, "Wave Mechanics and the Normal State of the Hydrogen Molecule," *Proc. Nat. Acad. Sci.* 13 (1927), 466–70; Friedrich Hund, "Zur Deutung der Molekülspektren," in 4 pts., *Zeit. Physik* 40 (November 1926), 742–64, 42 (February 1927), 93–120, 43 (May 1927), 805–26, and 51 (October 1928), 759–95; Hund, "Chemical Binding," *Trans. Faraday Soc.* 25 (1929), 646–48; John E. Lennard-Jones, "The Electronic Structure of Some Diatomic Molecules," *Trans. Faraday Soc.* 25 (1929), 668–86; Robert Mulliken, "The Assignment of Quantum Numbers for Electrons in Molecules: I," *Physical Review* 32 (1928), 186–222.

[9] Erich Hückel, "Zur Quantentheorie der Doppelbindung," in 3 pts., *Zeit. Physik* 60 (1930), 423–56, 70 (1931), 204–86, and 72 (1931), 310–37.

[10] Linus Pauling, *The Nature of the Chemical Bond*, p. vii.

Bibliography

Unpublished Material

Berkeley, California. University of California. Office of the Dean, College of Chemistry. G. N. Lewis Papers.

Fry, Harry S. "An Hypothesis Relative to the Constitution of Benzene Nucleus: An Application of the Corpuscular Atomic Conception of Positive and Negative Valencies to the Constituent Atoms of Benzene" (paper delivered at meeting of Cincinnati section of American Chemical Society, Cincinnati, Ohio, January 1908).

Hildebrand, Joel H. "Chemistry, Education, and the University of California." Interview conducted by Edna Tartaul Daniel. Berkeley: Regional Cultural History Project, General Library, University of California, 1962.

Published Material

Abegg, Richard. "Attempt at a Theory of Valency and of Molecular Compounds." *J. Chem. Soc. Abstracts* 84 (1903), 536.

―――. "Die Valenz und das periodische System: Versuch einer Theorie der Molekülarverbindungen." *Zeit. anorg. Chemie* 39 (1904), 330–80.

―――. *The Electrolytic Dissociation Theory.* Translated by Carl L. von Ende from the first German edition, 1903. New York: John Wiley & Sons, 1907.

―――. "Valency." *B.A.A.S. Report* 77 (1907), 481.

―――. "Versuch einer Theorie der Valenz und der Molekülarverbindungen." *Christiania Videnskabs-Selskabet Skrifter* 12 (1902). Reprinted with same title as pamphlet, Christiana: J. Dybwad, 1902.

Abegg, Richard, and Bodländer, Guido. "Die Elektroaffinität, ein neues Prinzip der chemischen Systematik." *Zeit. anorg. Chemie* 20 (1899), 453–99.

――― and ―――. "Electro-affinity as a Basis for the Systematization of Inorganic Compounds." *Amer. Chem. Journal* 28 (1902), 220–28.

Abegg, Richard, and Herz, W[alther]. *Practical Chemistry.* Translated by Harry T. Calvert from the German edition of 1900. London: Macmillan & Co., 1901.

Allen, Herbert Stanley. "The Case for a Ring Electron." *Chemical News* 118 (21 March 1919), 137–39, and (28 March 1919), 149–51.

Andrade, Edward N. da C. "The Birth of the Nuclear Atom." *Scientific American* 195 (November 1956), 93–104.

264   Bibliography

————. *The Structure of the Atom*. 3rd ed. London: George Bell and Sons, 1927.

Armstrong, Henry. "An Explanation of the Laws Which Govern Substitution in the Case of Benzenoid Compounds." *J. Chem. Soc.* 51 (1887), 258–68, 583–90.

————. "Valency." *Encyclopaedia Britannica*, 11th ed., Vol. 27, pp. 847–50.

Arrhenius, Svante. "Development of the Theory of Electrolytic Dissociation." *Nobel Lectures: Chemistry 1901–1912*. Amsterdam: Elsevier Publishing Company, 1966.

————. "Investigations of the Galvanic Conductivity of Electrolytes: I. Determination of the Conductivity of Extremely Dilute Solutions by Means of the Depolariser"; "II. Chemical Theory of Electrolytes." *Bihang till Kongliga Svenska Vetenskaps-Akademiens Handlingar* 8, nos. 13 and 14 (1884).

————. *Theories of Chemistry*. London: Longmans, Green, and Co., 1907.

————. *Theories of Solutions*. New Haven: Yale University Press, 1912.

————. "*Über die Dissociation der in Wasser gelösten Stoffe.*" *Zeit. phys. Chemie* 1 (1887), 631–48.

Arsem, William C. "A Theory of Valency and Molecular Structure." *J.A.C.S.* 36 (1914), 1655–75.

Baeyer, Adolf von. "Über das Isatin." *Berichte* 15 (1882), 2093–2102.

————. "Über die Konstitution des Benzols." *Annalen der Chemie* 245 (1888), 103–90; 269 (1892), 145–206, 403.

————. "Über die Verbindungen der Indigogruppe." *Berichte* 16 (1883), 2188–2204.

————. "Über Polyacetylenverbindungen." *Berichte* 18 (1885), 2269–81.

Baker, Herbert B. "Influence of Moisture on Chemical Change." *J. Chem. Soc.* 65 (1894), 611–24.

Baly, Edward C. C., and Desch, Cecil H. "The Ultra-violet Absorption Spectra of Certain Enol-keto Tautomerides: Part II." *J. Chem. Soc.* 87 (1905), 766–84.

Barkla, Charles G. "Energy of Secondary Röntgen Radiation." *Phil. Mag.* 7 (1904), 543–60.

————. "Note on the Energy of Scattered X-radiation." *Phil. Mag.* 21 (1911), 648–52.

Bates, Stuart J. "The Electron Conception of Valence." *J.A.C.S.* 36 (1914), 789–93.

Becker, August. "Messungen an Kathodenstrahlen." *Annalen der Physik* 17 (1905), 381–480.

Becquerel, Jean. "Sur un phénomène attribuable à des électrons positifs, dans le spectre d'etincelle de l'yttrium." *Comptes rendus* 146 (1908), 683–85.

Berzelius, J. J. "Essay on the Cause of Chemical Proportions, and on Some Circumstances Relating to Them: Together with a Short and Easy Method of Expressing Them." *Thomson's Annals of Philosophy* 2 (December 1813), 443–54.

Bichowsky, F. Russell von. "The Color of Inorganic Compounds." *J.A.C.S.* 40 (1918), 500–508.

Bjerrum, Niels J. *Selected Papers*. Edited by Friends and Co-workers. Copenhagen, 1949.

Bloch, Léon. "Recherches sur les actions chimiques et l'ionisation par barbotage." *Annales de Chimie et de Physique* 22 (1911), 370–417, 441–95, and 23 (1911), 28–144.

Blomstrand, Christian W. *Die Chemie der Jetztzeit.* Heidelberg: C. Winter, 1869.

Bohr, Niels. "Atomic Structure." *Nature* 107 (24 March 1921), 104–7.

———. *On the Constitution of Atoms and Molecules.* Edited by Leon Rosenfeld. New York: W. A. Benjamin, 1963.

———. "On the Constitution of Atoms and Molecules." *Phil. Mag.* 26 (1913), 1–25, 476–502, 857–75.

———. "The Structure of the Atom." Supplement to *Nature* 112 (7 July 1923), 29–44.

Born, Max. "Bemerkungen über die Grösse der Atome." *Zeit. Physik* 2 (1920), 87–89.

———. *The Constitution of Matter.* Translated by E. W. Blair and T. S. Wheeler. New York: E. P. Dutton and Company, 1923.

———. "Dynamik der räumlichen Atomstruktur." *Zeit. Physik* 2 (1920), 83–86.

———. "Über die elektrische Natur der Kohäsionskräfte fester Körper." *Verh. deut. phys. Ges.* 21 (1919), 533–38.

———. "Über kubische Atommodelle." *Verh. deut. phys. Ges.* 20 (1918), 230–39.

———. "Würfelatome, periodisches System und Molekülbildung." *Zeit. Physik* 2 (1920), 380–404.

Born, Max, and Landé, Alfred. "Über die Berechnung der Kompressibilität regulärer Kristalle aus der Gittertheorie." *Verh. deut. phys. Ges.* 20 (1918), 210–16.

Bragg, William L. "The Analysis of Crystals by the X-Ray Spectrometer." *Proc. Roy. Soc.* A 89 (1913), 468–89.

———. "The Arrangement of Atoms in Crystals." *Phil. Mag.* 40 (1920), 169–89.

———. "The Dimensions of Atoms and Molecules." *Nature* 107 (24 March 1921), 107.

Bray, William C., and Branch, Gerald E. K. "Valence and Tautomerism." *J.A.C.S.* 35 (1913), 1440–47.

Bray, William C., and Dowell, Carr Thomas. "Experiments with Nitrogen Trichloride." *J.A.C.S.* 39 (1917), 896–913.

Bredig, Georg. "Über die Affinitätsgrössen der Basen." *Zeit. phys. Chemie* 13 (1894), 289–326.

Broek, Antonius van den. "Die Radioelemente, das periodische System und die Konstitution der Atome." *Phys. Zeit.* 14 (1913), 33–41.

———. "Intra-atomic Charge." *Nature* 92 (27 November 1913), 372–73.

———. "Intra-atomic Charge and the Structure of the Atom." *Nature* 92 (25 December 1913), 476–78.

———. "The Number of Possible Elements and Mendeleev's 'Cubic' Periodic System." *Nature* 87 (20 July 1911), 78.

Brown, Alexander Crum, and Gibson, John. "A Rule for Determining Whether a

Given Benzene Monoderivative Shall Give a Meta-di-derivative or a Mixture of Ortho- and Para-di-derivatives." *J. Chem. Soc.* 61 (1892), 367–69.

Brühl, Julius W. "Neue Beiträge zur Frage nach der Konstitution des Benzols." *J. prakt. Chemie* 49 (1894), 201–94.

Brunel, Roger F. "A Criticism of the Electron Conception of Valence." *J.A.C.S.* 37 (1915), 709–22.

Bykov, G. V. "Historical Sketch of the Electron Theories of Organic Chemistry." *Chymia* 10 (1965), 199–253.

Cady, Hamilton P. "The Electrolysis and Electrolytic Conductivity of Certain Substances Dissolved in Liquid Ammonia." *J. Phys. Chem.* 1 (1897), 707–13.

Campbell, Norman R. "Atomic Structure." *Nature* 106 (25 November 1920), 408.

———. *Modern Electrical Theory.* Cambridge: At the University Press, 1907.

———. "A Static or Dynamic Atom." *Nature* 111 (28 April 1923), 569.

Cardwell, Donald S. L., ed. *John Dalton and the Progress of Science.* Manchester: Manchester University Press, 1968.

Carothers, Wallace H. "The Double Bond." *J.A.C.S.* 46 (1924), 2226–36.

Clarke, F. W., and Dennis, L. M. *General Chemistry.* New York: American Book Company, 1902.

Claus, Adolf. *Theoretische Betrachtungen und deren Anwendung zur Systematik der organischen Chemie.* Freiburg: H. M. Poppen & Sohn, 1866.

———. "Zur Frage nach den Affinitätsgrössen des Kohlenstoffs." *Berichte* 14 (1881), 432–35.

Clausius, Rudolf. "Über die Elektrizitätsleitung in Elektrolyten." *Annalen der Physik* 101 (1857), 338–60.

Cohen, Julius B. *Organic Chemistry.* London: Longmans, Green, and Co., 1907.

Collie, John N. "A Space Formula for Benzene." *J. Chem. Soc.* 71 (1897), 1013–23.

Compton, Arthur Holly, and Rognley, Oswald. "Is the Atom the Ultimate Magnetic Particle?" *Physical Review* 16 (1920), 464–76.

Conant, James B. "Addition Reactions of the Carbonyl Group involving the Increase in Valence of a Single Atom." *J.A.C.S.* 43 (1921), 1705–14.

Condon, Edward U. "Wave Mechanics and the Normal State of the Hydrogen Molecule." *Proc. Nat. Acad. Sci.* 13 (1927), 466–70.

Conn, George K. T., and Turner, Henry D. *The Evolution of the Nuclear Atom.* New York: American Elsevier Publishing Company, 1965.

"Constitution of the Atom, The." *Chemical News* 96 (23 August 1907), 94–95.

Crocker, Ernest C. "Application of the Octet Theory to Single-ring Aromatic Compounds." *J.A.C.S.* 44 (1922), 1618–30.

Crookes, William. "Elements and Meta-elements." *J. Chem. Soc.* 53 (1888), 487–504.

———. "Spectroscopic Researches on the Rare Earths." *J. Chem. Soc.* 55 (1889), 255–85.

Crosland, Maurice P. "The Origins of Gay-Lussac's Law of Combining Volumes of Gases." *Annals of Science* 17 (1961), 1–26.

Cuy, Eustace. "The Electronic Constitution of Normal Carbon Chain Compounds, Saturated and Unsaturated." *J.A.C.S.* 40 (1920), 503–14.

Dalton, John. *A New System of Chemical Philosophy*. Vol. 1, pt. 1. 1808. Facsimile. London: William Dawson & Sons Ltd., n.d.

Daniell, John F. "On the Electrolysis of Secondary Compounds." *Phil. Trans.* 129 (1839), 97–112.

―――. *An Introduction to the Study of Chemical Philosophy*. 2d ed. London: John W. Carter, 1843.

―――. "Second Letter on the Electrolysis of Secondary Compounds." *Phil. Trans.* 130 (1840), 209–24.

Daniell, John F., and Miller, W. A. "Additional Researches on the Electrolysis of Secondary Compounds." *Phil. Trans.* 134 (1844), 1–20.

Davey, Wheeler P. "The Cubic Shapes of Certain Ions as Confirmed by X-ray Crystal Analysis." *Physical Review* 17 (1921), 402–3.

―――. "Radiation." *Journal of the Franklin Institute* 197 (1924), 439–78.

Debye, Peter. *Polar Molecules*. New York: Chemical Catalog Company, 1929.

―――. "Report on Conductivity of Strong Electrolytes in Dilute Solutions." *Trans. Faraday Soc.* 23 (1927), 334–40.

―――. "Zur Theorie der Elektrolyte: II. Das Grenzgesetz für die elektrische Leitfähigkeit." *Phys. Zeit.* 24 (1923), 305–25.

Debye, Peter, and Hückel, Erich. "Zur Theorie der Elektrolyte: I. Gefrierpunktserniedrigung und verwandte Erscheinungen." *Phys. Zeit.* 24 (1923), 185–206.

Debye, Peter, and Scherrer, Paul. "Atombau." *Annalen der Physik* 19 (1918), 474–83.

Dirac, Paul. "On the Theory of Quantum Mechanics." *Proc. Roy. Soc.* A 112 (1926), 661–77.

Drude, Paul. "Optische Eigenschaften und Elektronentheorie." *Annalen der Physik* 14 (1904), 677–725, 936–61.

Dumas, Jean Baptiste. "Considerations générales sur la composition théorique des matières organiques." *Journal de Pharmacie* 20 (May 1834), 261–94.

―――. *Traité de Chimie*. Vol. 5. Paris: Chez Béchet Jeune, 1835.

Eastman, Ermon D. "Double and Triple Bonds, and Electron Structures in Unsaturated Molecules." *J.A.C.S.* 44 (1922), 438–51.

"Electronic Theory of Valency: A General Discussion." *Trans. Faraday Soc.* 19 (1923), 451–543.

Fajans, Kasimir. "Die Stellung der Radioelemente im periodischen System." *Phys. Zeit.* 14 (1913), 136–42.

―――. "Über eine Beziehung zwischen der Art einer radioaktiven Umwandlung und dem elektrochemischen Verhalten der betreffenden Radioelemente." *Phys. Zeit.* 14 (1913), 131–36.

Falk, K. George. *Chemical Reactions: Their Theory and Mechanism*. New York: D. Van Nostrand Company, 1920.

―――. *The Chemistry of Enzyme Actions*. New York: Chemical Catalog Company, 1921.

————. *The Chemistry of Enzyme Actions*. 2d ed. New York: Chemical Catalog Company, 1924.

————. "The Electron Conception of Valence: II. The Organic Acids." *J.A.C.S.* 33 (1911), 1140–52.

Falk, K. George, and Nelson, John M. "Die Aufussung der Valenz als Elektronenwirkung." *J. prakt. Chemie* 88 (1913), 97–128.

———— and ————. "The Electron Conception of Valence." *J.A.C.S.* 32 (1910), 1637–54; "V. Polar and Non-Polar Valence." *J.A.C.S.* 36 (1914), 209–14; "VI. Inorganic Compounds." *J.A.C.S.* 37 (1915), 274–86; "VII. The Theory of Electrolytic Dissociation and Chemical Change." *J.A.C.S.* 37 (1915), 1732–48.

———— and ————. "The Electron Conception of Valency in Organic Chemistry." *School of Mines Quarterly* 30, no. 3 (April 1909), 179–98.

———— and ————. "Some Comments on the Theories of the Structure of Matter." *Science* 46 (7 December 1917), 551–53.

Faraday, Michael. "Experimental Researches in Electricity—Seventh Series." *Phil. Trans.* 124 (1834), 77–122.

————. "On New Compounds of Carbon and Hydrogen and on Certain Other Products Obtained during the Decomposition of Oil by Heat." *Phil. Trans.* 115 (1825), 440–66.

Findlay, Alexander. *A Hundred Years of Chemistry*. 3rd ed. London: Gerald Duckworth & Co., 1965.

Frankland, Edward. "On a New Series of Organic Compounds Containing Metals." *Phil. Trans.* 142 (1852), 417–44.

Frankland, Percy. "Residual Affinity." *Nature* 70 (7 July 1904), 222–23.

Franklin, Edward C., and Kraus, Charles A. "The Electrical Conductivity of Liquid Ammonia Solutions." *Amer. Chem. Journal* 23 (1900), 277–313.

Friend, John Newton. "Electrochemical Conceptions of Valency." *J. Chem. Soc.* 119 (1921), 1040–47.

————. *The Theory of Valency*. London: Longmans, Green, and Co., 1909.

————. "Valency." *J. Chem. Soc.* 93 (1908), 262–64.

Fry, Harry S. "A Critical Survey of Some Recent Applications of the Electron Conception of Valence." *J.A.C.S.* 34 (1912), 664–73.

————. "Die Konstitution des Benzols vom Standpunkte des Korpuscularatomistischen Begriffs der positiven und negativen Wertigkeit." *Zeit. phys. Chemie* 76 (1911), 385–412.

————. "The Electronic Conception of Positive and Negative Valences." *J.A.C.S.* 37 (1915), 2368–73.

————. *The Electronic Conception of Valence and the Constitution of Benzene*. London: Longmans, Green, and Co., 1921.

————. "Interpretations of Some Stereochemical Problems in Terms of the Electronic Conception of Positive and Negative Valences: Part I. Anomalous Behavior of Certain Derivatives of Benzene." *J.A.C.S.* 36 (1914), 248–62; "Part II. Halogen Substitution in the Benzene Nucleus and in the Side Chain." *J.A.C.S.* 36 (1914), 1035–46; "Part III. A Continuation of the Interpretation of the Brown and Gibson Rule." *J.A.C.S.* 37 (1915), 855–63; "Part IV. The Simultaneous Formation of Ortho-, Meta- and Para-Substituted Derivatives of Benzene." *J.A.C.S.* 37 (1915), 863–83; "Part V. A Reply to A. F. Holleman." *J.A.C.S.* 37 (1915), 883–92.

———. "Positive and Negative Hydrogen, the Electronic Formula of Benzene, and the Nascent State." *J.A.C.S.* 36 (1914), 262–72.

———. "A Pragmatic System of Notation for Electronic Valence Conceptions in Chemical Formulas." *Chemical Reviews* 5 (1928), 557–69.

Garrett, Albert E. *The Periodic Law.* New York: D. Appleton & Co., 1909.

Gay-Lussac, Joseph Louis. "Memoir on the Combination of Gaseous Substances with Each Other." *Alembic Club Reprint* no. 4. Edinburgh: W. F. Clay, 1950.

Geiger, Hans, and Marsden, Ernest. "On a Diffuse Reflexion of the α-Particles." *Proc. Roy. Soc.* A 82 (1909), 495–500.

——— and ———. "The Laws of Deflexion of α-Particles through Large Angles." *Phil. Mag.* 27 (1913), 604–23.

Getman, Frederick, and Daniels, Farrington. *Outlines of Theoretical Chemistry.* 5th ed. New York: John Wiley & Sons, 1931.

Giauque, William F. "Gilbert Newton Lewis." *The American Philosophical Yearbook 1946.* Philadelphia: American Philosophical Society, 1947.

Gomberg, Moses. "An Instance of Trivalent Carbon: Triphenylmethyl." *J.A.C.S.* 22 (1900), 757–71.

———. "Triphenylmethyl, ein Fall von dreiwertigem Kohlenstoff." *Berichte* 33 (1900), 3150–63.

———. "On Trivalent Carbon." *Amer. Chem. Journal* 25 (1901), 317–35.

———. "Über das Triphenylmethyl." *Berichte* 34 (1901), 2726–33; 35 (1902), 2397–408.

———. "Über Triphenylmethyl." *Berichte* 36 (1903), 376–88; 37 (1904), 1626–44.

———. "Über Triphenylmethyl Acetat." *Berichte* 36 (1903), 3924–30.

Gomberg, Moses, and Cone, Lee H. "Über Triphenylmethyl." *Berichte* 37 (1904), 2033–51.

Grondahl, Lars O. "Experimental Evidence for the Parson Magneton." *Physical Review* 10 (1917), 586–89.

Hahn, Dorothy A., and Holmes, Mary E. "The Valence Theory of J. Stark from a Chemical Standpoint." *J.A.C.S.* 37 (1915), 2611–26.

Hanke, Milton T., and Koessler, Karl K. "The Electronic Constitution of Acetoacetic and Citric Acids and Some of Their Derivatives." *J.A.C.S.* 40 (1918), 1726–32.

Hartley, Harold. *Studies in the History of Chemistry.* Oxford: Clarendon Press, 1971.

Heilbron, John L. "J. J. Thomson and the Bohr Atom." *Physics Today* 30 (April 1977), 23–30.

Heisenberg, Werner. "Mehrkörperproblem und Resonanz in der Quantenmechanik." *Zeit. Physik* 38 (1926), 411–26.

Heitler, Walther, and London, Fritz. "Wechselwirkung neutraler Atome und homöopolare Bindung nach der Quantenmechanik." *Zeit. Physik* 44 (1927), 455–72.

Helmholtz, Hermann von. "The Modern Development of Faraday's Conception of Electricity." *J. Chem. Soc.* 38 (1881), 277–304.

Henrich, Ferdinand. *Theories of Organic Chemistry.* Translated by Treat B. Johnson and Dorothy A. Hahn. New York: John Wiley and Sons, 1922.

———. "Über die negative Natur ungesättigter Atomgruppen." *Berichte* 32 (1899), 668–76.

Hentschel, Willibald. "Über Chlorstickstoff." *Berichte* 30 (1897), 1434–37.

Hertz, Heinrich. "Die Kräfte elektrischer Schwingungen behandelt nach der Maxwell'schen Theorie." *Annalen der Physik* 36 (1899), 1–22.

———. *Electric Waves.* Translated by D. E. Jones. London: Macmillan & Co., 1893.

———. "Nachtrag zu der Abhandlung über sehr schnelle elektrische Schwingungen." *Annalen der Physik* 31 (1887), 543–44.

———. "Über die Ausbreitungsgeschwindigkeit der elektrodynamischen Wirkungen." *Annalen der Physik* 34 (1888), 551–69.

———. "Über elektrodynamische Wellen im Luftraume und deren Reflexion." *Annalen der Physik* 34 (1888), 609–23.

———. "Über Inductionserscheinungen hervorgerufen durch die elektrischen Vorgänge in Isolatoren." *Annalen der Physik* 34 (1888), 273–85.

———. "Über sehr schnelle elektrische Schwingungen." *Annalen der Physik* 31 (1887), 421–88.

Hess, Germain H. "Thermochemische Untersuchungen." *Annalen der Physik* 52 (1841), 97–113.

Hildebrand, Joel H. "Gilbert Newton Lewis." *National Academy of Sciences, Biographical Memoirs.* Vol. 21. New York: Columbia University Press, 1958.

———. "Irving Langmuir's Philosophy of Science with an Introduction by J. H. Hildebrand." In *Langmuir: The Man and the Scientist.* New York: Pergamon Press, 1962.

———. *Principles of Chemistry.* New York: Macmillan Company, 1919.

Holleman, Arnold F. "Substitution in the Benzene Nucleus." *J.A.C.S.* 36 (1914), 2495–98.

Hückel, Erich. "Zur Quantentheorie der Doppelbindung." *Zeit. Physik* 60 (1930), 423–56; 70 (1931), 204–86; 72 (1931), 310–37.

Huggins, Maurice L. "Electronic Structures of Atoms." *J. Phys. Chem.* 26 (1922), 601–24.

Hull, Albert W. "The Crystal Structure of Iron." *Physical Review* 9 (1917), 84–87.

Hund, Friedrich. "Chemical Binding." *Trans. Faraday Soc.* 25 (1929), 646–48.

———. "Zur Deutung der Molekülspektren." *Zeit. Physik* 40 (1926), 742–64; 42 (1927), 93–120; 43 (1927), 805–26; 51 (1928), 759–95.

Ihde, Aaron J. *The Development of Modern Chemistry.* New York: Harper & Row, 1964.

Jacobson, Paul Heinrich. "Zur Kenntnis der orthoamidinten aromatischen Mercaptane." *Berichte* 20 (1887), 1895–903; "III." *Berichte* 21 (1888), 2624–31.

Jakowkin, Alexander A. "Über die Dissociation des Chlorohydrates in wässeriger Lösung bei 0°." *Berichte* 30 (1897), 518–21.

———. "Über die Hydrolyse des Chlors." *Zeit. phys. Chemie* 29 (1899), 613–57.

James, Hubert M., and Coolidge, Albert S. "The Ground State of the Hydrogen Molecule." *J. Chem. Physics* 1 (1933), 825–35.

Jeans, James. "The Mechanism of Radiation." *Phil. Mag.* 2 (1901), 421–55.

Johnson, Frederick M. G. "Der Dampfdruck von trocknem Salmiak." *Zeit. phys. Chemie* 61 (1908), 457–63.

Jones, Lauder W. "Applications of the Electronic Conception of Valence." *Amer. Chem. Journal* 50 (1913), 414–43.

———. "The Beckmann Rearrangement of Hydroxamic Acids." *Amer. Chem. Journal* 48 (1912), 1–28.

———. "Electromers and Stereomers with Positive and Negative Hydroxyl." *J.A.C.S.* 36 (1914), 1268–90.

Kahl, Russell. *Selected Writings of Hermann von Helmholtz*. Middleton: Wesleyan University Press, 1971.

Kahlenberg, Louis. *Outlines of Chemistry*. New York: Macmillan Company, 1909.

———. *Outlines of Chemistry*. 2d ed. New York: Macmillan Company, 1915.

Kauffman, George B. "Alfred Werner's Early View of Valence." *Journal of Chemical Education* 56 (1979), 496–99.

———, ed. *Classics in Coordination Chemistry: Part 1. The Selected Papers of Alfred Werner*. New York: Dover Publications, 1968.

Kauffman, Hugo. *Die Valenzlehre*. Stuttgart: Ferdinand Enke, 1911.

———. "Elektronen und Valenzlehre." *Phys. Zeit.* 9 (1908), 311–14.

Kaufmann, Walther. "Die magnetische Ablenkbarkeit der Kathodenstrahlen und ihre Abhängigkeit vom Entladungspotential." *Annalen der Physik* 62 (1897), 596–98.

Kekulé, August. "On the Existence of Chemical Atoms." *American Journal of Science* 44 (1867), 270–73.

———. "Untersuchungen über aromatische Verbindungen." *Annalen der Chemie* 137 (1866), 129–77.

Kelvin, Lord (William Thomson). "Aepinus Atomized." *Phil. Mag.* 3 (1902), 257–83.

———. "Contact Electricity and Electricity According to Father Boscovich." *Nature* 56 (27 May 1897), 84.

Kermack, William O., and Robinson, Robert. "An Explanation of the Property of Induced Polarity of Atoms and an Interpretation of the Theory of Partial Valences on an Electronic Basis." *J. Chem. Soc.* 121 (1922), 427–40.

Ketteler, Eduard. "Constanz des Refraktions." *Annalen der Physik* 30 (1887), 285–316.

———. "On the Dispersion of Light in Gases." *Phil. Mag.* 32 (1866), 336–45.

———. *Theoretische Optik*. Braunschweig: F. Vieweg und Sohn, 1885.

———. "Über die Dispersion des Lichts in den Gasen." *Annalen der Physik* 124 (1865), 390–406.

Kharasch, Morris, and Darkis, Frederick R. "The Theory of Partial Polarity of the Ethylene Bond and the Existence of Electro-isomerism." *Chemical Reviews* 5 (1928), 571–602.

Kirkby, Paul J. "A Theory of Chemical Action of Electrical Discharge in Electrolytic Gas." *Proc. Roy. Soc.* A 85 (1910), 151–74.

Knorr, Carl A. "Eigenschaften chemischer Verbindungen und die Anordnung der Elektronenbahnen in ihren Molekülen." *Zeit. anorg. Chemie* 129 (1923), 109–40.

Knorr, Ludwig. "Studien über Tautomeric." *Annalen der Chemie* 293 (1897), 70–120; 303 (1898), 133–49; 306 (1899), 332–93.

Kohler, Robert E., Jr. "G. N. Lewis's Views on Bond Theory." *British Journal for the History of Science* 8 (1975), 233–39.

――――. "Irving Langmuir and the 'Octet' Theory of Valence." *Historical Studies in the Physical Sciences* 4 (1974), 39–87.

――――. "The Lewis-Langmuir Theory of Valence and the Chemical Community." *Historical Studies in the Physical Sciences* 6 (1975), 431–68.

――――. "The Origin of G. N. Lewis's Theory of the Shared Pair Bond." *Historical Studies in the Physical Sciences* 3 (1971), 343–76.

Kohlrausch, Friedrich W. "Das elektrische Leitungsvermögen der wässerigen Lösungen von Hydraten und Salzen des leichten Metalle, sowie von Kupfervitriol, Zinkvitriol, und Silbersalpeter." *Annalen der Physik* 6 (1879), 145–210.

Kossel, Walther. "Über Molekülbildung als Frage des Atombaus." *Annalen der Physik* 49 (1916), 229–362.

Laar, Peter Conrad. "Über die Hypothese der wechselnden Bindung." *Berichte* 19 (1886), 730–41.

――――. "Über die Möglichkeit mehrerer Strukturformeln für dieselbe chemische Verbindung." *Berichte* 18 (1885), 648–57.

Lachman, Arthur. *Borderland of the Unknown: The Life Story of Gilbert Newton Lewis, One of the World's Great Scientists.* New York: Pageant Press, 1955.

Ladenburg, Albert. "Bemerkungen zur aromatischen Theorie." *Berichte* 2 (1869), 140–42.

Lagowski, J. J. *The Chemical Bond.* Boston: Houghton Mifflin Company, 1966.

Landé, Alfred. "Dynamik der räumlichen Atomstruktur." *Verh. deut. phys. Ges.* 21 (1919), 2–12, 644–62.

Langmuir, Irving. "The Arrangement of Electrons in Atoms and Molecules." *J.A.C.S.* 41 (1919), 868–934.

――――. "The Arrangement of Electrons in Atoms and Molecules." *Journal of the Franklin Institute* 187 (1919), 359–62.

――――. "Future Developments of Theoretical Chemistry." *Chemical and Metallurgical Engineering* 24 (1921), 553–57.

――――. "Isomorphism, Isosterism and Covalence." *J.A.C.S.* 41 (1919), 1543–59.

――――. *Langmuir: The Man and the Scientist.* Vol. 12. *The Collected Works of*

Irving Langmuir. Edited by Guy Suits and Harold E. Way. New York: Pergamon Press, 1962.

———. "The Properties of the Electron as Derived from the Chemical Properties of the Elements." *Physical Review* 13 (1919), 300.

———. "The Structure of Atoms and Its Bearing on Chemical Valence." *Journal of Industrial and Engineering Chemistry* 12 (1920), 386–89.

———. "The Structure of Atoms and the Octet Theory of Valence." *Proc. Nat. Acad. Sci.* 5 (1919), 252–59.

———. *The Structure of Matter.* Vol. 6. *The Collected Works of Irving Langmuir.* Edited by Guy Suits and Harold E. Way. New York: Pergamon Press, 1961.

———. "The Structure of the Static Atom." *Science* 53 (25 March 1921), 290–93.

———. "Theories of Atomic Structure." *Nature* 105 (29 April 1920), 261.

Lapworth, Arthur. "The Form of Change in Organic Compounds and the Function of the α-Meta-Orientating Groups." *J. Chem. Soc.* 79 (1901), 1265–84.

———. "A Theoretical Derivation of the Principle of Induced Alternate Polarities." *J. Chem. Soc.* 121 (1922), 416–27.

Larmor, Joseph. "A Dynamical Theory of the Electric and Luminiferous Medium." *Phil. Trans.* pt. 1, 185 (1894), 719–822.

———. "On the Theory of Moving Electrons and Electric Charges." *Phil. Mag.* 42 (1896), 201–4.

———. "On the Theory of the Magnetic Influence on Spectra; and on the Radiation from Moving Ions." *Phil. Mag.* 44 (1897), 503–12.

Latimer, Wendell M., and Rodebush, Worth H. "Polarity and Ionization from the Standpoint of the Lewis Theory of Valence." *J.A.C.S.* 42 (1920), 1430–32.

Lenard, Philipp. "Über die Absorption von Kathodenstrahlen verschiedener Geschwindigkeit." *Annalen der Physik* 12 (1903), 714–44.

Lennard-Jones, John E. "The Electronic Structure of Some Diatomic Molecules." *Trans. Faraday Soc.* 25 (1929), 668–86.

Lewis, G. N. "Acids and Bases." *Journal of the Franklin Institute* 226 (1938), 293–313.

———. "The Atom and the Molecule." *J.A.C.S.* 38 (1916), 762–85.

———. "Color and Chemical Constitution." *Chemical and Metallurgical Engineering* 24 (1921), 869–75.

———. "The Magnetochemical Theory." *Chemical Reviews* 1 (1924), 231–48.

———. "The Static Atom." *Science* 46 (28 September 1917), 297–302.

———. "Valence and Tautomerism." *J.A.C.S.* 35 (1913), 1448–55.

———. "Valence and the Electron." *Trans. Faraday Soc.* 19 (1923), 452–58.

———. *Valence and the Structure of Atoms and Molecules.* New York: Chemical Catalog Company, 1923.

Lewis, G. N., and Gibson, G. E. "The Third Law of Thermodynamics and the Entropy of Solutions and of Liquids." *J.A.C.S.* 42 (1920), 1529–33.

Lewis, G. N., and Randall, Merle. *Thermodynamics and the Free Energy of Chemical Substances.* New York: McGraw-Hill, 1923.

Locke, James. "Electro-Affinity as a Basis for the Systematization of Inorganic Compounds." *Amer. Chem. Journal* 27 (1902), 105–17.

Lodge, Oliver. *Electrons*. London: George Bell and Sons, 1906.

———. *Modern Views on Matter*. Oxford: Clarendon Press, 1903.

———. "Residual Affinity." *Nature* 70 (23 June 1904), 176.

———. Review of *Electricity and Matter* by J. J. Thomson. *Nature* 70 (26 May 1904), 73–76.

London, Fritz. "Zur Quantentheorie der homöopolaren Valenzzahlen." *Zeit. Physik* 46 (1928), 455–77.

Lossen, Wilhelm. "Über die Verteilung der Atome in der Molekül." *Annalen der Chemie* 204 (1880), 265–364.

Lowry, Thomas. "Dynamic Isomerism." *B.A.A.S. Report* 74 (1904), 193–224.

Lucas, Howard J. "Electron Displacement in Carbon Compounds: IV. Derivatives of Benzene." *J.A.C.S.* 48 (1926), 1827–38.

Lucas, Howard J., and Jameson, Archibald Y. "Electron Displacement in Carbon Compounds: I. Electron Displacement versus Alternate Polarity in Aliphatic Compounds." *J.A.C.S.* 46 (1924), 2475–82.

Lucas, Howard J., and Moyse, Hollis W. " Electron Displacement in Carbon Compounds: II. Hydrogen Bromide and 2-Pentene." *J.A.C.S.* 47 (1925), 1459–62.

Lucas, Howard J.; Simpson, Thomas P.; and Carter, James M. "Electron Displacement in Carbon Compounds: III. Polarity Differences in Carbon-Hydrogen Unions." *J.A.C.S.* 47 (1925), 1462–69.

McPherson, William, and Henderson, William Edwards. *An Elementary Study of Chemistry*. Rev. ed. Boston: Ginn and Co., 1906.

Markownikov, Vladimir V. "Sur les lois qui régissent les réactions de l'addition directe." *Comptes rendus* 81 (1875), 668–71.

Maxwell, James Clerk. "Electrolysis." In *A Treatise on Electricity and Magnetism*. Vol. 1. Oxford: At the University Press, 1873.

Mayer, Alfred M. "Floating Magnets." *Nature* 17 (18 April 1878), 487–88; 18 (4 July 1878), 258–60.

———. "A Note on Experiments with Floating Magnets." *American Journal of Science* 15 (1878), 276–77.

———. "Note on Floating Magnets." *American Journal of Science* 15 (1878), 477–78.

Meisenheimer, Jakob. "Über die Ungleichartigkeit der fünf Valenzen des Stickstoffs." *Annalen der Chemie* 397 (1913), 273–84.

Mellor, Joseph W. *Modern Inorganic Chemistry*. London: Longmans, Green, and Co., 1916.

Mendeleev, Dmitri. "Das natürliche System der chemischen Elemente." *Ostwald's Klassiker der exakten Wissenschaften* 61–68. Leipzig: Wilhelm Engelmann, 1895. This consists of a German translation of Medeleev's first memoir of 1869, "Die Beziehungen zwischen den Eigenschaften der Elemente und ihren Atomgewichten"; a short German abstract of his first memoir, "Über die Beziehungen der Eigenschaften zu den Atomgewichten der Elemente"; and a

German translation of his second memoir of 1871, "Die periodische Gesetz-mässigkeit der chemischen Elemente."

―――. "Die periodische Gesetzmässigkeit der chemischen Elemente." *Annalen der Chemie* Supplementband 8 (1871), 133–229.

―――. "Essai d'un système des éléments d'après leurs poids atomiques et propriétés chimiques." *Journal of the Russian Physical and Chemical Society* 1 (1869), 60–77.

―――. "The Periodic Law of the Chemical Elements." *Chemical News* 40 (November–December, 1879), and 41 (February–March, 1880).

―――. "The Periodic Law of the Chemical Elements." *J. Chem. Soc.* 55 (1889), 634–56.

―――. *The Principles of Chemistry.* Translated by George Kamensky. 5th ed. London: Longmans, Green, and Co., 1891 (1st ed. 1868).

―――. "Über die Beziehungen der Eigenschaften zu den Atomgewichten der Elemente." *Zeitschrift für Chemie* 5 (1869), 405–6.

Meyer, Lothar. *Die modernen Theorien der Chemie und ihre Bedeutung für die chemische Statik.* Breslau: Maruschke & Berendt, 1864.

―――. "Die Natur der chemischen Elemente als Funktion ihrer Atomgewichte." *Annalen der Chemie* Supplementband 7 (1870), 354–64.

―――. "Die Natur der chemischen Elemente als Funktion ihrer Atomgewichte." *Ostwald's Klassiker der exakten Wissenshaften* 61–68. Leipzig: Wilhelm Engelmann, 1895.

Meyer, Victor, and Lecco, Marco T. "Über die Konstitution der Ammoniumverbindungen und des Salmiaks." *Berichte* 8 (1875), 233–42.

――― and ―――. "Untersuchungen über die Konstitution der Ammoniumverbindungen und des Salmiaks." *Annalen der Chemie* 180 (1876), 173–91.

Meyer, Victor, and Stuber, Otto. "Über die Nitroverbindungen der Fettreihe." *Berichte* 5 (1872), 399–406.

Michael, Arthur. "On the Nonexistence of Valence and Electronic Isomerism in Hydroxylammonium Derivatives." *J.A.C.S.* 42 (1920), 1232–45.

―――. "Über einige Gesetze und deren Anwendung in der organischen Chemie." *J. prakt. Chemie* 60 (1899), 286–384, 409–86.

Millikan, Robert. "Atomism in Modern Physics." *J. Chem. Soc.* 125 (1924), 1405–17.

―――. *The Electron.* 1917. Facsimile reprint. Chicago: University of Chicago Press, 1963.

Moissan, Henri. "Étude de la combinaison de l'acide carbonique et de l'hydrure de potassium." *Comptes rendus* 136 (1903), 723–27.

―――. "Sur la non-conductibilité électrique des hydrures métalliques." *Comptes rendus* 136 (1903), 591–92.

Moseley, Henry Gwyn-Jeffreys. "The High Frequency Spectra of the Elements." Pt. 1. *Phil. Mag.* 26 (1913), 1024–34; Pt. 2. *Phil. Mag.* 27 (1914), 703–13.

Muir, M. M. Pattison. *A History of Chemical Theories and Laws.* New York: John Wiley and Sons, 1906.

Mulliken, Robert. "The Assignment of Quantum Numbers for Electrons in Molecules; I." *Physical Review* 32 (1928), 186–222.

Nagaoko, Hantaro. "Kinetics of a System of Particles Illustrating the Line and Band Spectrum and the Phenomena of Radioactivity." *Phil. Mag.* 7 (1904), 445–55.

Nef, John U. "Über das Phenylacetylen, seine Salze und seine Halogensubstitutionsprodukte." *Annalen der Chemie* 308 (1899), 264–328.

———. "Über das Verhalten der tri- und tetrahalogen substituierten Methane." *Annalen der Chemie* 308 (1899), 329–33.

Nelson, John M.; Beans, Hal T.; and Falk, K. George. "The Electron Conception of Valence: IV. The Classification of Chemical Reactions." *J.A.C.S.* 35 (1913), 1810–21.

Nelson, John M., and Falk, K. George. "The Electron Conception of Valence: III. Oxygen Compounds." *Original Communications to the Eighth International Congress of Applied Chemistry* 6 (September 1912), 212–21.

Nernst, Walther. "Die elektrolytische Zersetzung wässiger Lösungen." *Berichte* 30 (1897), 1547–63.

———. *Theoretical Chemistry.* Translated by Robert A. Lehfeldt from the 4th German edition of 1903. London: Macmillan & Co., 1904.

———. *Theoretical Chemistry.* Translated by H. T. Tizard from the 7th German edition of 1913. London: Macmillan & Co., 1916.

———. *Theoretical Chemistry.* Translated by L. W. Codd from the 8th–10th German edition of 1921. London: Macmillan & Co., 1923.

———. *Theoretische Chemie.* 2d ed. Stuttgart: Ferdinand Enke, 1898.

Newth, George S. *A Textbook of Inorganic Chemistry.* 11th ed. London: Longmans, Green, and Co., 1905.

Noyes, Arthur A. "The Conductivity and Ionization of Salts, Acids, and Bases in Aqueous Solutions at High Temperatures." *J.A.C.S.* 30 (1908), 335–53.

———. "Die Reaktion zwischen Chlor und Ammoniak." *Zeit. phys. Chemie* 41 (1902), 378.

Noyes, William A. "An Attempt to Prepare Nitro-Nitrogen Trichloride, an Electromer of Ammono-Nitrogen Trichloride." *J.A.C.S.* 35 (1913), 767–75.

———. "An Attempt to Prepare Nitro-Nitrogen Trichloride: II. The Conduct of Mixtures of Nitrogen and Chlorine in a Flaming Arc." *J.A.C.S.* 43 (1921), 1774–82.

———. "The Electronic Interpretation of Oxidation and Reduction." *J.A.C.S.* 51 (1929), 2391–96.

———. "Electronic Theories." *Chemical Reviews* 17 (1935), 1–26.

———. "The Electron Theory." *Proceedings: Illinois Academy of Science* 5 (1912), 20–28.

———. "The Electron Theory." *Journal of the Franklin Institute* 185 (January 1918), 59–84.

———. "The Interaction between Nitrogen Trichloride and Nitric Oxide: Reactions of Compounds with Odd Electrons." *J.A.C.S.* 50 (1928), 2902–10.

————. "A Kinetic Hypothesis to Explain the Function of Electrons in the Chemical Combination of Atoms." *J.A.C.S.* 39 (1917), 879–82.

————. "Molecular Rearrangements." *J.A.C.S.* 31 (1909), 1368–74.

————. "The Nature of the Forces Holding Atoms in Combination." *J.A.C.S.* 36 (1914), 214.

————. "A Possible Explanation of Some Phenomena of Ionization by the Electron Theory." *J.A.C.S.* 34 (1912), 663–64.

————. "A Possible Reconciliation of the Octet and Positive-Negative Theories of Chemical Combination." *J.A.C.S.* 45 (1923), 2959–61.

————. "Present Problems of Organic Chemistry." *Science* 20 (14 October 1904), 490–501.

————. "The Reaction between Chlorine and Ammonia: III. Probable Formation of Trichloro-Ammonium Chloride." *J.A.C.S.* 42 (1920), 2173–79.

————. "The Relation of Shared Electrons to Potential and Absolute Polar Valences." *Chemical Reviews* 5 (1928), 552–55.

Noyes, William A., and Haw, Arthur B. "The Reaction between Chlorine and Ammonia: II." *J.A.C.S.* 42 (1920), 2167–73.

Noyes, William A., and Lyon, Albert C. "The Reaction between Chlorine and Ammonia." *J.A.C.S.* 23 (1901), 460–63.

Noyes, William A., and Wilson, Thomas A. "The Ionization Constant of Hypochlorous Acid: Evidence for Amphoteric Ionization." *J.A.C.S.* 44 (1922), 1630–37.

Onsager, Lars. "Zur Theorie der Elektrolyte." *Phys. Zeit.* 27 (1926), 388–92; 28 (1927), 277–98.

Ostwald, Wilhelm. "Elektrochemische Studien: Das Verdünnungsgesetz." *J. prakt. Chemie* 31 (1885), 433–62.

————. "Notiz über das elektrische Leitungsvermögen der Säuren." *J. prakt. Chemie* 30 (1884), 93–95.

————. *Outlines of General Chemistry.* Translated by William W. Taylor. 3rd ed. London: Macmillan & Co., 1908.

————. "Über die Affinitätsgrössen organischer Säuren und ihre Beziehungen zur Zusammensetzung und Konstitution derselben." *Zeit. phys. Chemie* 3 (1889), 170–97, 241–88, 369–422.

————. "Über die Dissociationstheorie der Elektrolyte." *Zeit. phys. Chemie* 2 (1888), 270–83; 3 (1889), 588–602.

————. *Wissenschaftlichen Grundlagen der analytischen Chemie.* Leipzig: W. Engelmann, 1901.

Palmer, W. G. *A History of the Concept of Valency to 1930.* Cambridge: At the University Press, 1965.

Parson, Alfred L. "A Magneton Theory of the Atom." *Smithsonian Miscellaneous Publication* 65 (1915), 1–80.

Partington, J. R. *A History of Chemistry.* Vol. 4. London: Macmillan & Co., 1864.

————. *A Short History of Chemistry.* 3rd ed. New York: Harper & Brothers, 1960.

Pauli, Wolfgang. "Über den Zusammenhang des Abschlusses der Elektronengruppen in Atom mit der Komplexstruktur der Spektren." *Zeit. Physik* 31 (1925), 765–83.

———. "Zur Quantenmechanik des magnetischen Elektrons." *Zeit. Physik* 43 (1927), 601–23.

Pauling, Linus. *The Nature of the Chemical Bond.* 3rd ed. Ithaca: Cornell University Press, 1960.

———. "The Nature of the Chemical Bond: I. Application of Results Obtained from the Quantum Mechanics and from a Theory of Paramagnetic Susceptibility to the Structure of Molecules." *J.A.C.S.* 53 (1931), 1366–1400.

———. "The Shared-Electron Chemical Bond." *Proc. Nat. Acad. Sci.* 14 (1928), 359–62.

Pease, Robert N. "An Analysis of Molecular Volumes from the Point of View of the Lewis-Langmuir Theory of Molecular Structure." *J.A.C.S.* 43 (1921), 991–1004.

Perkin, William H. "Tautomerism." Presidential address. *J. Chem. Soc.* 105 (1914), 1176–89.

Planck, Max. "Das chemische Gleichgewicht in verdünnten Lösungen." *Annalen der Physik* 34 (1888), 139–54.

Prescott, Albert B., and Johnson, Otis C. *Qualitative Chemical Analysis.* 5th ed. New York: D. Van Nostrand Co., 1903.

Ramsay, William. "Compounds of Electrons." *Rice Institute Pamphlet* 1 (July 1915), 410–24.

———. "The Electron as an Element." *J. Chem. Soc.* 93 (1908), 774–88.

———. "Elements and Electrons." *J. Chem. Soc.* 95 (1909), 624–37.

———. *Essays, Biographical and Chemical.* London: Constable & Co., 1908.

———. "A Hypothesis of Molecular Configuration in Three Dimensions of Space." *Proc. Roy. Soc.* A 92 (1916), 451–62.

———. *Modern Chemistry: Part II. Systematic Chemistry.* 5th ed. London: Macmillan & Co., 1912.

———. *Modern Chemistry: Part I. Theoretical Chemistry.* 4th ed. London: Macmillan & Co., 1907.

Ramsay, William, and Soddy, Frederick. "Experiments in Radioactivity and the Production of Helium from Radium." *Proc. Roy. Soc.* 72 (1903), 204–7.

Raoult, François Marie. "Sur le point de congélation des dissolutions salines." *Annales de Chimie et de Physique* 4 (1885), 401–30.

Rayleigh, Lord (John William Strutt). "On Some Physical Properties of Argon and Helium." *Proc. Roy. Soc.* 59 (1896), 198–208.

Remick, Arthur E. *Electronic Interpretations of Organic Chemistry.* New York: John Wiley & Sons, 1943.

Richards, Theodore W. "Die Bedeutung der Änderung der Atomvolume." *Zeit. phys. Chemie* 40 (1902), 597–610.

———. "Die mögliche Bedeutung der Änderung des Atomvolums." *Zeit. phys. Chemie* 40 (1902), 169–84.

Romer, Alfred. *Radiochemistry and the Discovery of Isotopes.* New York: Dover Publications, 1970.

Ruggli, Paul. *Die Valenzhypothese von J. Stark vom chemischen Standpunkt.* Stuttgart: F. Enke, 1912.

Runge, Carl, and Kayser, Heinrich. "Über die Spektren der Alkalien." *Annalen der Physik* 41 (1890), 302–20.

Russell, Colin A. "Berzelius and the Development of the Atomic Theory." *John Dalton and the Progress of Science.* Edited by Donald S. L. Cardwell. Manchester: Manchester University Press, 1968.

———. *The History of Valency.* Leicester: Leicester University Press, 1971.

Rutherford, Ernest. "The Scattering of α and β Particles by Matter and the Structure of the Atom." *Phil. Mag.* 21 (1911), 669–88.

———. "The Structure of the Atom." *Nature* 92 (11 December 1913), 423.

Rutherford, Ernest, and Royds, Thomas, "The nature of the α-Particle from Radioactive Substances." *Phil. Mag.* 17 (1909), 281–86.

Rydberg, Johannes R. "Über den Bau der Linienspektren der chemischen Grundstoffe." *Zeit. phys. Chemie* 5 (1890), 227–32.

Saltzman, Martin D. "The Robinson-Ingold Controversy." *Journal of Chemical Education* 57 (1980), 484–88.

Schott, George A. "On the Electron Theory of Matter and on Radiation." *Phil. Mag.* 13 (1907), 189–213.

Schützenberger, Paul. "Über die Substitution electronegativer Körper an die Stelle der Metalle in Sauerstoffsalzen." *Annalen der Chemie* 120 (1861), 113–18.

———. "On the Substitution of Electro-negative Bodies (Chlorine, Bromine, Iodine, Cyanogen, Sulphur, etc.) for the Metals in Oxygenised Salts: Production of a New Class of Salts in Which the Electro-negative Bodies Replace the Basic Hydrogen." *Chemical News* 3 (13 April 1861), 225–26.

Selivanov, Feodor. "Beitrag zur Kenntnis der gemischten Anhydride der unterchlorigen Säure und analoger Säuren." *Berichte* 25 (1892), 3617–23.

Sidgwick, Nevil V. "Coordination Compounds and the Bohr Atom." *J Chem. Soc.* 123 (1923), 725–30.

———. *The Electronic Theory of Valency.* Oxford: Clarendon Press, 1927.

———. "The Nature of the Nonpolar Link." *Trans. Faraday Soc.* 19 (1923), 469–75.

Simon, William, and Base, Daniel. *Manual of Chemistry.* 9th ed. Philadelphia and New York: Lea and Febiger, 1909.

Slater, John D. "Directed Valence in Polyatomic Molecules." *Physical Review* 37 (1931), 481–89.

———. "Note on Hartree's Method." *Physical Review* 35 (1930), 210–11.

Smith, Alexander. *General Chemistry for Colleges.* New York: Century Company, 1908.

Smith, J. D. Main. "Friend's Theory of Valency." *Chemical News* 124 (17 February 1922), 84–86.

Soddy, Frederick. "Die Radioelemente und das periodische Gesetz." *Jahrbuch der Radioaktivität und Elektronik* 10 (1913), 188–97.

———. "Intra-Atomic Charge." *Nature* 92 (4 December 1913), 399–400.

———. "The Radioelements and the Periodic Law." *B.A.A.S. Report* 83 (1913), 445–47.

———. "The Radio-Elements and the Periodic Law." *Chemical News* 107 (28 February 1913), 97–99.

———. "The Structure of the Atom." *Nature* 92 (18 December 1913), 452.

Spiegel, Leopold. "Über Neutralaffinitäten." *Zeit. anorg. Chemie* 29 (1902), 365–70.

Stark, Johannes. "Anwendung einer Valenzhypothese auf Erscheinungen der Fluoreszenz." *Zeit. Elektrochemie* 17 (1911), 514–17.

———. "Die Valenzlehre auf atomistisch elektrischer Basis." *Jahrbuch der Radioaktivität und Elektronik* 5 (1908), 124–53.

———. *Dissoziierung und Umwandlung chemischer Atome.* Braunschweig: F. Vieweg und Sohn, 1903.

———. "Folgerung aus einer Valenzhypothese: I. Bandenspektrum und Valenzenergie." *Jahrbuch der Radioaktivität und Elektronik* 9 (1912), 15–27.

———. *Prinzipien der Atomdynamik: Part III. Die Elektrizität in chemischen Atom.* Leipzig: S. Hirzel, 1915.

———. *Prinzipien der Atomdynamik: Part II. Die elementare Strahlung.* Leipzig: S. Hirzel, 1911.

———. "Über das dreiatomige Wasserstoffmolekül nach Hypothese und Erfahrung." *Zeit. Elektrochemie* 19 (1913), 862–63.

———. "Zur Energetik und Chemie der Bandenspektra." *Phys. Zeit.* 9 (1908), 85–94.

Stewart, Alfred W. *Stereochemistry.* London: Longmans, Green, and Co., 1907.

Stieglitz, Julius. "On the Beckmann Rearrangement: I. Chlorimidoesters." *Amer. Chem. Journal* 18 (1896), 751–61.

———. "The Electron Theory of Valence as Applied to Organic Compounds." *J.A.C.S.* 44 (1922), 1293–1313, 1833–34.

———. *The Elements of Qualitative Analysis with Special Consideration of the Application of the Laws of Equilibrium and of the Modern Theories of Solution.* New York: Century Club, 1911.

———. "Molecular Rearrangements of Triphenlymethane Derivatives." *Proc. Nat. Acad. Sci.* 1 (1915), 196–210.

———. "On Positive and Negative Halogen Ions." *J.A.C.S.* 23 (1901), 797–99.

Stieglitz, Julius, and Leech, Paul N. "The Molecular Rearrangement of Triarylmethylhydroxyl Amines and the Beckmann Rearrangement of Ketoximes." *J.A.C.S.* 36 (1914), 272–301.

Stieglitz, Julius, and Peterson, Peter P. "Über Stereoisomere Chlorimido-Ketone." *Berichte* 43 (1910), 782–87.

Stieglitz, Julius, and Stagner, Bert A. "Molecular Rearrangements of β-Triphenylmethyl–β-Methylhydroxylamines and the Theory of Molecular Rearrangements." *J.A.C.S.* 38 (1916), 2046–69.

Stock, Alfred. *The Structure of Atoms*. Translated by S. Sugden. New York: E. P. Dutton and Co., n.d.

Stoddard, John T. *Introduction to General Chemistry*. New York: Macmillan Company, 1910.

Stoney, G. Johnstone. "On the Cause of Double Lines and of Equidistant Satellites in the Spectra of Gases." *Scientific Proceedings of the Royal Dublin Society* 4 (1891), 563–608.

————. "Of the 'Electron,' or Atom of Electricity." *Phil. Mag.* 38 (1894), 418–20.

————. "On the Physical Units of Nature." *Phil. Mag.* 11 (1881), 381–90.

Sugiura, Yoshikatsu. "Über die Eigenschaften des Wasserstoffmoleküls im Grundzustande." *Zeit. Physik* 45 (1927), 484–92.

"Symposium on Atomic Structure and Valence." *Chemical Reviews* 5 (1928), 361–617.

Taylor, Richard, ed. *Scientific Memoirs*. Vol. 5. London: R. and J. E. Taylor, 1852.

Thiele, Johannes. "Zur Kenntnis der ungesättigten Verbindungen: I. Theorie der ungesättigten und aromatischen Verbindungen." *Annalen der Chemie* 306 (1899), 87–142.

Thomson, J. J. "Application of the Electron Theory of Chemistry to Solids." *Phil. Mag.* 43 (1922), 721–57.

————. "Cathode Rays." *Phil. Mag.* 44 (1897), 293–316.

————. "On the Charge of Electricity Carried by Ions Produced by Röntgen Rays." *Phil. Mag.* 46 (1898), 528–45.

————. *Conduction of Electricity through Gases*. 2d ed. Cambridge: At the University Press, 1906.

————. *The Corpuscular Theory of Matter*. London: A. Constable & Co., 1907.

————. *Electricity and Matter*. New York: Charles Scribner's Sons, 1904.

————. *The Electron in Chemistry*. Philadelphia: Franklin Institute, 1923.

————. "The Forces between Atoms and Chemical Affinity." *Phil. Mag.* 27 (1914), 757–89.

————. "Further Experiments on Positive Rays." *Phil. Mag.* 24 (1912), 209–53.

————. "On the Masses of the Ions in Gases at Low Pressures." *Phil. Mag.* 48 (1899), 547–67.

————. "On the Number of Corpuscles in an Atom." *Phil. Mag.* 11 (1906), 769–81.

————. "On the Origin of Spectra and Planck's Law." *Phil. Mag.* 37 (1919), 419–66.

————. *Rays of Positive Electricity*. London: Longmans, Green, and Co., 1913.

————. "Rays of Positive Electricity." *Phil. Mag.* 21 (1911), 225–49.

————. "Rays of Positive Electricity." *Proc. Roy. Soc.* A 89 (1913), 1–20.

————. "On the Scattering of Light by Unsymmetrical Atoms." *Phil. Mag.* 40 (1920), 393–413.

————. "The Structure of the Atom." *Engineering* 95 (1913), 328–30, 346–47, 397–98.

————. "On the Structure of the Atom." *Phil. Mag.* 26 (1913), 792–99.

————. "On the Structure of the Atom: An Investigation of the Stability and Periods of Oscillation of a Number of Corpuscles Arranged at Equal Intervals around the Circumference of a Circle, with Application of the Results to the Theory of Atomic Structure." *Phil. Mag.* 7 (1904), 237–65.

————. "On the Structure of the Molecule and Chemical Combination." *Phil. Mag.* 41 (1921), 510–44.

van't Hoff, Jacobus H. "Über die Menge und die Natur des Sogen. Ozons, das sich bei langsamer Oxydation des Phosphors bildet." *Zeit. phys. Chemie* 16 (1895), 411–16.

Van Vleck, John H. "The New Quantum Mechanics." *Chemical Reviews* 5 (1928), 467–507.

Walden, Paul. "Über abnorme Elektrolyte." *Zeit. phys. Chemie* 43 (1903), 385–464.

————. "Über einige anorganische Losungs- und Ionisierungsmittel." *Zeit. anorg. Chemie* 25 (1900), 209–26.

Walker, James. "The Arrhenius Memorial Lecture." *Memorial Lectures Delivered Before the Chemical Society.* Vol. 3. London: Chemical Society, 1933.

Watson, Herbert B. *Modern Theories of Organic Chemistry.* 2d ed. London: Oxford University Press, 1941.

Weber, Wilhelm. "Elektrodynamische Massbestimmungen." *Leipzig Abhandlung Jablonowskische Gesellschaft der Wissenschaften* 1 (1846), 209–278.

————. "Elektrodynamische Massbestimmungen." *Annalen der Physik* 73 (1848), 193–240. English translation in Richard Taylor, ed. *Scientific Memoirs.* Vol. 5. London: R. and J. E. Taylor, 1852.

————. "Elektrodynamische Massbestimmungen inbesondere über das Prinzip der Erhaltung der Energie." *Leipzig Abhandlung Mathematische-Physikalische* 10 (1871), 1–62. English translation in *Phil. Mag.* 43 (1872), 1–20, 119–49.

————. *Galvinismus und Elektrodynamik. Wilhelm Weber's Werke.* Vol. 4. Berlin: J. Springer, 1894.

Webster, David L. "Parson's Magneton Theory of Atomic Structure." *Physical Review* 6 (1915), 54.

————. "The Scattering of α-Particles as Evidence of the Parson Magneton Hypothesis." *J.A.C.S.* 40 (1918), 375–79.

————. "The Theory of Electromagnetic Mass of the Parson Magneton and Other Non-spherical Systems." *Physical Review* 9 (1917), 484–99.

Werner, Alfred. "Beiträge zum Theorie der Affinität und Valenz." *Naturforschung Gesellschaft, Zürich* 36 (1891), 129–69.

————. "Beiträge zur Konstitution anorganischer Verbindungen." *Zeit. anorg. Chemie* 3 (1893), 267–330; 8 (1895), 153–97.

————. *New Ideas on Inorganic Chemistry.* Translated by Edgar P. Hedley. London: Longmans, Green, and Co., 1911.

————. "Über Spiegelbildisomerie bei Chromverbindungen: I." *Berichte* 44 (1911), 3132–40.

————. "Über Spiegelbildisomerie bei Eisenverbindungen." *Berichte* 45 (1912), 433–36.

————. "Über Spiegelbildisomerie bei Rhodiumverbindungen. I." *Berichte* 45 (1912), 1228–36.

————. "Valency." *Chemical News* 96 (13 September 1907), 128–31.

————. "Zur Kenntnis des asymmetrischen Kobaltatoms: I." *Berichte* 44 (1911), 1887–98; "II." 44 (1911), 2445–55; "IV." 44 (1911), 3279–84.

Werner, Alfred, and Miolati, Arturo. "Beiträge zur Konstitution anorganischer Verbindungen." *Zeit. phys. Chemie* 12 (1893), 35–55; 14 (1894), 506–21.

Wiechert, J. Emil. "I. Über das Wesen der Elektrizität. II. Experimentelles über Kathodenstrahlen." *Annalen der Physik* 21 (1897), 443–44.

Wien, Wilhelm. "Untersuchungen über die elektrische Entladung in verdünnten Gasen." *Annalen der Physik* 65 (1898), 440–52.

Williamson, Alexander W. "Über die Theorie der Aetherbildung." *Annalen der Chemie* 77 (1851), 37–49.

Wood, Robert W. "On the Existence of Positive Electrons in the Sodium Atom." *Phil. Mag.* 15 (1908), 274–79.

Wright, Stephen, ed. *Classical Scientific Papers: Physics.* New York: American Elsevier Publishing Company, 1964.

Zeeman, Pieter. "Doublets and Triplets in the Spectrum Produced by External Magnetic Forces." *Phil. Mag.* 44 (1897), 55–60, 255–59.

————. "On the Influence of Magnetism on the Nature of the Light Emitted by a Substance." *Phil. Mag.* 43 (1897), 226–39.

Index